Home Satellite TV Installation & Troubleshooting Manual

Fourth Revised Edition

Dr. Frank Baylin

contributions by

Brent Gale and Ron Long

Baylin Publications

Boulder/London

ACKNOWLEDGMENTS

Since first published in 1985, this book has been thoroughly revised for its second and third editions. Now due to rapid developments in satellite communications technology, we have created an essentially new fourth edition. Thanks to Brent Gale of Echosphere Corporation whose participation was instrumental in preparing the first two editions and to Ron Long who contributed to the third edition.

DEDICATION

This book is dedicated to a hope, a hope that satellite communications technology can serve to educate people of the world quickly enough to reverse the pervasive damage to our environment that is occurring this very moment.

Dr. Frank Baylin, founder and president of Baylin Publications, provides technical and marketing consulting services. He has authored or co-authored more than ten books about satellite television, MMDS and SMATV systems.

He received a PhD in Physics from the University of Pennsylvania in 1975 and an MBA from the Wharton School of Business in 1977. He has worked in technical areas including biophysics, solar energy and communications as a researcher and analyst. Before founding Baylin Publications he was a Senior Scientist at the Solar Energy Research Institute.

For speaking engagements or technical consulting services contact:

Dr. Frank Baylin
1905 Mariposa
Boulder, Colorado 80302, U.S.A.
Tel: (303) 449-4551
FAX: (303) 939-8720

First Edition - September 1985
Second Edition - August 1987
Third Edition - January 1991
Fourth Edition - October 1995

ISBN: 0-917893-27-1
LCCCN: 85 - 71662

Table of Contents

CHAPTER 1

INTRODUCTION

SATELLITE TV COMES of AGE

Today the satellite TV revolution is entering a bold new era. While receive dishes, sometimes as large as 15 feet in diameter or more, that originally sprouted in North America in the late 1970s have become prevalent throughout the world, the transition from "analog" to "digital" technology is under way. Today many "large-dish" C-band systems in some countries already receive CD-quality digital audio and laser-disk quality digital video signals.

Large and small dish home satellite TV systems, all described as direct-to-home (DTH) systems, now thrive in both urban and rural areas. The larger dishes typically have 3 to 10 foot (1 to 3.5 meter) diameters and detect signals in the 3.7 to 4.2 billion cycles/second C-band frequency range while "small" dishes usually fall in the 18 inch to 3 foot (45 cm to 1 meter) range and receive Ku-band signals in the 10.95 to 12.75 billion cycles/second range.

While small-dish, Ku-band "MPEG-2 DVB" (digital video broadcasts) systems have captured much publicity lately, "analog" C-band systems are still the mainstay of satellite television delivery in most parts of the world including North America. In late-1995 approximately 3.5 million homes in the United States sported large-dish satellite television systems while 600,000 Ku-band PrimeStar and 700,000 DIRECTV®/USSB digital small-dish systems had been installed.

Experts believe that "full-view" analog C-band systems will continue to thrive in North America since they have access to a wider variety of programming than the new small-dish Ku-band services. Since the installed base of analog C-band system is so large this industry will continue for some time to come. Furthermore, the same digital technologies used in the new small-dish systems can be applied to large-dish C-band systems. For example, the African company, MultiChoice, already offers an MPEG-2 DVB service on both C- and Ku-bands.

In the midst of this alphabet soup of buzzwords, the objective of this book is to paint a clear picture of the operation, installation and maintenance of large-dish C-band systems. With some exception, the techniques are basically the same for both analog and digital technologies. Our intent the intent is to make many seemingly complicated ideas and terms such as compressed digital video, MPEG-2, error correction and others clear and understandable. The transition to digital broadcasts that are received by C-band systems is also explored. Readers wishing to learn more about Ku-band small dish systems can consult *Miniature Satellite Dishes - The New Digital Television* that is listed in the reference section at the end of this book.

Television Programming

Since the advent of television in the early 1940s, numerous methods have been used to bring news, sports and other forms of entertainment from the TV studio and the broadcast site to consumers. At first, over-the-air VHF channels, the familiar 2 through 12, and then higher frequency UHF channels were the only choices. The large American networks, ABC, CBS and NBC, were born in this era, the late 1940s and 1950s.

Cable television (CATV) began its evolution in the early 1950s with systems designed to bring over-the-air network programming to rural areas. CATV plants featured very large aerials to detect weak broadcast signals from distant cities and a distribution system cabled directly to each subscriber's residence. This industry reached full maturity in the 1980s with the creation and uplinking of specialized and popular cable TV networks such as CNN (Cable News Network), HBO (Home Box Office), TBS (Turner Broadcast Systems) and The Movie Channel. While conventional television broadcasters have seen CATV as a form of potentially lethal competition, the evolution of both industries has continued at a rapid pace.

Satellite communications, ironically the driving force behind the rapid growth and evolution of cable TV, became the spawning ground of the home TVRO (television receive only) industry, as characterized by the familiar "dish" ranging from 5 to 10 feet in diameter. During the mid-to late-1970s programmers such as HBO, Bravo, The Discovery Channel, WTBS and many others were delivered via satellite to cable networks. Before long electronic experimenters discovered how to inexpensively intercept these signals for private use. This development is evidenced today by the millions of home TVROs in operation today. This technology has spread around the world from America to Europe to Asia.

Today a consumer can choose TV entertainment from a variety of sources including the local cable television company, an over-the-air network station, or MMDS (multipoint multichannel distribution system) when available, as well as satellite broadcasts. The new technologies usher in an era of competition, one in which programming that is fairly available to all broadcasters can be offered via numerous alternative systems.

As digital TV and digital compression technologies evolve, all program distribution systems, including over-the-air, satellite, cable and MMDS are developing the capability of not only providing tens or hundreds of additional programs but also two-way links between viewer and sender known as interactive entertainment. We are entering a dramatic era in which the differences and distinctions between the television, telephone and computer are simply vanishing.

WHAT IS SATELLITE BROADCASTING?

Since the 1974 landmark broadcast of the Mohammed Ali heavyweight boxing match, the "Thrilla in Manilla" carried by HBO in the U.S., satellite TV has captured the imagination and linked people throughout the world. It has had dramatic effects. Wars and famine can be made so real and present that public pressure to eliminate them has mushroomed. Ethnic programming is now available to groups far from their homelands.

Satellite TV is playing a major role in the revolution occurring in man-made communications. Satellites and computers have altered everything from business structures to the way our children learn and entertain themselves. In particular, dishes that receive audio and video entertainment and information are having a dramatic, global effect from our largest cities to the most sparsely populated rural areas.

Satellite communication devices have uses that extend far beyond the familiar urban backyard, farm or mountain retreat. These technologies now allow business people in cities scattered all over the world to be linked together by television. A whole new breed of broadcasters is now able to relay hundreds of channels of entertainment and information via satellite from the largest cities to the most remote corners of our globe. "Narrowcasting" to "small-market" special interest groups such as gem enthusiasts or gold miners is now possible with the combination of "digital video compression" techniques and small-dish technologies.

The concept underlying modern satellite television is simple (see Figure 1-1). Television signals are focused and "uplinked" by a large dish, typically larger than 15 feet (5 meters) in diameter, to an orbiting satellite. This spacecraft lies in the geosynchronous orbit at 22,300 miles (35,786 kilometers) above the equator. In this location its rotational speed precisely matches that of the earth below. It therefore appears stationary to an observer on earth below. This eliminates the need for "tracking" so that fixed receive dishes can be used. The communication satellite receives the uplinked signal, lowers its frequency and then beams it to dishes within its "footprint" below. A receive dish and associated electronics detect and process the downlinked signal.

The principal advantage of satellite broadcasting lies in its broad coverage. Microwave signals carrying television, telephone or computer messages can be received at any place within its footprint.

In recent years large "C-band" satellite dishes have become a common sight throughout the world. These receive signals in the 3.7 to 4.2 GHz (gigahertz or billion cycles per second) range. The more common large-dishes are now being joined by smaller-dish Ku-band systems. These typically feature dishes from about 18 inches (45 cm) to perhaps 3 feet (1 meter) in diameter and operate in a portion of the higher frequency "Ku-band" that spans 10.95 to 12.75 GHz.

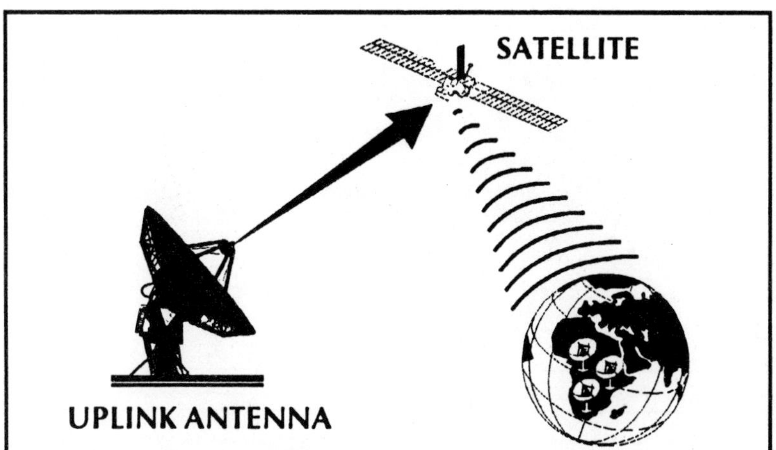

Figure 2-1. The Satellite Link. *Signals are transmitted from an uplink site to a satellite in the geostationary orbit. They are received, amplified, lowered in frequency and fed into television sets in individual homes or in small to large-scale cable systems. (Courtesy of MultiChoice)*

CHAPTER 2

SATELLITE COMMUNICATION FUNDAMENTALS

A satellite link is used to relay radio waves that carry some form of communications, video, data or audio, from one or more uplinks via a satellite to an unlimited number of receive stations. The fundamental operational principles of this link are explored in this chapter.

THE SATELLITE CIRCUIT

The concepts underlying satellite broadcasting, as first stated in Arthur C. Clarke's ground-breaking article in the October 1945 edition of *Wireless World*, are straightforward: signals are beamed into space by an "uplink" antenna, received by an orbiting satellite, electronically processed, broadcast back to earth by a "downlink" antenna and detected by a receive dish and associated electronics. The receive station or TVRO (television receive-only) can be situated anywhere within the satellite's "footprint." The overwhelming strength of satellite broadcasting lies in its ability to reach any number of customers within the downlink "footprint" regardless of their geographic location, without the need for any physical connection.

Nearly all communication satellites destined for commercial use are positioned or "parked" in the "Clarke" orbit, also known as the "geostationary" arc. The Clarke orbit lies in the equatorial plane of the Earth at 22,247 miles (35,786 kilometers) directly above the equator of the earth. This orbit about the earth is unique because in this circle the velocity of a spacecraft matches that of the rotational velocity of the surface of the earth below. Therefore, to an observer on the surface of the earth each satellite appears in one stationary orbital slot in the sky above. This allows a fixed receive dish to be permanently aimed towards a target satellite (see Figure 2-1).

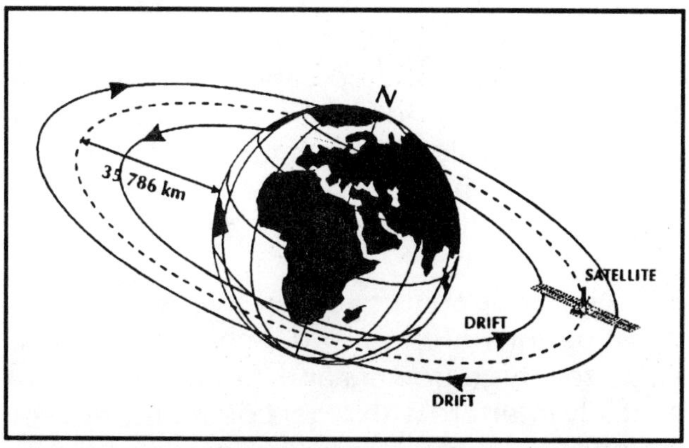

Figure 2-1. The Geostationary Orbit. *Satellites in higher or lower circular orbits in the equatorial plane of the earth rotate more slowly or quickly, respectively, than those in the geostationary orbit at 22,247 mile (35,786 kilometers). Only a spacecraft in the geostationary orbit appears to stay in a fixed position relative to an observer on the surface of the earth below. (Courtesy of MultiChoice)*

In contrast, pioneer communication satellites were placed into lower, more complex elliptical orbits because sufficiently powerful launching vehicles were not available to lift them into the distant geostationary arc. Thus, both transmitting and receiving antennas had to track the spacecraft. Telstar, one of the first communication satellites launched by the United States in 1962, had to be targeted by very costly and bulky mobile equipment that was mounted on rails. It relayed communications between Europe and the U.S. during a 36 minute window while it was visible from both sides of the Atlantic. Today some communication satellites are still boosted into various types of elliptical orbits. These non-geosynchronous systems include a few military and experimental spacecraft, satellites used in the global positioning (GPS) network as well as the Soviet Molniya television broadcast satellites.

Non-geostationary satellites are making a comeback in the mid-1990s with the launch of new low earth orbit (LEO), global communication systems such as Motorola's Iridium. These could be the subject of another book.

The Uplink

Uplinks are relatively complex systems designed to relay a beam of microwaves to a pinpoint target in space. In principle, the operation of an uplink antenna is similar to that of a car headlight which has a small but powerful light bulb at its center and a parabolic reflector to direct the illumination to the road in front. Microwave transmissions having powers of hundreds of watts are reflected from the antenna surface and beamed into outer space (see Figure 2-2).

Uplink operators are carefully regulated in most nations of the world. Their signals have relatively high powers and can therefore potentially interfere with other communicators or harm people and other living organisms. Furthermore, each uplink antenna relays messages which can be detected at millions of receive sites. So an improperly aimed uplink could easily disturb communications relayed by an adjacent satellite, thus interfering with thousands or millions of receiving stations.

Uplinks are increasingly being used in many segments of the business community. Today countless television and radio stations, telephone systems and data networks are reliant on satellite links. Numerous businesses with multiple locations have established privately owned satellite networks for managing internal communications. This transformation of our communication system has now been speeded up by the introduction of Ku-band systems because smaller and less expensive receive dishes can be used compared to those employed in conjunction with the lower frequency C-band technology.

Figure 2-2. Two Uplink Dishes. *These 9.1 meter (30 foot) uplink dishes at MultiChoice headquarters in Johannesburg, South Africa, relay signals to two different communication satellites. (Courtesy MultiChoice)*

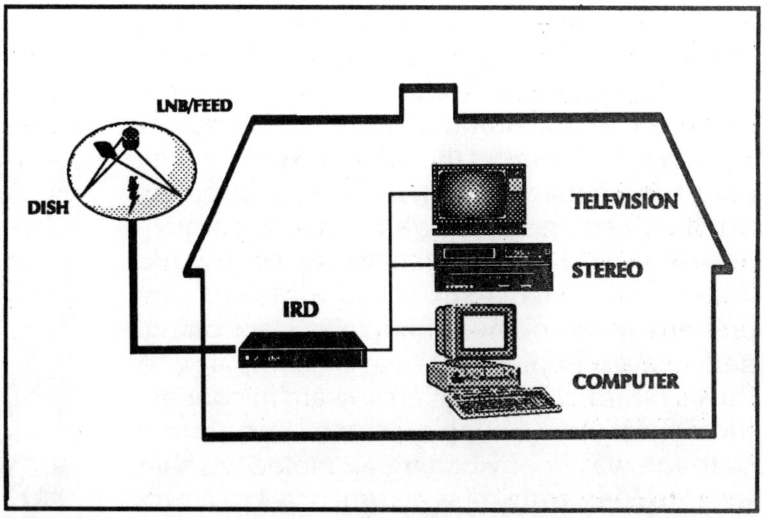

Figure 2-3. The TVRO Receive Site. *Satellite broadcasts are intercepted by a dish, essentially a large signal reflector, focused to a feed and amplifier (LNB) and then relayed by cable to an indoors satellite receiver. The receiver output can be connected to a television, a stereo or a computer. A typical 8 foot dish operating in the middle of the C-band (3.95 GHz) concentrates the signals by a factor of about 6,000. An LNB increases the power of the signals by an additional factor of more than 100,000. (Courtesy of Multi-Choice)*

A fixed uplink can be fed by fiber optic or coaxial cable links from broadcast studios or from other sources such as telephone switching centers or mainframe computers. Many television stations also relay signals by conventional, over-the-air methods to distant uplinks and subsequently to communication satellites for rebroadcast. Organizations needing uplink facilities on an occasional basis for teleconferencing, one-time events such as concerts or sporting matches or for news gathering activities have another option, that of "taking Mohammed to the mountain," i.e. of bringing the uplink equipment directly to the action. This is a viable economical alternative to over-the-air broadcasting to a fixed uplink site, particularly over Ku-band satellite links. Portable uplinks complete with all equipment and operating crews are now a mainstay of news gathering activities.

The Communication Satellite

Geosynchronous broadcast satellites receive the uplinked signal, lower the frequency of the transmission and then beam it to any chosen geographic area on the surface of the earth below. Downlink antennas can target up to 40% of the earth's surface with global beams, can broadcast to selected countries or continents via zone beams, or can pinpoint smaller areas with spot beams. Thus, for example, many American broadcast satellites blanket the continental United States with one beam and direct a separate beam to the Hawaiian islands. Similarly, PanAmSat-4 positioned over the Indian ocean covers the African continent with one C-band beam and relays a more localized Ku-band spot beam to southern Africa. Spacecraft that relay higher frequency Ku-band beams have the attractive capability of easily downlinking more localized spot beams than their C-band counterparts. This is because dishes operating at higher frequencies can focus signals into a more narrow beam and have a higher "gain". European satellites such as Astra or the Eutelsat series transmit three or more beams that cover various regions.

The Receive Station

A system for receiving satellite signals typically uses a parabolic reflector, the dish, designed to collect and concentrate as much of the very weak downlinked signal as possible by reflection to its focus (see Figure 2-3). The feed, located precisely at this focus, channels microwave signals reflected and concentrated by the dish into the first electronic component, the low noise block converter or LNB. The signal is downconverted or lowered in frequency in the LNB and then cabled indoors to the satellite receiver where it is processed into a form understandable by a television, stereo or computer. A receive station is essentially an uplink operating in reverse.

RADIO WAVES and COMMUNICATIONS

The transmission of extremely low power radio waves underlies the mechanism by which entertainment and information are relayed via satellite. The agents by which radios, conventional TVs and other man-made devices work, radio waves are a subset of a more comprehensive phenomena known as electromagnetic waves. Radio waves can be compared to the disturbances that travel outward in concentric circles when a pebble is tossed into a pond or to the waves of vibrating air molecules that we know as sound (see Figure 2-4). While sound travels at a plodding 760 miles per hour (1223 kilometers per hour) at sea level, radio waves and all other electromagnetic waves travel 15,000 times faster at the speed of light, 186,000 miles per second (299,339 kilometers per second). At this rate, a signal travels from an uplink antenna to a satellite and back again to earth in about 0.4 second, or over-the-air from a television station or along a cable network to a customer virtually instantaneously.

Figure 2-4. Wave Motion in a Pond. *Radio waves can be visualized by considering the wave motion in a pond caused when a pebble disturbs the water.*

Signal Frequency

The frequency of a radio wave or electrical signal transmitted by wire is the number of vibrations that occur every second as the energy passes any point in space (see Figure 2-5). Just as the frequency of sound vibrations determines whether a musical note is either a soprano or a bass, so the frequency of a radio wave determines whether it can be used to transmit regular AM radio broadcasts or satellite television broadcasts. Microwaves, a relatively higher frequency radio wave, have frequencies in excess of one billion cycles per second (known as one gigaHertz and abbreviated 1 GHz). By comparison, the electricity from a wall outlet has a frequency of 60 Hertz in America or 50 Hertz in Europe. This means that the voltage changes from positive to negative 50 times each second.

Many seemingly different phenomena encountered in nature including light, X-rays, infrared heat rays, microwaves used for both cooking and communicating and gamma rays from the cosmos are electromagnetic waves. The only difference among them is their frequency. Since all electromagnetic waves travel at the same speed, the speed of light, as the frequency increases the wavelength must decrease. For example, the wavelength of visible light is comparable to the dimensions of atoms

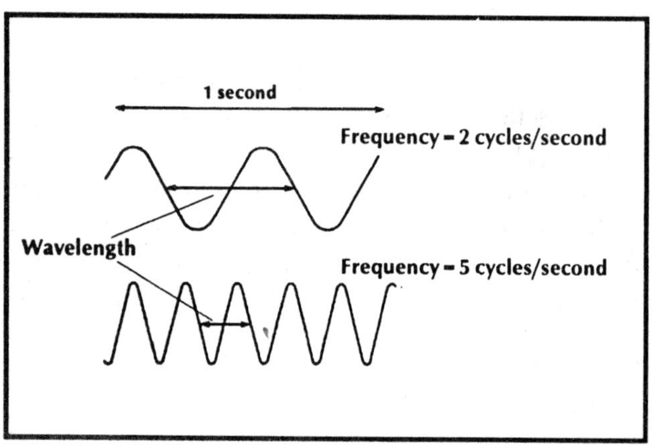

Figure 2-5. Frequency. *All electromagnetic waves are defined by how many vibrations occur each second, the frequency. Higher frequency waves have shorter wavelengths. For example, a 4 GHz microwave signal has a wavelength of 7.5 cm..*

and molecules, and their frequencies are many billions of billions of cycles per second. In contrast, microwaves have lower frequencies of one to fifty gigaHertz and wavelengths ranging from one foot to fractions of an inch (30 centimeters to a few millimeters). Radio waves have wavelengths which can be kilometers long and frequencies in the millions of cycles per second (megaHertz, abbreviated MHz) range.

Phase

The simplest form of electromagnetic or electrical wave is a pure sine wave. For example, the line power from a wall outlet is either a 60-cycle or 50-cycle per second sine wave, depending upon the convention adapted in any particular country. The phase of a sine wave is measured relative to another reference wave or starting point (see Figure 2-6). Not knowing the reference point would be comparable to attempting to time a 100 meter dash without knowing when it began.

Each full vibration encompassing positive and negative voltages around the baseline is considered to occupy 360 degrees which equals the number of degrees in one complete circle. Therefore, for example, if the peaks of two sine waves are one quarter of a wavelength apart, they have a 90 degree phase difference.

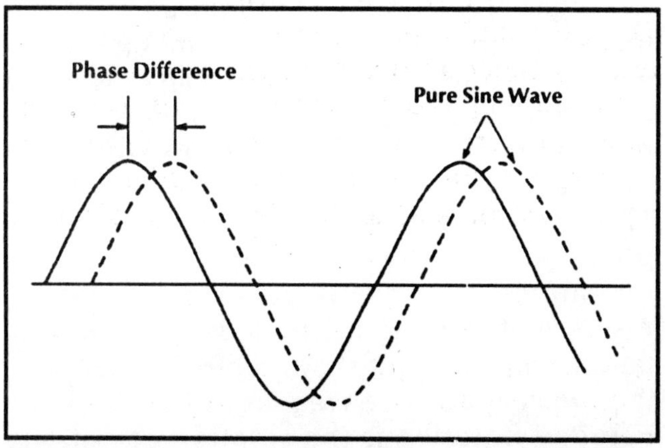

Figure 2-6. Phase. *Phase defines the relationship between two signals. The signals shown here are approximately one eighth of a wavelength or 45° (one eighth of 360°) out-of-phase with each other. One wave is said to lead and the other to lag by 45°.*

Transmission Power

The power or power density of radio waves is commonly measured in watts or watts per square meter, respectively. For example, 10 watts/meter^2 means that the power passing through each square meter is ten watts. Satellite television broadcasts are typically received by a dish at power densities of less than 10^{-17} watts per square meter!

In contrast, the power of a signal relayed over a conductor is commonly measured in watts. An electronic circuit might, for example, transform a 1 milliwatt signal into a more powerful one of 10 watts.

Wave Polarity

Electromagnetic waves can be linearly or and circularly polarized. With vertically and horizontally polarized signals, a form of linear polarization, the electric and magnetic fields of the signal (that are perpendicular to each other) remain in the same planes in which they were originally transmitted (see Figure 2-7a and b). To visualize linear polarization, imagine a car driving along a highway. It can travel to the same destination by following a curving road along flat ground or by following a straight road over rolling hills. Horizontally polarized waves vibrate in a horizontal plane like the car traveling along a curving road. Vertically polarized waves in a vertical plane.

Feeds that are designed to receive linearly polarized signals have to be properly oriented. Skew is the term used to describe how the probe in a feed must be rotated to match the plane of polarity of the incoming satellite signal.

Until recently, most domestic satellites used the linear polarization format. In contrast, Intelsat spacecraft have often relayed C-band circularly polarized signals. However, circular polarization is now used on the new American DirecTv/USSB DBS broadcasts as well as on Intelsat telephony links. Soviet C-Band transmissions as well as the French TDF satellite also use circular polarization.

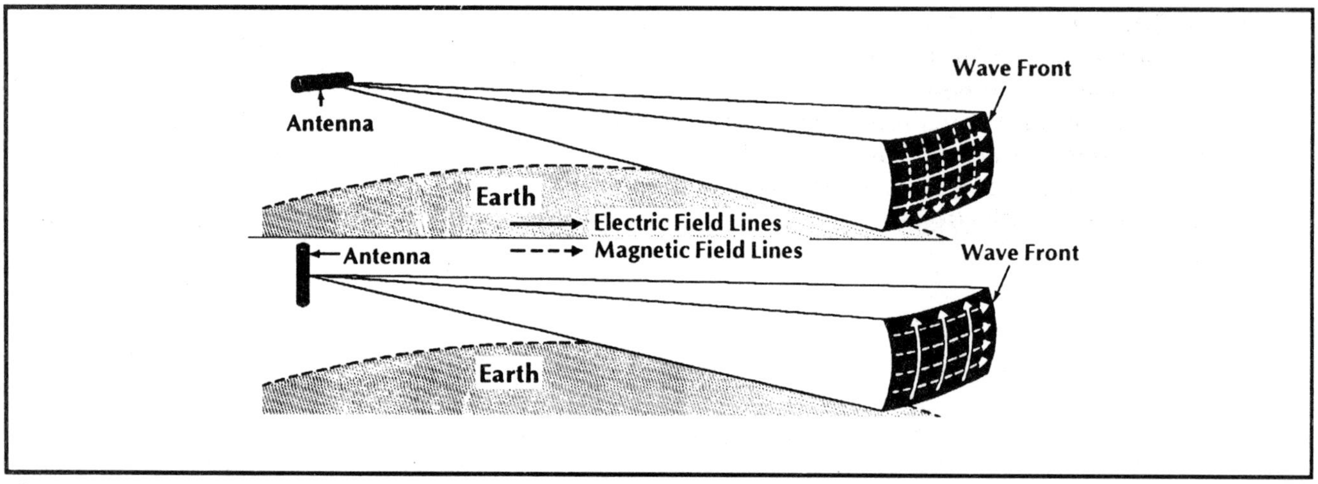

Figure 2-7a. Generating Polarized Waves. *A vertically or horizontally oriented antenna generates electromagnetic waves that are vertically and horizontally polarized, respectively. The direction of polarization is defined by the electric field (solid lines). The magnetic field lines (dotted) are always at right angles to the electric field.*

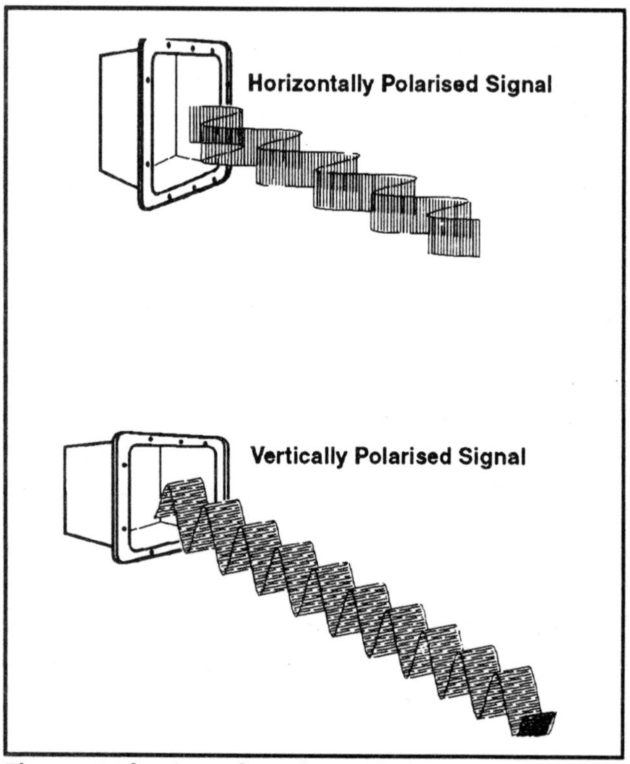

Figure 2-7b. Linearly Polarized Waves.
Polarization is the orientation of the electric field radiated from the transmitting antenna. When oriented parallel to the ground, it is defined as horizontal polarization. When the direction of the radiated electric field is perpendicular to the ground it is defined as vertical polarization. The reason for using horizontal and vertical polarization on the same satellite is known as frequency re-use. This allows twice the number of channels to be transmitted in a given bandwidth, one vertical and one horizontal.

In circularly polarized signals, the electrical and magnetic fields rotate in a circular screw-like motion as they travel through space, a motion similar to a spiral (see Figure 2-8). The direction of rotation determines the sense of circular polarization. A signal rotating in a right-hand direction as viewed from the satellite is right-hand circularly polarized (RHCP); a signal rotating in the left-hand direction is left-hand circularly polarized (LHCP).

Circular polarization was transmitted by pioneer satellites because the need for skew adjustment was eliminated. These spacecraft were generally not in geostationary orbit so that both the receiving and transmitting antennas had to track the satellite. If a linearly polarized signal had been transmitted, both the transmitting and the receiving antenna would have required a continual skewing adjustment. This need would have been further compounded since the scattering of the signal by rain affects the linearly polarity orientation. Also, with the transient problem of Faraday Rotation, some form of skew control would have theoretically also been required on the satellite. Faraday Rotation is the rotation of polarity caused by the Earth's magnetic field and/or magnetic storms when a signal travels through the atmosphere. This transparency to Faraday rotation makes circular polarization eminently suitable for transmitting voice and data telephony via satellite.

Figure 2-8. Circular Polarization. *Circular polarization can also be used for satellite transmissions. Instead of positioning the microwave energy in either a vertical or horizontal plane, circularly polarized signals are transmitted in a spiral pattern like a coiled spring. The direction of rotation of the electric field vibrations can follow either a clockwise or counterclockwise motion.*

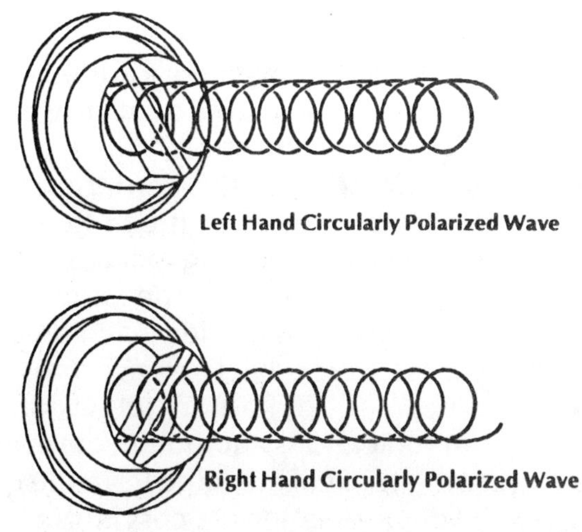

COMMUNICATION FUNDAMENTALS

The same principles underlie all forms of man-made communication. The first step is creating and coding the message. Next, this information must be modulated or added onto the medium, the carrier wave, that relays the signal. At the receiving end, the signal is demodulated, a process whereby the original information is extracted from the carrier wave. The amount of information that can be relayed is determined by the "bandwidth" used. The signal can be amplified or attenuated so that its power is increased or decreased. In addition, communication engineers must always account for unwanted signals or noise which are always present and hinder perfect communication.

Coding the Message

Any message, whether it be the image and voice of an entertainer or details of stock-market transactions, must first be changed into a form that can be transmitted by radio waves. Analog coding methods mimic the pattern of a message by changes in electrical voltages. For example, a voice can be changed into an analog signal by a microphone that creates a voltage pattern determined by the loudness and frequency of the sound. The louder the sound, the higher the voltage. The higher the sound frequency, the more rapid are the changes that occur in this voltage.

In contrast, a digital coding method uses just the numbers 0 and 1, represented by any two different voltage levels, to convey all information about these signal powers and frequencies. Thus, a voice could be transformed from analog to digital form if the analog voltage waveform representing this voice were "sampled" or measured at regular intervals in time and each of these voltages levels were then expressed by a binary number, a sequence of 0s and 1s. A photograph could similarly be described in digital form. Each "pixel" or dot on the picture would be coded so that binary numbers would impart information both about the location of the dots composing the picture as well as the brightness and color of these dots. Computers are exclusively digital devices.

Satellite uplinks can relay either digital or analog forms of the same message. Analog-to-digital (A/D) or digital-to-analog (D/A) converters, in reality mass-produced voltage-to-number or number-to-voltage microchips, translate between these two languages or forms of representation. For example, compact disc players convert digital streams detected by a laser from the surface of a revolving plastic disc via a D/A chip into analog waveforms that can drive loudspeakers. Until recently, nearly all TV broadcasts have been in analog form for transmission over satellite circuits. However, we are now plunging into the new age of digitally compressed broadcasts of video, voice and data.

Modulation

Analog or digital signals are added onto or impressed upon radio waves or microwaves by a process called "modulation." Once the message is modulated onto the carrier wave of a predetermined frequency it can be relayed from a sending to a receiving location whether it is via a satellite link, over-the-air or along a cable route. The term carrier wave describes precisely what the modulated wave does; it carries the information. Radios, televisions and other communications equipment demodulate or extract the original message from the carrier wave. The demodulation process is also known as detection.

Analog Modulation

Traditional analog modulation techniques, developed in the early days of radio communications, are amplitude modulation (AM) and frequency modulation (FM). These are used extensively in terrestrial services where bandwidth is limited, for example, in AM and FM radio broadcasts. In amplitude modulation, the power of a carrier wave is varied in accordance with the voltage level of the message being relayed, while in frequency modulation the frequency of the carrier wave is varied (see Figure 2-9).

Each of these types of modulation has distinct advantages and disadvantages. On one hand, AM messages must have relatively high powers to be capable of traveling long distances without being weakened too severely by atmospheric disturbances and other forms of noise. They are also more prone to picking up static than are FM messages. On the other hand, FM signals must use a substantially wider range of frequencies than AM messages would to carry the same amount of information but have the advantage of needing lower power for successful, long distance transmission.

Analog satellite services are always frequency modulated. As signals cover the long distances between uplinks, satellites and earth receive stations, their power becomes so low that standard AM transmissions would be unusable. Also, because extremely high frequencies are used in satellite communication, the large bandwidth required by FM broadcasts is available. In contrast, terrestrial television broadcasts as well as cable network signals use an AM video signal in order to conserve space in their limited frequency spectrum.

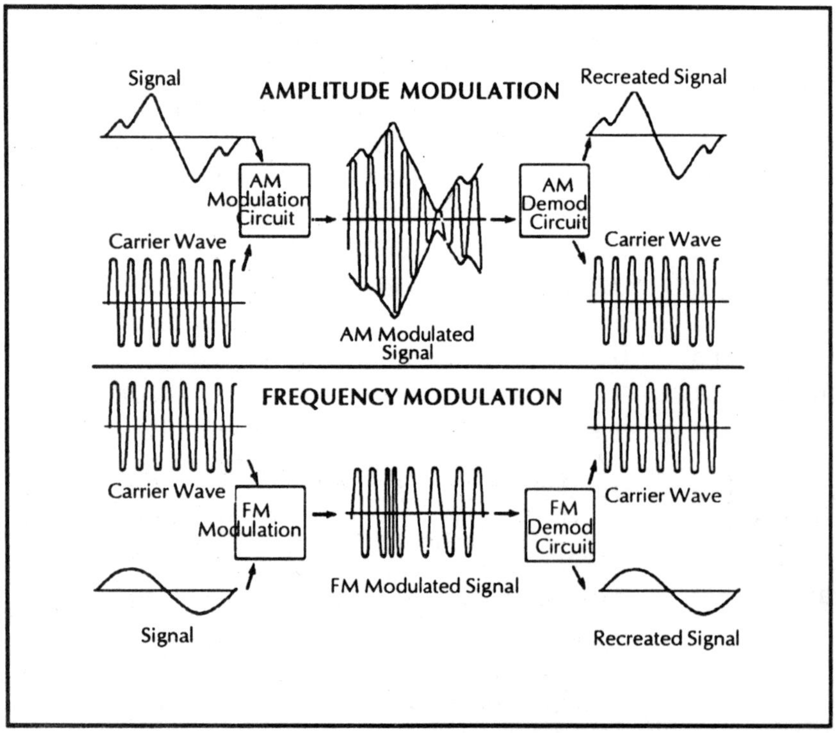

Figure 2-9. AM and FM Modulation. *Two formats used to impress an analog audio, video or data signal onto a carrier wave are amplitude and frequency modulation. The modulated signal, the carrier wave, can then be relayed by cables, over-the-air, by satellite or any other transmission media to its final destination.*

I Modulation

Digital modulation techniques, derived from the original Morse code, are geared to the language used by computers, wherein a pattern of 0s and 1s relay all information. Digital transmissions are quite tolerant to noise because a receiver must only determine whether the signal is either "on" or "off." This tolerance can be improved by using forward error correction (FEC) techniques that reconstruct bits that are missing or are in error at the receive site.

The simplest method to modulate a carrier wave is to switch it on and off. For example, Morse code can be relayed as a series of dots and dashes by turning the carrier wave on and off. In one of the simpler forms of digital modulation, binary phase shift keyed (BPSK) modulation, a shift in the phase of the carrier wave determines whether a 0 or 1 is transmitted (see Figure 2-10a). In QPSK (quadrature phase shift keyed modulation) two BPSK modulated bit streams are combined 90 degrees out of phase with each other. As shown in figure 2-10b this results in four states, 00, 01, 10 or 11, being transmitted in each sine wave cycle. While more information is thus carried, this method is somewhat more sensitive to interference from noise than BPSK.

Digital satellite links are characterized by the number of bits transmitted per second. For example, a typical broadcast TV signal requires in excess of 200 Mbs (200 million bits per second) for high quality reproduction. Telephone channels have traditionally required circuits that transmit 64 kilobits per second.

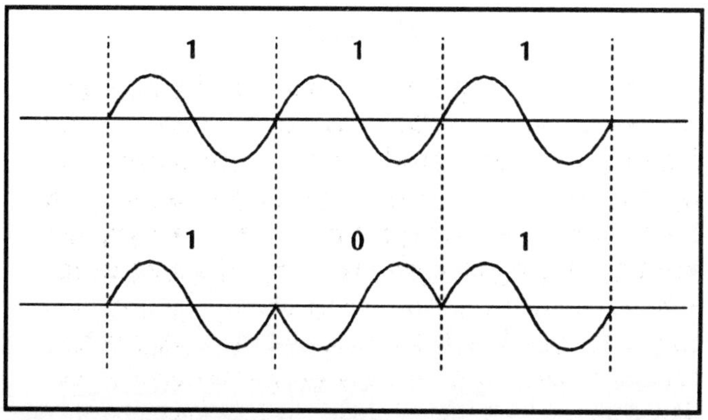

Figure 2-10a. BPSK Modulation. *BPSK modulation is one of the simplest forms of phase shifting digital modulation. The phase of the carrier wave is shifted once each complete cycle. Each shift in phase determines whether or not a 1 or a 0 is transmitted. (Courtesy of MultiChoice)*

Digital communication systems have a number of distinct advantages in addition to their noise tolerance. Computers can easily manipulate the video, audio or data bit streams to create secure encrypted transmissions. Compression techniques can also be used to squeeze more information into each satellite channel. Today methods have been developed to use as low as 2 to 3 megabits per second to transmit a full motion color picture or 8 kilobits per second for a two-way voice circuit. In addition, forward error correction methods can be used to reconstruct bit streams that have been corrupted during transmission and can therefore improve the quality of communications. (More details on signal compression as well as error correction techniques are presented in Chapter 3).

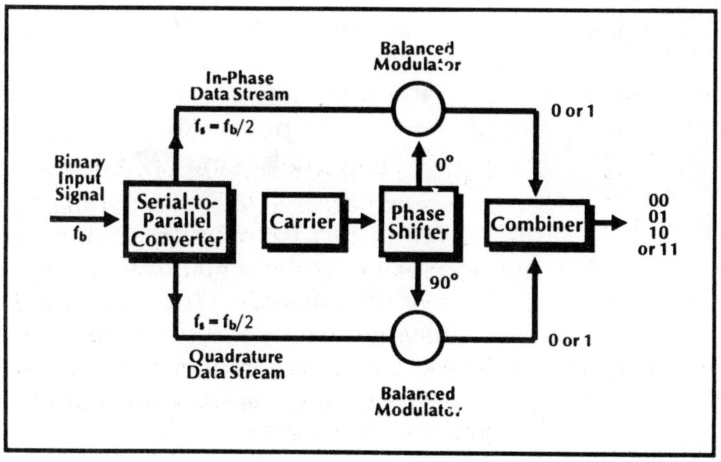

Figure 2-10b. BPSK and QPSK Modulation. *QPSK modulation is the most typical method used on satellite links. A QPSK modulated signal is created from two BPSK signals. As a result, the phase of the carrier wave is shifted four times each complete cycle. The net result is that four states, 00, 01, 10 and 11, can be relayed each full sine wave cycle.*

Bandwidth

Just as a large-diameter pipe can carry more water than a small one, so a signal spanning a wide band of frequencies can carry more information than can one using a narrow band. This range of frequencies is termed the "bandwidth." For example, TV channel 4 in North America is broadcast in the frequency range from 66 to 72 MHz and thus has an AM bandwidth of 6 MHz. When transmitted over satellite, this signal is spread over an FM band spanning up to 36 MHz.

Each type of communication medium requires a characteristic bandwidth. TV links need a substantially wider bandwidth than does radio or telephone because much more information is necessary to recreate a picture than music or a voice. To illustrate, a non-compressed PAL-I television signal broadcast by satellite typically requires a video bandwidth of 5.5 MHz. However, voice channels normally require a bandwidth of only 3 to 4 kHz for quality sound reproduction.

The bandwidth requirements in MHz or kHz as well as the modulation technique used determine the number of bits per second required for digital transmissions. "Better" modulation methods can relay higher bit rates in a given bandwidth.

Amplification and Attenuation

Man-made communications must often be amplified during their voyage from sender to receiver to keep the information intact because their power is usually weakened or attenuated. In the same fashion that a photograph is enlarged but not changed, a perfect amplifier would simply increase the voltage but not change the waveform of the original message (see Figure 2-11). All receiving devices amplify a signal before demodulation occurs. For example, messages beamed into space from an uplink antenna are weakened on their voyage to a satellite as the signal becomes spread out and is absorbed by water vapor, clouds and other materials. The purpose of a satellite receive dish is to collect and concentrate these weak signals much as a magnifying glass focuses light waves.

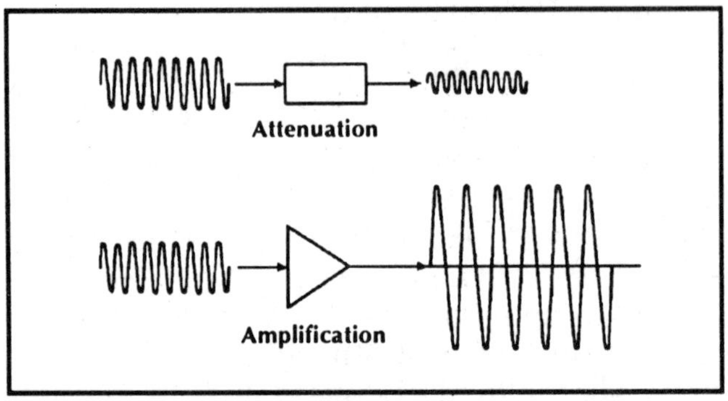

Figure 2-11. Amplification and Attenuation. *Amplification increases a signal's peak-to-peak voltage, while attenuation reduces its strength.*

Occasionally, a signal is intentionally attenuated. For example, a cable TV headend may deliver an excessively powerful signal to a feeder line that could over-drive connected TVs and cause distortion. Pads or line attenuators are then inserted to reduce the power. Some satellite receivers when installed very close to the receive dish might also be overdriven. Power-passing attenuators are then required to avoid distortion.

Noise

In an ideal communication system transmitted signals would be free of interference or noise. However, TV or radio broadcasts are never perfect but are "noisy." All transmissions are accompanied by some noise, which is present in all matter at temperatures above absolute zero, zero degress Kelvin (0K), the temperature at which all molecular motion ceases. There is no temperature colder than absolute zero! Kelvin degrees have the same magnitude as degrees Centigrade, they are just measured from a different starting point. Absolute zero, zero degrees Kelvin, equals minus 273.16 °C.

Noise is caused by the endless motion of the molecules that compose all matter. These small, vibrating charged particles generate radio waves that can mask the organized signal sent by man-made devices. Noise from the environment becomes stronger as the temperature increases.

Satellite dishes intercept noise from the warm ground. The temperature on an average 17°C day equals 290K. At the frequencies employed in Ku-band satellite relays, outer space also generates noise at about 3K, as a result of the "big bang." So, for example, a satellite downlink antenna pointed directly towards earth detects an additional 290K of noise; a receive dish on earth pointed directly into space sees on the order of 50K of noise, about 3K from space and the rest leaking into the dish from the surrounding warm ground as well as surrounding buildings, trees and other objects. Receive systems detect more of this environmental noise as the signal bandwidth increases. Additional noise is generated by internal heat in power supplies, amplifiers, receivers and other electronics.

Man-made signals not intended for the receive station are also considered a type of noise called interference. Terrestrial interference, known as TI, results when communications other than those from the targeted satellite broadcaster are inadvertently received by a TVRO. This type of interference, while of an annoying familiarity in receiving C-band broadcasts because of terrestrial line-of-sight microwave relays, is less common in Ku-band transmissions. However, both C and Ku-band systems are susceptible to interference from adjacent satellites operating in similar frequency ranges as well as from signals of opposite polarity. Cross-polarity signals, also operate in similar frequency ranges.

Analog Link Quality and Noise

The quality of all communication links is determined by the relative levels of carrier power versus noise power, the carrier-to-noise (C/N) ratio (see Figure 2-12). Random noise is always present in satellite communication systems and cannot be easily reduced and certainly cannot be avoided. However, the carrier power can be more easily controlled.

The C/N ratio defines analog transmission quality. For example, if a carrier signal of 1 watt (1000 milliwatts) power rides on 1 milliwatt of noise power, the C/N ratio equals 1000 in that bandwidth. In this case, picture quality would surpass that of a carrier of 2 watts power that is detected with 4 milliwatts of noise, a C/N ratio of 500.

Digital Signal Quality and Noise

Noise also affects the quality of a digital signal. The parameter used to measure this quality is the ratio of the energy density per bit in the carrier wave to the noise power in 1 Hertz of bandwidth. The C/N ratio is then derived from this quantity. This process is briefly outlined in more detail in Chapter 13.

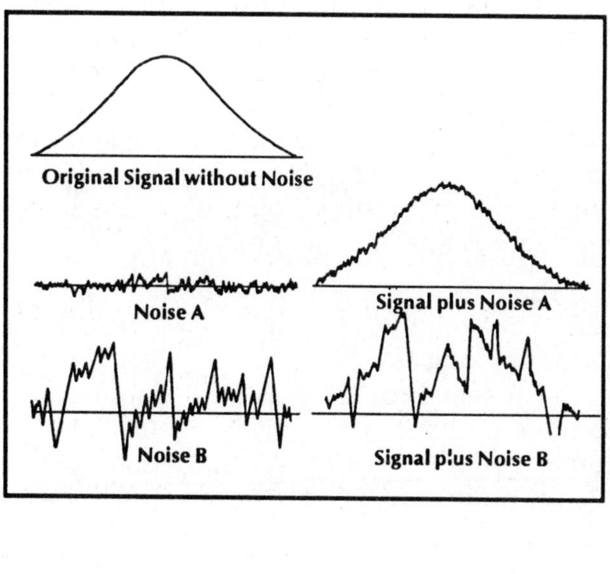

Figure 2-12. Noise. *The ratio of carrier power to noise power, the carrier-to-noise (C/N) ratio, defines the quality of a communication link. Excessive amounts of noise can completely garble an audio or video transmission so as to make it unintelligible.*

The Decibel Notation

The decibel scale, named in honor of Alexander Graham Bell, was created to describe the enormous changes in power that occur at various stages of the communication chain by relatively small numbers. This was necessary because components like amplifiers and dishes often increase power levels many hundreds of thousands of times. Constantly writing out such numbers could become very cumbersome. The equation to calculate decibel changes is presented in Appendix A. The most useful fact to remember is that small changes in decibels mean much larger changes in power levels. Thus, for example, a change of 3 dB means a doubling of power; a change of 30 dB means an increase of power by a factor of 1,000. Note that dBw means decibels relative to one watt; dBm means decibels relative to one milliwatt. So, by comparison with Table 2-1, 3 dBw means 2 watts and 0 dBm means 1 milliwatt.

TABLE 2-1. CHANGES in POWER EXPRESSED in dB

dB FACTOR	MULTIPLIER
0	1
1	1.26
2	1.58
3	2.00
5	3.16
10	10
15	31.6
20	100
30	1,000

This table provides a simple method to determine the multiplier for any dB factor. For example, 7 dB equals 5 dB (3.16 multiplier) plus 2 dB (1.58 multiplier). The 7 dB multiplier of 4.99 is found by multiplying 3.16 by 1.58. Similarly, the 11 dB multiplier of 12.6 is calculated by multiplying the 10 dB by the 1 dB multiplier.

WHY MICROWAVES?

Microwaves have been used in satellite communication for a number of well-defined reasons. Higher frequency electromagnetic waves can relay larger quantities of information because as frequency increases more bandwidth is available. Microwaves can relay more information per satellite than lower frequency signals and thus can pay off the expensive investment in satellite launching, operation and maintenance more quickly.

Uplink antennas must aim a highly directional beam towards an extremely small target in space. Physics dictates that electromagnetic waves can be better focused by an antenna that is substantially larger than the wavelength of radiation it is managing. For example, sending a directional beam of an AM radio signal having a 100 meter wavelength would require an extremely large, cumbersome and expensive antenna. Since 6 GHz microwaves have wavelengths of approximately 2 inches (5 cm), a 15-foot uplink dish can aim most of its radiation into a very narrow beam, and relatively low power can be used.

Microwave transmissions to satellites or between line-of-sight relay stations on earth are not as susceptible to noise from atmospheric disturbances as are lower frequency transmissions. For example, several times each year, for periods as long as two or three days, short-wave radio is useless for long-distance communication because sun-spot activity disturbs the required reflection of these relatively low frequency radio waves by the upper atmosphere.

Microwaves can be used for satellite communication since they can pass through the upper atmosphere into outer space. Below frequencies of approximately 30 MHz, a radio wave is reflected back from the ionosphere layer in the atmosphere towards earth. Since microwave frequencies are far above the 30 MHz range, they easily pass through the ionosphere shield.

The microwave region of the electromagnetic spectrum was a virgin territory during the late 1950s and 1960s when frequencies were being allocated by the FCC and the International Telecommunications Union. Lower frequency space was already occupied by many different communication media and users.

As geosynchronous orbital space has become increasingly populated, progressively higher microwave frequencies have been allocated to satellite communications. Until the early 1980s most satellite broadcasters used C-band frequencies. Today, the Ku-band is extensively employed and higher frequency bands such as the Ka-band are being eyed by numerous potential users.

A technical difficulty does lie in this path since microwaves are depolarized and more strongly absorbed by water vapor in the atmosphere at higher frequencies. So higher power transmissions must be used to counteract this effect.

WORLDWIDE FREQUENCY ALLOCATIONS and DTH SATELLITE BROADCASTS

Nations around the world have worked together in allocating portions of the electromagnetic frequency spectrum to the various communication media and user groups. The International Telecommunication Union, the ITU, an agency of the United Nations, coordinates meetings every four years, known as World Radio Conferences (WRC), to plot the paths for future technologies and to develop standards for the worldwide communication system. Organizations such as the Federal Communications Commission (FCC) in the United States, the Independent Broadcast Authority (the IBA) in South Africa, the Japanese Ministry of Post and Telecommunications, the Independent Broadcast Authority (IBA) in Great Britain and the Department of Communications (DOC) in Australia have sent representatives to these WARCs. Their decisions affect other international organizations such as INTELSAT, the International Telecommunications Satellite Consortium and INMARSAT, the International Maritime Satellite Organization. The frequency spectrum was allocated up to only 30 MHz at the first World Administrative Radio Conference (WARC) in 1927. The latest WARC was held in the fall of 1995. Today, worldwide frequency allocations range up to 300 GHz.

The history of communication can be unfolded by an exploration of these frequency assignments. As technology has progressed, man has become capable of utilizing ever higher frequencies (see Table 2-2 and Figure 2-14). Wire transmissions were first relayed on relatively low frequency carriers because the pioneers were limited by the rather primitive electronics available at that time. Ironically, when radio waves above 1.5 MHz frequency were first produced, they were relegated to the "hams" because regulators could not see a use for this region of the spectrum. Ham radio operators have contributed numerous innovations since then. For example, the first home satellite receiving stations were built by hams. As technology developed, coaxial cable transmission, microwave relays and subsequently satellite communication were allocated successively higher frequencies.

TABLE 2-2. U.S. ALLOCATIONS of COMMUNICATIONS FREQUENCIES

Frequency	FCC User Assignment
3-30	Wire Service, CB, Ham
30-54	Mobile Radio
54-72	VHF TV Channels 2-4
72-76	Radio Services
76-88	VHF TV Channels 5 &6
88-108	FM Radio
108-120	Aeronautical
120-136	Aeronautical
136-144	Government
144-148	Amateur Radio
148-151	Radio Navigation
151-174	Land, Mobile, Maritime
174-216	VHF TV Channels 7-13
216-329	Government
329-420	UHF TV Channels
420-450	Ham Radio
450-470	Mobile Radio
470-806	UHV TV Channels 14-69

TABLE 2-3. MICROWAVE FREQUENCY BANDS

Band Name	Bandwidth (GHz)
L-band	1 - 2
S-band	2 - 4
C-band	4 -8
X-band	8 - 12
Ku-band	12 - 18
K-band	18 - 27
Ka-band	27 - 40
V-band	40 - 75
W-band	75 - 110
mm-band	110 - 300

Satellite broadcasts are transmitted at frequencies ranging from about 1 GHz to above 30 GHz. The designations for these bands are outlined in Table 2-3. Today most geostationary communication satellites operate in the C- and Ku-bands.

In spite of the rapid development in man-made forms of communication, a relatively small portion of the vast electromagnetic frequency spectrum is presently being used. However, in those ranges where frequency space is available, it is in heavy use and competition for allocated space can be fierce. As a result, innovative methods have been developed to "re-use" or to have more than one user simultaneously share the same portion of this spectrum. Re-use methods are outlined later in this book.

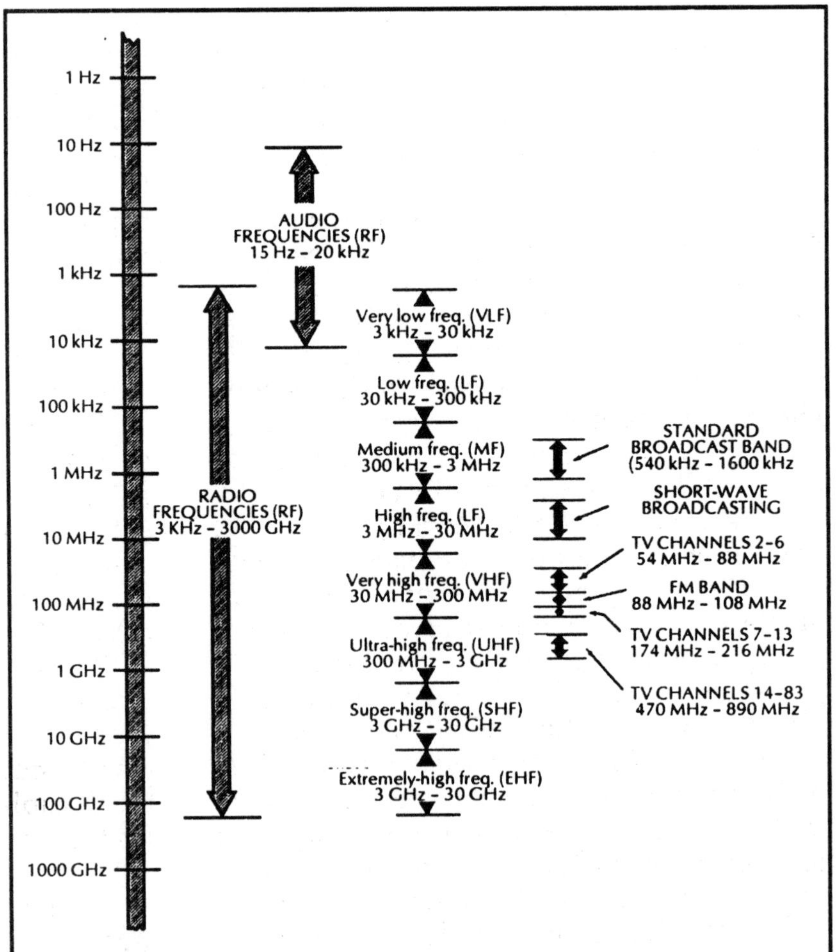

Figure 2-13. Allocation of Audio and Radio Frequencies. *The allocation of frequencies for conventional radio wave communications is illustrated here.*

CHAPTER 3

MPEG-2 DIGITAL VIDEO BROADCASTS (DVB)

Digital television has been a revolution anxiously waiting to happen since the early 1970s. While the basic methods have been well understood for years, affordable technology to handle the vast quantities of data and the rate at which this data had to be transmitted were simply not available. Nevertheless, high cost professional TV products such as standards converters and editing consoles that used digital methods were slowly introduced.

The light at the end of the digital tunnel began to brighten as engineers and researchers came to more fully understand how data compression methods could be used to reduce bit rates for a TV signal down to manageable proportions. However, it was this knowledge combined with parallel developments in a number of other areas that has finally brought us into the full light of day. Communication engineers have been hard at work perfecting advanced modulation schemes that can pack more data into a given amount of signal bandwidth. During the same period, powerful mass-produced microprocessors and memory chips that would have seemed incredible just ten years ago have been developed. The final element has been the creation of an international standard, the MPEG or Motion Picture Experts Group standard, that supports an environment in which low-cost compression chips can be mass produced. As a result, a transition from analog to digital communications is under way. The unification of communications, computers and display media into a "multimedia" industry is upon us.

How will this new technology affect the C-band home satellite TV industry? It is clear that in twenty years' time all broadcasting will be digital. However, in markets such as the United States where vast investments by both cable companies and private individuals in analog C- and Ku-band uplink and receive equipment have been made, the transition will be relatively slow. Other areas of the world, for example, such as Africa where MultiChoice has recently launched a compressed digital video broadcast service will likely take the market by force since there have been few other options.

In North America already-launched or soon-to-be-launched broadcasters such as DirecTV, USSB, PrimeStar Partners, EchoStar and AlphaStar are offering programming via small-dish, Ku-band systems. However, these still do not provide the same wide variety of programming as the "full-view" C-band analog systems. Furthermore, as more C-band broadcasters decide to make the transition to digital many will likely simulcast on both analog and digital. The new GI Digicipher II module expected to replace the VideoCipher II+ in analog C-band home satellite receivers will be backwards compatible. It will process both VideoCipher II+ encrypted analog signals as well as digital CVB.

The concepts underlying MPEG-2 digital video broadcasts are examined in this Chapter. This includes an outline of digital compression methods, the MPEG-2 standard and the role of error correction and modulation methods

DIGITAL TV METHODS

Digital television broadcasts have two key advantages. They are inherently less noisy to transmit since a receiver must simply determine whether or not a bit, a 0 or a 1, is present as opposed to discriminating between small differences in signal voltage as is required in analog methods. A satellite link using a digital decoder/receiver can therefore be relayed at up to 4 dB below the power level of an analog signal to achieve a similar quality picture! In addition, digital signals are more amenable than analog to the use of very secure and reasonably priced encryption methods.

There are other advantages in using digital methods to process and transmit TV signals. Just a few bytes of information can be used for vertical and horizontal synchronization. In contrast, an analog signal repeats sync information each line and frame. Large chunks of time thus need not be allocated. Furthermore, the luminance and the color signals can be relayed independently thus avoiding picture distortions that can occur when these signals are multiplexed together as in analog broadcasts.

However, before the advent of compression methods in the late 1980s, digital television had a serious limitation. It required an enormous transmission bandwidth a whopping 216 Mbits/second (million bits per second). But data transfer rates have technical limitations. For example, the fastest computer CD-ROM available today transfers data at just 4.5 Mbits/second while a typical voice-grade channel on the switched telephone network transfers data at a maximum of 20 kilobits/second. A device for the consumer electronics market that would read an incoming 216 Mbits/second data rate would be prohibitively expensive, if not impossible with existing technology.

Furthermore, using a relatively simple form of modulation a 216 Mbits/second rate requires a 100 MHz wide bandwidth, nearly four times the typical 27 MHz that is used to transmit one analog video signal via satellite. Digital satellite links were certainly not economical without the development of digital compression techniques.

WHAT IS COMPRESSED DIGITAL VIDEO?

Compression methods are used to reduce the data rate in a digital video signal to a fraction of its original value. Once this has been accomplished with modern techniques, transmission of DVB signals becomes not only feasible but quite attractive. This fact lies at the heart of the birth of, for example, the successful DIRECTV/USSB and PrimeStar services in the United States.

The reason for using any type of data compression is that fewer resources are needed to transmit or store a compressed signal. Power, bandwidth, telephone budgets, computer memory, diskette space or other resources, can be "stretched" by compressing data. In the case of expensive television transmission and relay facilities, as with costly satellites positioned in a limited number of orbital slots, using these resources better becomes very advantageous.

Video Compression Methods

Data compression is achieved by pre-arranging data handling methods between the "transmitter" and "receiver" of the data. When a compressed message is received, these methods are used to restore the signal to its original form, i.e. the signal is decompressed. A classic example is the "thumbs-up or thumbs-down" signal in ancient Rome that told whether to save or put gladiator combatants to death.

Morse Code first demonstrated the usefulness of data compression by speeding up telegraph message transmission and so increasing the capacity and revenues of the telegraph system. In this code, a form of "Huffman Coding", the most often used letters, such as "e", are assigned the shortest code words ("dot"), and the unlikely one, such as "q", as assigned the longest code words (dash, dash, dot, dash).

Pre-Processing

The first step in compressing a video signal, pre-processing, is both an art and a science that stems from advances in the understanding of the human visual process and its limitations. Information that is the most difficult to encode but least important to visual quality is removed from the analog video signal.

Removing Spatial or Temporal Redundancy

After the video signal has been digitized, unnecessary or redundant information is removed. Repeated parts of the picture can fall within a single frame, as, for example, when the entire background consists of identical blue pixels. Thus instead of transmitting information to describe each "pixel" along the TV line, the information can be sent once with a message informing the decoder how often it is repeated on this line. Repetitions can also occur from frame to frame – a temporal or inter- frame redundancy – wherein the same background is repeated in frame after frame as, for example, in a "talking head" picture.

Intra- and inter-frame compression combined with replacing other forms of redundant information, such as the highly repetitive horizontal and vertical sync pulses used in an analog transmission, with simple codes can reduce the overall bit rate by dramatic amounts. In some video scenes the amount of redundant information removed can be as high as 98 percent, quite an astonishing amount!

Motion compensation prediction is another important compression method. During motion, a portion of any picture can be predicted from frame to frame. For example, a falling object appears almost identical in each frame; only its position changes. If a predictor block can be identified from a scan of previous frames, i.e. the falling object, then the transmitted digital video signal must contain information only about the position of the predictor block and any relatively small differences between the appearance of this block and the current block, i.e. any changes in color, shape or shadowing that occur in this block during its fall. In essence, a command can be sent which says: "move this block of pixels over by m and up by n." This saves having to re-transmit the entire pixel block. Since each frames occurs every 1/25th of second, the actual differences between frames are small.

DCT Transform Coding

Once redundant information has been removed and the signal has been expressed as a series of numbers, a mathematical formula known as the discrete cosine transformation (DCT) is applied. This method converts the digital signal into one that represents its frequency structure. The "transform coding" is, in essence, similar to the method used by a spectrum analyzer to display the frequency composition of a communication signal.

Transform coding has two advantages. First, the signal generally has most of its information concentrated in lower frequency components so not all the co-efficients must be relayed and those that are needed do not have to be relayed with perfect accuracy. Second, this process mimics the perception of errors in the human visual system, and permits the next quantization step to be tailored to the sensitivity of the viewer.

Quantization

Quantization is the process of selecting only a portion of the information, i.e. reducing the number of bits transmitted and thus the required data rate. If done properly, this does not significantly degrade visual quality but does dramatically reduce the bit rate. In essence, bits that fall within a set of values are replaced by a single value, compressing the range of many numbers to just a single one. The greater the amount of compression, i.e. the number of quantizing levels, the more accurately the original analog picture is represented and the higher the bit rate required.

The use of quantization is based on the fact that the human visual system can perceive only a limited number of brightness levels and can respond to varying signals that only fall below a threshold frequency value. Thus just 256 brightness levels, represented by a 8-bit word, need be relayed and rapidly changing video levels can be more coarsely quantized with little loss in perceived visual quality.

Variable Length or Entropy Coding

In the next step, variable length or entropy coding, the signal is scanned for short, frequently occurring digital words sequences which are then represented by a set of smaller words. This is a form of "Huffman Coding" that is similar to the method use in Morse Code.

Both quantization and entropy coding can be applied in varying degrees. The result is that a slow motion scene, as for example in a teleconference, may support relatively high levels of quantization and entropy coding. This translates into a higher level of compression.

Post-Processing

Some DVB designs incorporate post-processing at the end-user decoder. This is an attempt to identify the most annoying artifacts resulting from compressing the video signal and to reduce their noticeable effects. This technique is typically used only in situations were fairly high levels of distortion are present.

THE MPEG-2 STANDARD

The MPEG standard grew out of the development of video conferencing compression systems. When this technology had advanced to the point where high volume applications were possible, a clear standard was needed to insure compatibility between systems from various manufacturers as well as economies of scale and thus lower prices.

In the late 1980s The International Standards Organization began the process of drafting a series of worldwide standards that would underlie the use of digital compression in both videoconferencing and digital video applications. The Joint Photographic Experts Group (JPEG) leads the effort for still photos, graphics and X-rays, and the Motion Picture Experts Group (MPEG) for multimedia applications.

The MPEG group had four major goals:

- to produce flexible and generic world standards for video and audio coding
- to define features suitable for use in transmission and recording media
- to create procedures for evaluating systems
- to define a bit stream structure to enable coder and decoder designs

MPEG I, completed in 1988, set the stage for rapid growth in videoconferencing applications, the oldest and most evolved use of compressed digital video. It defined the system to provide VHS quality pictures at data rates up to 1.5 Mbit/second.

MPEG-2, created to be reverse compatible to MPEG-1, defines a fluid architecture to allow a true multi-standard environment where one system can output video and audio signals in NTSC, PAL, SECAM, RGB or any other video format so desired. It is a layered structure in which the outer layers provide the system-level functions, synchronization and multiplexing, necessary for using one or more data streams in the system. The inner layers define the image and audio compression mechanisms, many of which have been described above.

MODULATION, MULTIPLEXING and ERROR CORRECTION

Modulation Methods for Satellite and Terrestrial Links

Once a program data stream data has been compressed, one or more signals must be modulated onto a carrier wave. The typical goal in choosing the modulation system is to fit as much data onto the transponder as possible.

The downlink signal from a geostationary satellite has severe power limitations. This situation is aggravated when small receive dishes that collect less signal are used. However, a satellite transponder has a good deal of available bandwidth, typically 30 to 40 MHz. It is thus power but not bandwidth limited. The modulation method chosen for relaying DVB signals via the typical satellite transponder must thus be very power efficient but not extremely bandwidth efficient. This is the same reason that FM analog signals, not AM signals, are used in analog satellite links.

Two modulation techniques are used for digital satellite systems: bi-phase shift keyed (BPSK) and quadrature-phase shift keyed (QPSK) as briefly described in Chapter 2. These signals have optimum power efficiencies, thus making them suitable for power-limited, satellite channels. The QPSK modulation scheme can fit twice as much data into a bandwidth as the BPSK method without a need to increase power.

Multiplexing DVB Streams

When multiple television programs are transmitted over a transponder that previously contained just a single channel, the data from the programs has to be combined, or "multiplexed," in some fashion. The two simplest multiplexing methods used are frequency-division multiplexing (FDM) and time-division multiplexing (TDM).

Figures 3-1a and 3-1b illustrate the differences between frequency- and time-division multiplexing. In an FDM system each program is first modulated onto the carrier wave, then all the modulated signals are either relayed in a single-channel-per-carrier (SCPC) format or are all multiplexed together and transmitted via a single carrier. In a TDM system, the data streams from each program are first sequentially combined into a single one and then the entire stream is modulated onto a carrier wave and transmitted. The entire band is used for just one carrier, not broken into individual carriers as is the case with FDM.

An FDM satellite receiver must demodulate just a single program at one time and so does not have to manage the higher data rate of the combined signals. However, guard band space between the individual carriers must be used to eliminate interference between signals. This guardbanding "wastes" spectrum and decreases the overall data rate that can be put through the channel bandwidth. The TDM approach eliminates the need for guardbands except at the edges of the entire transponder and so makes more complete use of the spectrum. TDM can fit more data into the available bandwidth.

However, a TDM satellite receive must demodulate the entire data stream, a feat that may be more difficult than demodulating the lower rate individual FDM data streams. For multiple NTSC video programs relayed over a single bandwidth, the combined TDM data rate is about 24 Mbits/second for the current systems, well within the capability of cost-effective circuitry. Thus most DVB transmission systems are expected to use TDM.

Error Correction

An uncompressed digital video signal tolerates errors well. An error in one bit might change the color or brightness of just one pixel but many errors must occur in transmission before they become noticeable to the eye. With DVB bit streams, however, a single bit error can cause a noticeable image degradation since each bit encodes more than a single point of the image. For example, if the wrong bit were sent for a motion estimation vector, a whole pixel block might be placed in the wrong location.

Why compress a digital signal in the first place if the tolerance for errors is lost? If forward error correction (FEC) bits are added to the compressed data stream, for a very small amount of "overhead" nearly all of the expected errors can be corrected. For example, "Reed/Solomon FEC" can correct a bit error rate (BER) of 10^{-5} (one error in every 10^5 or 100,000 bits) to a BER of 10^{-9} (one error in a billion bits), with just a small increase in data rate.

How Does FEC Work in General?

It may seem surprising that errors in a digital message can be corrected. But, one mentally correct errors in the morning newspaper. When a misspelled word is seen, it can usually be determined what the word has to be for the sentence to make sense. That is, the correct word can be determined from the context. FEC codes create an artificial "context" that allows determination of where the error must lie.

If the context changes, the error might be corrected in a different way. As a demonstration of this concept, consider that the misspelled word "COT" would be corrected in the middle letter to "CAT" if the context were "COT-DOG-PARAKEET." It will be corrected in the third letter to "COW" if the context were "COT-HORSE-PIG." FEC methods accomplish this same function, but in a very well-planned, efficient manner that gets the most correction from the smallest number of added FEC bits.

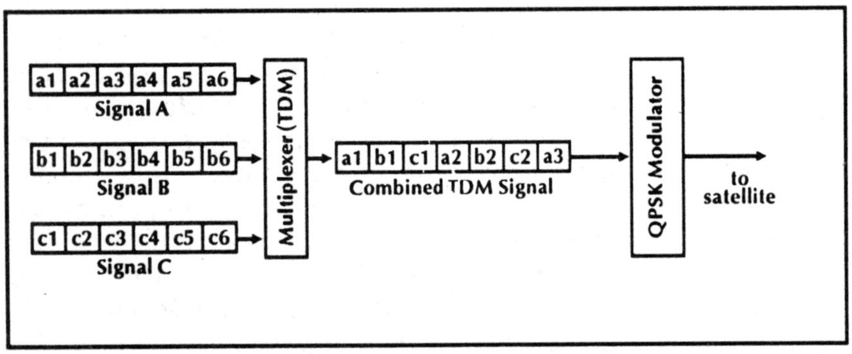

Figure 3-1a. Time Division Multiplexing. *Digitally compressed signals are time division multiplexed by first combining them in sequence. The time-interwoven signal, a1..b1..c1..a2..b2..c3..a3..b3..etc... is then multiplexed, usually via QPSK,and transmitted to satellite.*

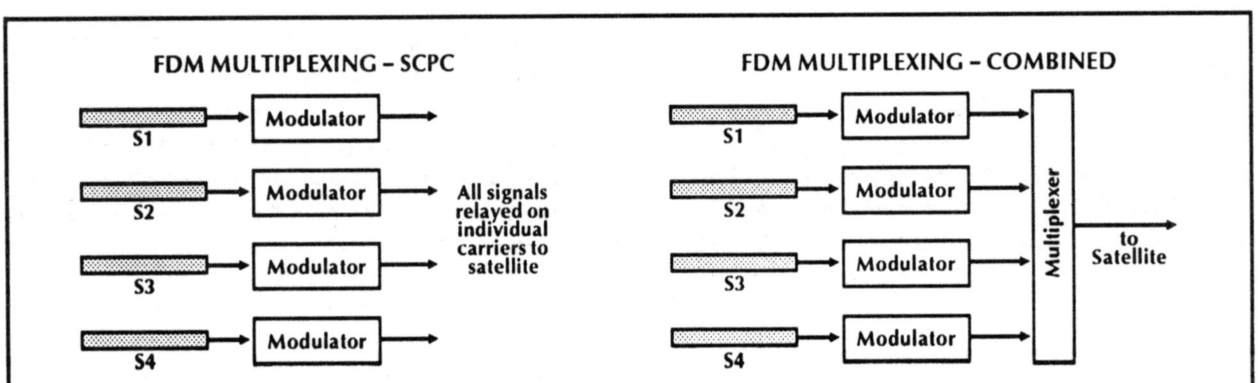

Figure 3-1b. Frequency Division Multiplexing –SCPC and Combined. *Digitally compressed signals can be frequency multiplexed and transmitted to satellite on separate carriers or all multiplexed signals can re-multiplexed onto one carrier.*

CHAPTER 4

THE COMMUNICATION SATELLITE

SATELLITE DESIGN and OPERATION

The satellite is the key element in the communication circuit. Transmitting signals from deep space is a potent technology because any location within a satellite's "view" can be linked without the use of cables or line-of-sight relay towers. So a single satellite can simultaneously serve vast areas of our globe.

Communication spacecraft that broadcast audio, data and video to vast regions of the earth below have been developed during the past forty to fifty years. Satellites that originally served only as passive reflectors of man-made communication signals today are high power, complex switching and frequency translation relay devices. This rapid technical evolution is extraordinary. For example, while the first Intelsat satellite, "Early Bird," was capable of relaying just 480 voice channels, the current Intelsat 600 series of spacecraft can transmit a full 80,000 voice conversations!

A modern communication satellite receives transmissions, typically in the 1 to 40 GHz range, amplifies and processes these signals, and then lowers their frequency for broadcast to a target area on the earth below. The downlinked signals are characterized by their power levels as well as their frequency composition.

The Downlink Path and Satellite Footprints

Downlink antennas can broadcast microwave signals to any chosen geographic region within view of a communication satellite. The satellite footprint defines this geographic coverage and is determined by precise details of the downlink antenna and feed design as well as by the power levels generated by each onboard communication channel, or transponder (see Figure 4-1). While each transponder is a physically separate circuit of a satellite, signals from more than one transponder can be downlinked by just one antenna. Power levels are usually higher near the center of the downlink beam and weaker in off-target areas. A larger antenna is therefore required to receive off-axis signals than would be needed at the center of the footprint.

More advanced satellites now employ "shaped" downlink antennas that transmit power to cover just the area desired so it is not "wasted" in non-targeted areas. These permit the distribution of downlink power to be modified so that higher powers may be directed to off-boresight areas.

Understanding Footprint Maps

Footprint maps provide crucial information in sizing receive dishes. Power levels on these maps are measured by a quantity known as the "effective isotropic radiated power" (the EIRP) which is expressed in units of decibels relative to one watt (dBw). A footprint map is constructed by joining all those points on the map having an equal EIRP by continuous lines. Thus a distinctive "footprint" can be created for any orbiting satellite as a series of contour lines superimposed upon the map of the region served (see Figure 4-1).

The apparently small variation in EIRPs on footprint maps can be deceptive. They translate into large differences in signal power. For example, the Galaxy 5 signal in Atlanta, Georgia at 40.5 dBw is roughly double that received in Corpus Christi, Texas at 37.5 dBw because of the 3 dB difference in EIRP. Everything else being held the same, a receive dish in Atlanta would need twice the surface area as one in Corpus to receive an equivalent amount of power.

Satellites that broadcast to more limited geographical regions have footprints covering a smaller area. These "spot beams" have proportionately higher EIRPs than those of a satellite targeting a much larger area with "hemispherical," "zone" or "global" beams because the same power is more concentrated. Global beams cover 42.4% of the earth's surface, the maximum view of any one geosynchronous satellite. Hemispheric and zone downlink antennas target 20% and 10%, respectively, of the globe. Spot beams are focused on even smaller areas. Clearly both the shape and orientation of the downlink antenna as well as the power generated by each transponder determine geographic coverage and EIRPs.

The term effective isotropic radiated power (EIRP) is derived from the word isotropic which means equal in all directions. Effective isotropic radiated power means the power levels that would be received at any location if an antenna were radiating equally in all directions. Therefore, a 37 dBw isotropic radiated power reading means that a perfect antenna would direct 37 dBw or 5012 watts per square meter in

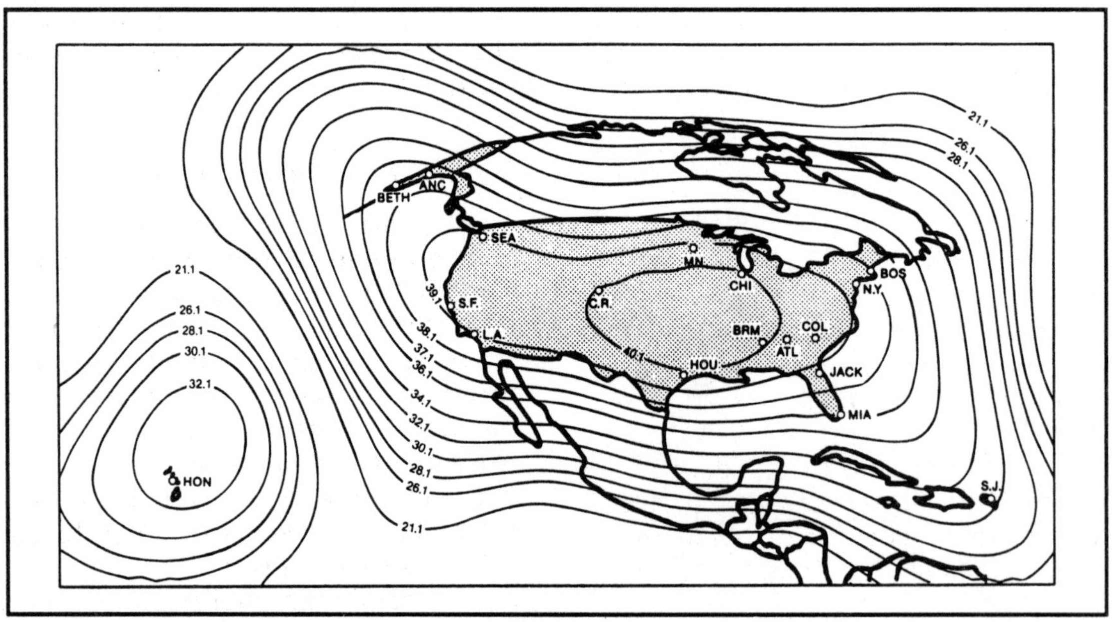

Figure 4-1. Galaxy IV C-Band Footprint Map – Vertical Transponders. *Galaxy IV directs a spot beam to the Hawaiian Islands with a beam center EIRP of 33.6 dBw as well as a zone beam to the continental United States with a beam center EIRP of 41.6 dBw*

all directions. The reason that a transponder having rather limited power can apparently have such a high EIRP stems from the fact that this power is not radiated equally in all directions but is concentrated in a narrow beam aimed at the earth below. Ku-band transponders having a total power of 50 watts have EIRPs as high as 50 dBw or higher when this power is directed into a zone beam.

EIRP levels refer to the power of signals measured at the downlink antenna on the satellite. In the example given above, 5012 watts/m^2 would be directed towards a selected location on earth below. The 50 watts total power would be concentrated into an area much less than one square meter at the downlink antenna. The downlinked signal then spreads out in a cone-like beam as it travels to the earth below. This weakening or dilution in power as it moves away from the satellite is called the "free space path loss." The greater the distance to the target, called the "slant range," the greater are the free space path losses.

Also contributing to signal losses incurred on the homeward voyage is absorption by molecules of the atmosphere. Water vapor is the main culprit in the attenuation of downlinked signals. In fact, during a severe rainstorm, the power received on the surface of the earth can be reduced by as much as 10 dB, a factor of 10, at higher Ku-band frequencies.

Free space path loss and atmospheric absorption explain why a 50 watt Ku-band transponder is detected on earth at a strength of less than roughly one billionth of a billionth of a watt per square meter! Detecting a transmission from a 50 watt transponder is like clearly seeing a typical light bulb from a distance of 22,300 miles. Or it is like receiving a relatively powerful CB radio transmission, designed to have a 10 to 20 mile range, from the same distance.

Satellite Operation

Uplinked microwaves have powers of fractions of a millionth of a watt/meter^2 when received by a geosynchronous satellite. In the heart of the spacecraft, typically about the dimensions of a medium sized truck, these signals are amplified many thousands of times, shifted to a lower frequency range and then broadcast back to earth. The frequency downconversion that occurs in a satellite also reduces interference at the receive site. If the same frequency had been used, some of the uplinked signal could mistakenly be detected by receiving stations near uplink sites along with the desired signal. This would cause feedback between the uplinked and downlinked signals like the loud ringing caused by feedback in a sound system when a microphone is too close to a speaker. This problem could be more severe at uplinks, typically designed to both receive and transmit satellite signals.

All the power to carry out these electronic functions is provided by the sun's energy which is captured by arrays of solar cells, also known as photovoltaic cells. The illustration in Figure 4-1 shows these arrays as large wing-like structures. The smaller central features mounted on the spacecraft "bus" are the uplink and downlink antennas.

Television Channel Formats

The number of television channels, telephone conversations or the amount of data transmitted is related to the electronic design of a satellite. Early American C-band Western Union spacecraft such as Westar I and II simultaneously relayed 12 TV programs on signals having one sense of linear polarity. The 500 MHz in the C-band range from 3.7 to 4.2 GHz was subdivided into twelve segments of 40 MHz plus a remainder of 20 MHz. Since a 36 MHz bandwidth is more than sufficient to broadcast a high quality analog TV picture, Western Union designed its pioneer satellites to relay 12 channels each having a 36 MHz bandwidth with 4 MHz guard bands between each channel to eliminate the possibility of crosstalk. Each channel was handled separately on-board the satellite by a "transponder."

Engineers who designed the Satcom I spacecraft went further. They doubled the number of channels that could be relayed within this 500 MHz total bandwidth by a technique called frequency re-use. All even channels were transmitted earthward with horizontal polarity, all odd channels with vertical polarity (see Figure 4-2). Then the frequency centers of these cross-polarized channels were offset from each other for further security against crosstalk. Since each receive site was equipped to detect just vertical or horizontal polarity at one time, there could be some overlap between the frequencies used for the odd or even channels without interference occurring between channels of opposite polarity. The center frequencies of each channel is listed in Table 4-1.

Although this 500 MHz C-band channel format has become a standard for most American satellites and many domestic spacecraft from other nations, it is by no means the only one used. Designers are free to chose any combination of center frequencies and channel bandwidth desired to meet their particular needs.

Video signals have traditionally been transmitted in roughly a 28 MHz portion of a 36 MHz wide transponder. This was sufficient to provide a high quality video signal at the receive site. The remaining bandwidth was then used for audio or data signals. While this is still

TABLE 4-1. C-BAND CHANNEL CENTER FREQUENCIES & BANDWIDTHS

Transponder Number	Downlink Frequency (MHz)	Frequency Band (MHz)
1	3720	3702-3738
2	3740	3722-3758
3	3760	3742-3778
4	3780	3762-3798
5	3800	3782-3818
6	3820	3802-3838
7	3840	3822-3858
8	3860	3842-3878
9	3880	3862-3898
10	3900	3882-4008
11	3920	3902-3938
12	3940	3922-3958
13	3960	3942-3978
14	3980	3962-3998
15	4000	3982-4018
16	4020	4002-4038
17	4040	4022-4058
18	4060	4042-4078
19	4080	4062-4098
20	4100	4082-4118
21	4120	4102-4138
22	4140	4122-4158
23	4160	4142-4178
24	4180	4162-4198

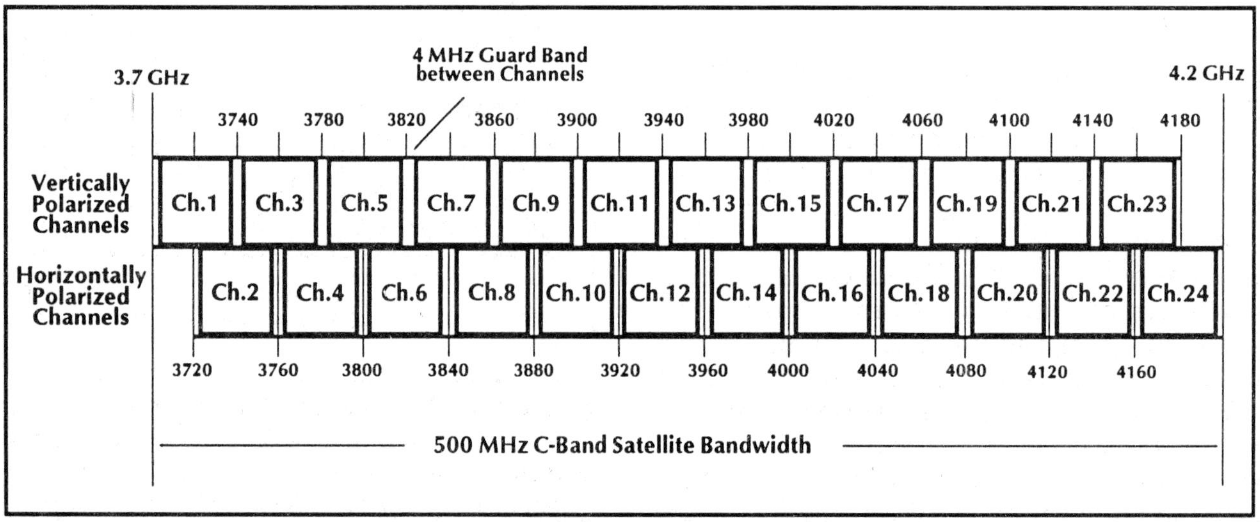

Figure 4-2. The Standard C-Band Frequency Layout. *Most C-band broadcast satellites are designed to transmit 24 channels each having a maximum bandwidth of 36 MHz. Twelve channels each are relayed on horizontal and vertical polarity with 4 MHz guard bands between each transponder circuit.*

the case with many analog satellite broadcasts, bandwidth can be reduced and still carry "reasonable" quality video signals. In order to increase the carrying capacity of a satellite "half-transponder" broadcasts that each use as little as 14 MHz of bandwidth have been employed, with some degradation in video quality. Of course, the actual bandwidth of a half-transponder transmission depends upon the full transponder bandwidth. On 40 MHz Intelsat transponders, this translates to a reasonable 20 MHz while on 54 MHz PanAmSat-4 transponder, a half-transponder is a full 27 MHz wide.

When two video signals are carried via a single transponder the total available power must be cut in half and split between the two channels. Uplink power must then be "backed off" by an additional 3 or 4 dB to avoid interference between these signals.

Audio and Data Signals

A variety of methods has been used to relay audio and data signals via satellite. Traditionally, such signals have either been included as "subcarriers" attached to a video signal or have been relayed separately on non-video transponders as single-channel-per-carrier signals. More advanced systems have transmitted digital audio signals either as sound-in-sync or actually woven into a MPEG-2 digital video/audio/data broadcast.

Audio Subcarriers

Following the reception and processing of a conventional analog satellite transmission in a receiver, the recovered raw video and audio signal is contained in a band of frequencies ranging from near zero to about 10 MHz. Video signals occupy only a portion of this region. For example, an NTSC video signal occupies the region between zero and 4.5 MHz. The remaining space can be used for audio "audio subcarriers", usually analog signals, some of which relay stereo or monaural sound that matches the television picture. Many other audio messages such as radio stations not having an identity related to the video signal are often present (see Figure 4-3).

Most analog C-band television broadcasts relay audio subcarriers in the 5.5 to 9.0 MHz range, with 6.2, 6.6 and 6.8 MHz being the most common mono audio frequencies. Stereo sound is often modulated onto two separate subcarriers, for example at 7.20/7.38 or 7.02/7.20 MHz, via systems such as Wegener Panda stereo.

SCPC Single-Channel-per-Carrier

Audio messages can also be broadcast independently of video carriers in the single-channel-per-carrier (SCPC) mode. In this case each audio channel is transmitted on its own carrier independent of the video signal. An SCPC carrier can occupy from as little as 5 kHz of bandwidth for voice-grade messages to as much as 250 kHz for high fidelity audio. Thus hundreds of single channel per carrier (SCPC) audio or data signals or telephone relays with individually assigned carriers and frequencies can be transmitted over each satellite transponder. For example, a single 36-MHz wide transponder can transmit nine hundred 40-kHz wide signals, each having a separate carrier. Today more than 300 SCPC audio channels are transmitted via North American satellites with

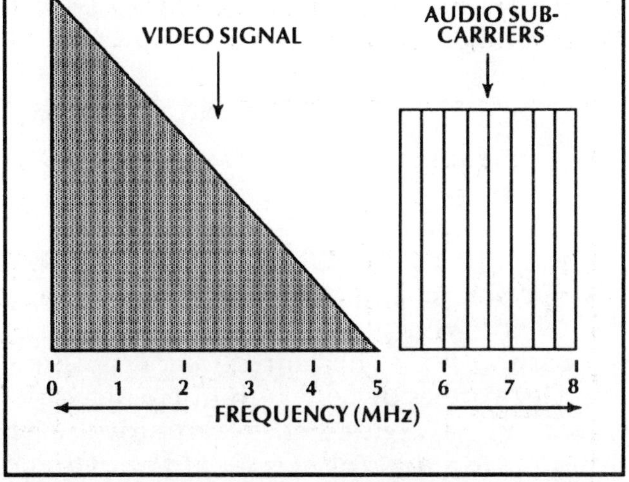

Figure 4-3. NTSC Audio Channel Format. *The audio information in an NTSC television broadcast is typically relayed on subcarriers having center frequencies ranging from 5.0 to 9.0 MHz. Monophonic sound accompanying many satellite TV broadcasts is often located on either a 6.2 or 6.8 MHz subcarrier.*

more than 90 channels on Galaxy IV at 90°W alone. In contrast, a single 36 MHz wide transponder can manage at the very most 10 to 15 audio subcarriers because these are relayed at the expense of the accompanying video power and bandwidth. SCPC signals can be detected and processed by a device known as an FM scanner, basically a beefed up FM radio.

Sound-in-Sync

An audio signal may also be relayed within the structure of the analog video signal, a technique known as sound-in-sync. The digital audio signal is inserted into either the horizontal or vertical blanking intervals of the standard TV signal. (Blanking intervals are the time between line traces or frames often seen as the black line between frames when a picture "rolls"). This is feasible because these intervals, which occupy nearly 20% of the time taken to transmit a television picture, have little information content. Using sound-in-sync eliminates the need for separate audio subcarriers and allows the transponder power to be concentrated on the video information so that the video signal-to-noise power can be maximized.

Digital audio signals are transmitted within the blanking intervals in a number of hybrid analog/digital encryption systems including the familiar North American VideoCipher II+ and the Oak Orion as well as the European/African IRDETO. In these systems, encrypted digital audio and data messages are inserted into the vertical blanking interval of an analog video signal which itself is scrambled. In fact the first "broadcast" use of digital audio signals occurred in the United States when HBO began using the VideoCipher II encryption system.

Pure Digital Methods

Digital audio and data are an integral component of MPEG-2 DVB (digital video broadcasts). As previously described in Chapter 3, at the beginning of each 188 byte data packet is a 4-byte header that informs the receive system whether the information is an audio, video and data signal. An MPEG-2 broadcast makes very efficient use of transponder bandwidth. A mixture of high quality audio, data and video programs can be transmitted in space reserved for much less analog capacity.

SATELLITE MAINTENANCE

Stationkeeping

Communication satellites would remain stationary forever above a chosen location in the Clarke belt over the equator if there were no gravitational forces from the sun and moon, if solar winds did not sweep past our globe and if the earth were perfectly spherical. This is not the case, so these unbalanced forces cause any satellite to drift slowly away from its assigned location into an a figure-eight excursion with a 24-hour period that follows a path above and below the orbital plane. Ground controllers must periodically adjust satellite positions to counteract these forces. If not, the travel above and below the orbital plane would build up at a rate of about 0.9 degrees per year. A geostationary satellite would also drift in an east or west direction along the orbital arc. This process of maintaining a satellite within a pre-assigned "window" is known as "stationkeeping."

Communication spacecraft are equipped with small hydrazine gas charged thrusters that are remotely fired by ground controllers whenever stationkeeping is necessary. Most geosynchronous satellites are maintained within 0.1° of their assigned position, essentially in a box having dimension of about 90 miles per side. Thrusters are fired once every 3 to 4 weeks to adjust the east/west location and once every 30 to 90 days to maintain a position close to the arc. Ten times as much fuel is required to keep a satellite on the equatorial plane as that needed for east/west stationkeeping.

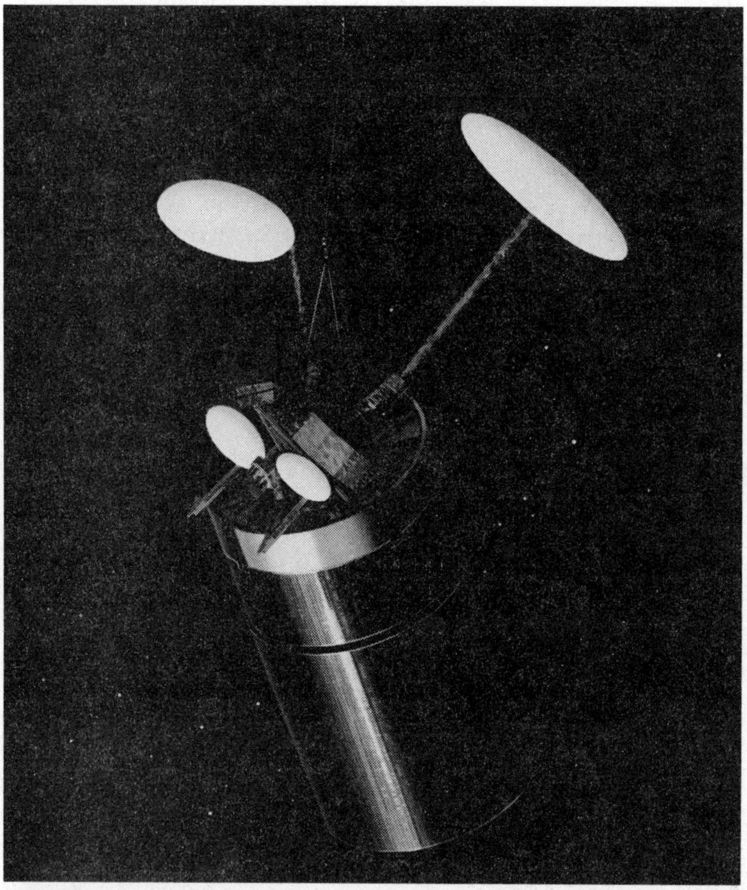

Figure 4-4. Intelsat VI. *The Intelsat VI satellite is equipped to handle both C- and Ku-band transmissions. It stands nearly 39 feet high with its antennas unfolded and its aft solar panel extended. (Courtesy Hughes Aircraft Company)*

There are two "zero-pull" locations at approximately 104.5° West and 75.5° East on the exact opposite side of our globe where these forces balance so that a geostationary body will remain stationary along the arc. However, the rotation of the earth around the sun and the subsequent change in its position relative to the equâtor causes satellites located even at these two points to have some vertical movement above or below the arc.

A satellite must also be maintained in a stable orientation with respect to the earth so its downlink antennas do not stray from their targets. Until recently, most spacecraft were aligned by spinning their whole body in order to create a stabilizing gyroscopic effect similar to those forces that keep a rapidly spinning top upright (see Figure 4-4). On such "spinners" or body-stabilized spacecraft the antenna platform was spun in the opposite direction so the reception and downlink antennas would be fixed on target. Most recent satellites, like the Hughes HS601 vehicles, are three-axis stabilized and use an internal three-axis gyroscope to perform this task. An example of an earlier RCA Americom satellite is shown in Figure 4-5.

Figure 4-5. A Three-Axis Stabilized Satellite. *The large wing-like structures on this satellite are solar cells which provide on-board power. The downlink and uplink antennas receive and re-transmit power, respectively. (Courtesy of Martin Marietta)*

In order to relay a global beam having a width of approximately 2 to 3 degrees, as is typically required to cover a land mass such as continental North America or Africa, downlink antennas must be pointed to within a 0.2 degree accuracy. This is crucial because most satellites, even when accurately maintained at their assigned station, still move in a small figure-eight pattern, usually about 15 miles across. Even this small movement can impair reception of signals by large, narrow beamwidth, Ku-band receiving dishes. Even small wobbles must be corrected on body-stabilized satellites. If not, this motion can evolve into a "flat" spin (as opposed to the proper vertical spin) resulting in a total loss of communications with the spacecraft.

Solar cell arrays on three-axis stabilized satellites, the large wing- like structures, track the sun to maximize the solar power received. Spinners, in contrast, have solar arrays covering their whole external surface so only 40% of the cells are illuminated at once. Spinners typically look like large cans having antennas at one end.

While a satellite is almost always in full view of the sun, during two periods each year, from either February 27th through April 12th or August 31st through October 16th, it passes through the shadow of the Earth. Receive sites then experience a series of 45 separate eclipses that gradually increase in length to a peak of about 1.2 hours per day on the equinox dates, March 21st and September 21st. Following the maximum at the equinoxes, the eclipse time decreases. Onboard nickel-hydrogen batteries are then used to maintain designed power levels.

Today, both the improved stability and stationkeeping of present-day vehicles permit the use of larger antennas, capable of focusing higher-power microwaves and of housing larger solar cell arrays to increase satellite power.

Satellites do not survive forever. Their life expectancy is determined by how long adequate power and stability can be maintained. Repositioning rockets usually run out of hydrazine gas before surviving 10 to at most 15 years in space. Solar cells under constant bombardment with micrometeorites and ultraviolet rays from the sun, also slowly deteriorate. Either a 30% reduction in solar cell power or expiration of the hydrazine fuel supply is taken as a signal to retire a satellite from active duty. Transponder powers also slowly decay during the final years of a satellite.

When ground controllers put a satellite to rest it is boosted into a slightly higher, unstable, non-geosynchronous orbit where, in time, it eventually re-enters the atmosphere and burns up. Some of the decommissioned pioneer communication vehicles are still resting in "eternal orbits."

TRENDS in SATELLITE DESIGN and OPERATION

Satellite technology is still in its infancy and developments have proceeded at a rapid pace during the last twenty to thirty years. The future will also certainly bring some extraordinary advances in design and operation of the worldwide satellite communication system. Designing, building, launching and operating a satellite communication system is at the forefront of the high technology business of the 1990s.

Some interesting recent developments include low and inclined orbit vehicles, high power satellites, frequency re-use and multiple frequency satellites, digital compression and other improvements in signal processing and inter-satellite communications.

Inclined Orbit Satellites

The geostationary orbit is just one of an infinite series of orbits that a spacecraft, meteor or any other object can assume. Orbits that lie within the equatorial or polar planes are known as equatorial or polar orbits, respectively. Any other tracks are known as inclined orbits.

While most geostationary satellites are rigorously maintained within their assigned "windows," some have been allowed to drift outside of their normal stationkeeping position. This strategy has typically been implemented to extend a spacecraft's lifetime. Since it takes relatively little fuel to maintain the longitude of a spacecraft, most of the drift allowed is in the north/south plane. For example, the failure of an apogee kick motor during the launch of the American satellite GSTAR III and the subsequent use of stationkeeping fuel to achieve a geostationary orbit forced controllers to allow it to drift more than 4.2° below and above the equatorial plane to extend its lifetime. However, it is being maintained within 0.1° of its assigned longitude in accordance with FCC requirements.

The difference between the angle that the plane of the spacecraft makes with respect to the equatorial plane is the measure of its inclination. Geostationary satellites typically lie within 0.1° inclination while inclined orbit spacecraft can have inclinations as high as 3.0° or more.

A spacecraft in an inclined orbit follows a figure-eight pattern about its assigned orbital location. As the inclination and thus the north/south motion increases so does the width of this figure-eight. Dishes that detect signals from inclined-orbit satellites must have adequate tracking ability built into their mounts. If the reception site longitude is within about 10° of the satellite longitude most of the figure-eight motion can be tracked by movements in just the elevation. However, outside this range tracking must be in both the elevation and azimuth. For inclined orbit satellites that lie near the ends of the arc, most of the required motion must be in the azimuth direction.

High Power Satellites

Satellite transponder levels have increased from a few watts to as high as 200 watts or more during the past twenty years. As a result, the diameter of receive dishes has plummeted from as high as nearly 100 feet (30 meters) to under one foot (30 centimeters) for some Ku-band systems. This trend has been quite evident in both C- and Ku-band reception.

Devices known as traveling wave tube amplifiers (TWTs), one associated with each transponder, have provided this power. While earlier TWTs were expensive, difficult to manufacture and subject to failure, higher power models have functioned well. In addition, a relatively new option, the solid state power amplifier (SSPA), has the advantages of being lightweight, long-lived and reliable without the bandwidth limitations characteristic of TWTs. While SSPAs have not yet achieved the power levels of TWTs, especially in the higher frequency Ku-band, this is rapidly changing. Intelsat satellites now feature a combination of TWTs and SSPAs in both C and Ku-band transponders.

Satellites have generally been classified as either low, medium or high power. Their power is determined by the TWT capability. Low power spacecraft with powers of 25 watts per transponder have employed hemispherical or global beams to transmit telephone or video programming for cable TV concerns. Medium powered satellites such as the European Ku-band spacecraft Astra which relays 55 watts per transponder with a beam center EIRP of 52 dBw have been used for direct-to-home television and radio broadcasts that can be received by dishes as small as 60 cms in diameter. Transmissions from high power satellites that generate signals of approximately 110 watts with beam center EIRPs exceeding 58 dBw can be received by dishes as small as 30 centimeters. The Ku-band DBS-1 satellite in the United States can run two 100-watt transponders in parallel to achieve a full 200 watt output.

Frequency Re-Use and Multiple Frequency Satellites

There are of persuasive reasons to use satellites and orbital space as efficiently as possible. First, launching a satellite is an expensive proposition. Clearly, the more communication traffic a spacecraft can manage, the more economical the end result. Second, there is simply limited space in the geosynchronous orbit. If satellites are spaced 2 degrees apart, a total of just 180 satellites can be in orbit at once. Third, competition for frequency allocations can be fierce so there is strong motivation to use the available bandwidth as efficiently as possible. A number of solutions has been developed in response to these concerns.

Regulators and innovators have been progressively using higher and higher frequencies for satellite communications. C- and Ku-band systems are now extensively used; Ka-band frequency satellites that operate in the 20/30 GHz range are being developed and, in some cases, being tested.

Satellites can be co-located in the geostationary arc if they operate in different frequency bands and are thus effectively invisible to each other. Intelsat spacecraft and numerous domestic satellites are dual-band, that is they have both C- and Ku-band transponders. Perhaps, in time, clusters of non-interfering, co-located satellites will receive information from just one uplink. In addition, numerous co-located spacecraft could further conserve resources by sharing a single platform, positioning system and power supply. This could result in increased pointing accuracy for the downlink antennas, thus enhancing circuit performance.

The "re-use" of frequency allocations is a common technique in extending the range of a communication system. The use of dual polarization signals is one example of frequency re-use; this has effectively doubled the channels a given satellite can transmit. Downlink antennas have developed to the point where localized spot beams can be downlinked to multiple locations re-using the same frequency band in each location. Ku-band downlink antennas are particularly suited to this "beam shaping" and localization. Using localized spot beams effectively increases satellite EIRP at the receiving site.

Digital Compression and Other Improvements in Signal Processing

Digital compression techniques that allow more information to be carried per transponder are now a reality. Use of a satellite circuit can be practically increased by a factor of 5 to 10, an impressive gain. Digital video compression systems have the potential to transmit eight or more programs in the bandwidth previously used to transmit one analog video broadcast.

Inter-Satellite Communication

Satellite communication circuits will be used more effectively by employing communication directly between satellites with microwave or laser beams. This is feasible in outer space because no atmosphere is present to weaken signals bouncing between satellites. In fact, the ITU has already allocated frequency bands for inter-satellite links. Using such relays could make communication between very distant countries less expensive by eliminating an intermediate link. A message uplinked in Seattle could be received by a domestic satellite, relayed directly to PAS-4 and subsequently downlinked to Africa. The alternative would be to use an intermediate Intelsat satellite earth station or a submarine fiber optic cable, at a potentially higher cost.

CHAPTER 5

THE RECEIVE SITE

INTRODUCTION

Direct-to-home (DTH) systems designed to receive satellite broadcasts have evolved rapidly since the early 1980s. This pace is now heating up as both analog and digital broadcasts are being transmitted. The underlying theory and operation of all the components required to assemble a viable satellite reception system, quite similar for both types of broadcasts, are explored in this chapter.

DISHES

A satellite dish intercepts the very weak signal coming from a targeted satellite and reflects it to a focal point where the feed and LNB are placed. In this process the signal is concentrated so that the necessary power is available for subsequent electronic components.

The quality of a dish is determined by how well it targets a satellite and concentrates the desired signal and by how well it ignores unwanted noise and interference. These seemingly simple design objectives, the subject of a unique scientific and engineering discipline for many years, must be mated with economic considerations. Therefore, a technically excellent dish must also be moderately priced, aesthetically pleasing, durable, easy to install and virtually maintenance-free.

Most dishes in use today for receiving satellite broadcasts have designs based upon variations of circular or parabolic surfaces. Signals reflected from surfaces incorporating such designs are concentrated to a "focal point" or to a series of focal points. Flat plate antennas have been recently introduced.

Single-Focus Dishes

Prime Focus Reflector

By far the most common dish, the prime-focus parabola, theoretically focuses all signals arriving parallel to its axis to a single "prime focus." Any incoming signals from off-axis directions are reflected away from this focus (see Figures 5-1 and 5-2).

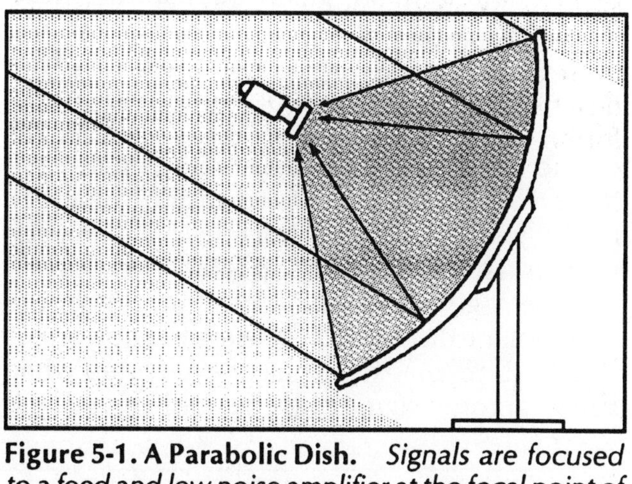

Figure 5-1. A Parabolic Dish. *Signals are focused to a feed and low noise amplifier at the focal point of the parabolic reflector.*

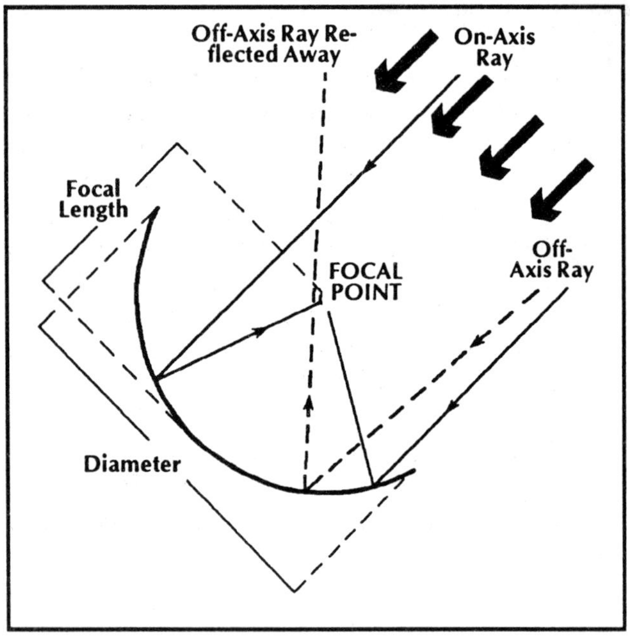

Figure 5-2. Parabolic Antenna Geometry. *In the prime-focus configuration all rays are directed via one reflection to a feed at the focal point.*

In practice, a prime-focus dish does not behave as dictated by theory for three reasons. First, the feed mounted at the dish focus and designed to collect and concentrate incoming signals occupies a finite area in the vicinity of this point. As a result, it also intercepts some radiation arriving from directions slightly off the main axis. Second, the reflective surface does not perform exactly as geometrical ray-tracing methods would suggest because the intercepted microwaves behave according to the principles of waves. As a result, a portion of the incoming signal spreads around the dish and feed edges. This effect is well known in optics as diffraction. Also, because satellite signals have a wave-like nature, the feed cannot be localized at just a single point but must have finite dimensions, resulting in "aperture blockage." Third, irregularities and imperfections in the shape of the dish surface cause reflection errors so that some off-axis signals are detected while some targeted signals pass by unobserved.

Offset Fed Reflector

The surface of an "offset fed" dish is a section of a larger prime-focus parabola (see Figures 5-3 and 5-4). The feed which is still located at the focus of the larger "parent" dish thus appears to be offset from that portion of the reflective surface in use. This design eliminates blockage of incoming signals by the feed/LNB structure (see Figure 5-5). Offset fed dishes are more commonly encountered in Ku- band than in C-band TVROs because the shorter focal length and smaller dimensions of both the feed and LNB permit the offset supporting arm to have more stability.

Offset dishes are aimed towards the arc of satellite at a steeper angle than the prime-focus variety. They therefore have the advantage of being better able to shed snow.

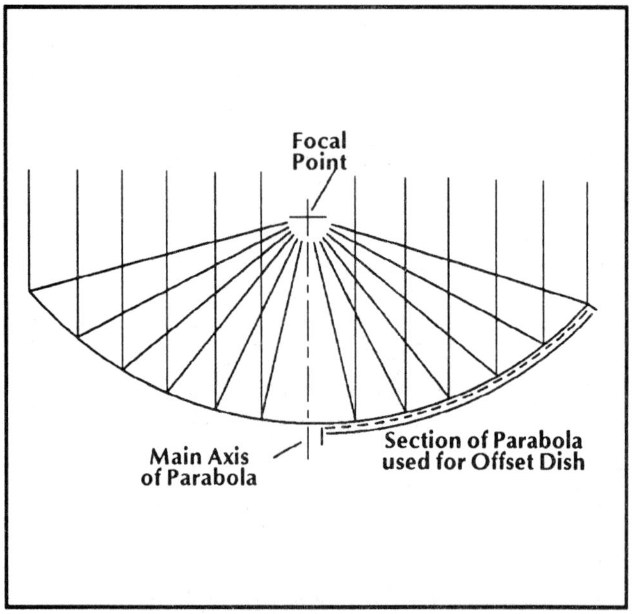

Figure 5-3. Offset Dish Geometry. *An offset dish is cut from a section of a larger "parent" parabolic prime focus dish. The feed is still located at the same place in relation to the parent dish and all signals are reflected to this focal point. The only difference is that a good sized portion of the larger dish is missing.*

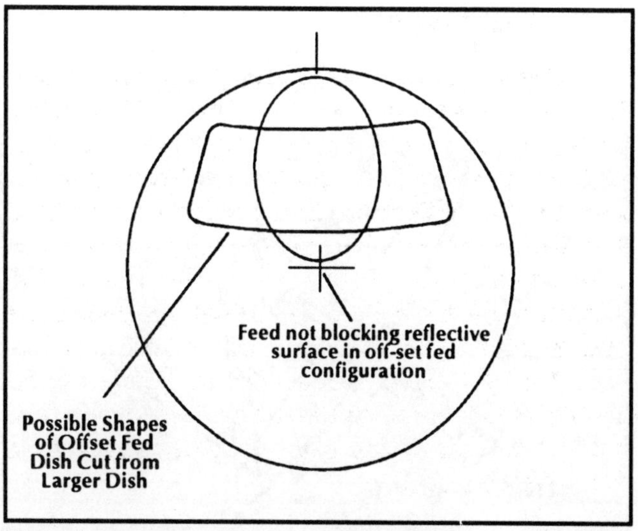

Figure 5-4. Front View of Offset Dish. *Various shapes can be cut from the "parent" prime focus reflector so an offset dish can assume an elliptical, round, rectangular or any other shape. Generally an elliptical shape is used to simplify feed design.*

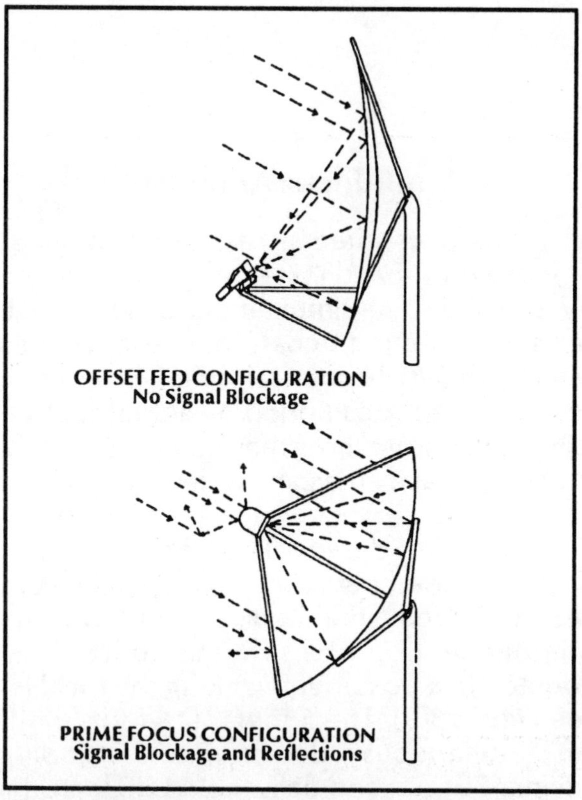

Figure 5-5. Aperture Blockage in Prime Focus versus Offset Fed Dishes. *The feed in an offset dish does not shadow the reflector surface as it does in a prime focus configuration. Dish noise temperature is also generally lower since the feed faces more into deep space and further away from the warm ground. The increased tilt also avoids snow load in colder climates.*

Cassegrain Parabolic

A Cassegrain dish also has a parabolic surface but redirects radiation via a second reflective surface, the hyperbolic subreflector, down a waveguide to an LNB mounted behind the dish. In extremely hot climates, earth stations using Cassegrains can perform more effectively than prime focus brands because the LNB is installed behind the antenna and is therefore protected from direct exposure to solar energy (see Figures 4-6 and 4-7).

Cassegrain antennas are generally more expensive than the ordinary prime-focus variety and are usually found in commercial installations.

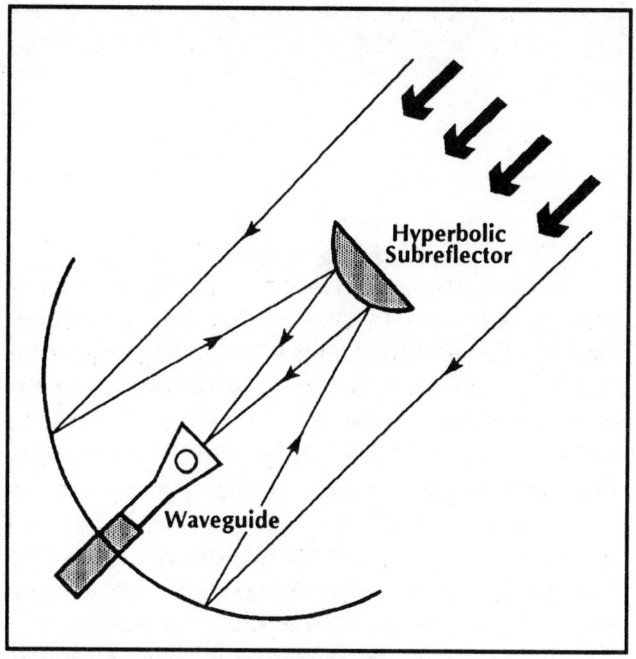

Figure 5-6. Cassegrain Antenna Geometry. *A Cassegrain parabolic dish employs a second reflector, the hyperbolic subreflector, to direct microwaves to a feed located behind the reflective surface.*

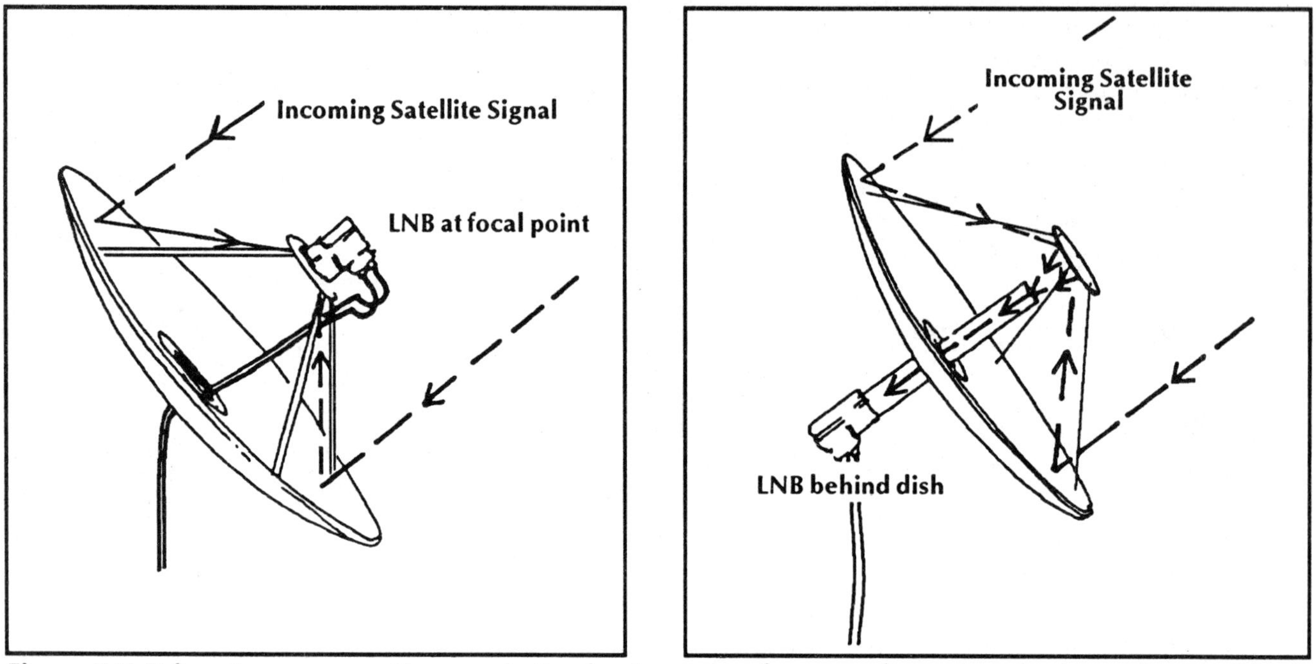

Figure 5-7. Prime Focus versus Cassegrain Feeds. *In a prime focus configuration, the low noise amplifier is located at the antenna focal point. In contrast, in a Cassegrain system, the amplifier is in a more protected rear position.*

The Flat Plate Antenna

Flat plate antennas have not yet been widely used in home satellite systems. The operation of passive flat plate antennas is based upon Fresnel reflection while that of active designs is based upon solid state circuit elements.

In a passive flat plate antenna a series of concentric rings overlaid onto a transparent sheet creates a lens that redirects and focuses radiation such as microwaves, light or X-rays. If the concentric rings are elongated into a set of ellipses, signals arriving from off-boresight can still be focused to a point behind the Fresnel pattern.

If a negative pattern is used whereby concentric rings are made transparent and the space between the rings is opaque, then a reflective sheet laid behind this assembly can reflect incoming satellite signals to a feed in front of the plate.

Multi-focus Antennas

Multi-focus antennas, developed originally for use with C-band systems, allow more than one satellite to be simultaneously detected by focusing incoming signals to a series of feeds (see Figure 5-8). In contrast, single-focus reflectors must be repositioned by actuators to receive signals from more than one satellite. Only one satellite at a time can be viewed.

Multi-focus designs incorporate variations of spherical and parabolic surfaces. For example, the C-band Simulsat dish is cut from a rectangular section of a sphere and has feeds mounted in a box oriented along the focal line (see Figure 5-9). The 5 meter (16-foot) model can simultaneously detect up to 35 satellites within a 70° arc and deliver a 44 dB gain. The Torus antenna, also designed to receive C-band broadcasts, is a dual curvature reflector having a circular cross section along the arc of satellites and a parabolic contour at right angles. This design also causes incoming signals to be reflected to a focal line.

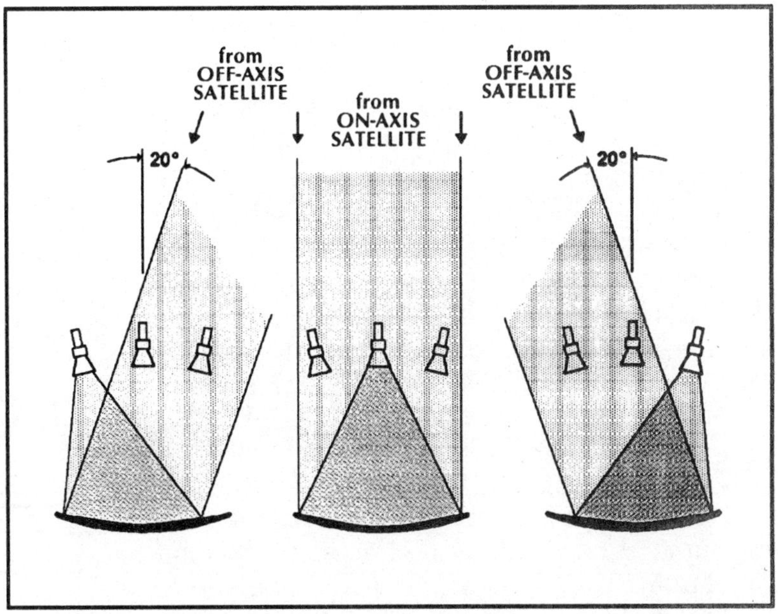

Figure 5-8. Spherical Dish Geometry. *A spherical reflective surface can be used to focus signals from neighboring satellites to a series of focal points. The feed is located further from the surface of the dish than in a prime focus configuration since a "shallow" reflector is used. In addition, some deterioration of signal, known as spherical aberration, occurs.*

A prime-focus parabolic dish can be used in a multi-focus configuration where one or two extra feeds and LNBs are offset on either side of the central feed. In this case, the reception of signals is not as "clear" because reflections from different parts of the reflective surface travel slightly different distances to reach these additional feeds. Multipath distortion also occurs in the operation of other multi-focus designs. The resulting decrease in performance can be partially overcome by using a slightly larger reflector. Two or more C-band signals can be simultaneously received with minimum degradation of performance by a 3.7 to 4.9 meter diameter dish.

The Horn Antenna

The horn reflector, shaped like a horn, sends intercepted signals entering via a large aperture down the reflective body of the horn to a focus. Thus the whole body of the horn performs the same function as the feedhorn and dish does for a prime-focus dish or Cassegrain antenna (see Figure 5-10). This type of antenna is almost always installed in line-of-sight relays used by common carriers terrestrial telephone, video and data transmissions.

Figure 5-9. The Simulsat Multi-focus Dish. *The multi-focus Simulsat dish can view up to 35 satellites at once. Either C- or Ku-band signals are directed to up to 35 feed and low noise amplifiers placed at the focal "line." (Courtesy of Antenna Technology Corporation)*

Figure 5-10. Horn Antennas on Line-of-Sight Microwave Relay. *Before the advent of satellite broadcasting, microwaves were relayed either via cables or between line-of-sight stations like the one shown here. Such towers are still widely used. (Courtesy of Microwave Filter Company)*

Benchmarks of Antenna Performance

A number of interrelated factors affects the overall performance quality of a dish. These include gain and efficiency, beamwidth, side lobe patterns and noise temperature. These performance criteria are determined by the overall design of a dish/feed system. Thus, for example, the "depth" of a dish, expressed by the distance between the feed and the reflective surface, and the type and extent of reflector surface irregularities all affect these parameters.

Dish Gain

Dish gain is the factor by which an incoming signal is concentrated when it reaches the focal point. Gain is measured in comparison to an "isotropic antenna" which radiates energy in all directions as a light bulb hanging by a wire does. A microwave reflector concentrates the signal into one direction like a car headlight directs a beam of light. Gain is measured in decibels relative to an isotropic antenna. For example, a gain of 1,000 is equivalent to 30 dBi.

Dish gain depends upon three parameters. First, as reflector surface area increases, more microwave radiation can be intercepted and therefore gain increases. If the reflective surface area is doubled, so is its gain. For example, a 10 foot dish has 4 times the area and thus 4 times the gain of a 5-footer. The surface area used to calculate gain is the total effective area aimed directly towards any particular satellite. Therefore, a dish with an offset feed which is angled perhaps as much as 30° off the satellite line of sight has less effective "intercept" area.

Second, gain increases with the square of the signal frequency. This can be conceptually explained because higher frequency signals do not spread out like waves in water but are more easily focused into straight lines like beams of light. To illustrate, Ku-band dishes receiving broadcasts with triple the frequency of C-band signals have gains 9 times higher.

Third, gain is determined by how accurately the surface of a reflective surface is manufactured to match its designed shape. Surface errors can be either random or systemic imperfections. Random errors, small distortions in the reflective surface, can cause substantial decreases in gain (see Table 5-1). Thus, a dish that has a large number of deep welts or ripples in its surface behaves more poorly than one that is smooth and more closely approximates its designed shape. Signals striking a dent are reflected away from the focal point. This loss of gain with surface imperfections becomes more important at higher frequencies. Therefore, a dish which may be perfectly adequate at C-band may not have a tight enough surface tolerance for quality performance at Ku-band.

TABLE 5-1. LOSS OF GAIN WITH SURFACE DISTORTION

Average Surface Distortion (cm)	Loss of Gain (dB) 4 GHz	12.5 GHz
0.01	0.001	0.012
0.05	0.03	0.29
0.10	0.12	1.19
0.25	0.76	7.44

Systemic surface errors are caused by design or assembly problems other than random surface irregularities. If reflector curvature is smooth but inaccurate or if a dish is incorrectly designed or assembled so that panels do not fit together properly, systemic errors occur. Systemic gain reductions can also be caused by "springback" or "metal memory" in metal dishes when ribs or panels "remember" their original flat shape and straighten out a little. In fiberglass dishes springback, as well as shrinkage, can cause displacement of the internal reflective surface. Mesh dishes can have gain reductions since there are often flat areas of reflector between supporting ribs. Systemic errors affect large portions of the reflective surface and can cause a double-focus to occur. Like random errors, the negative effect of such design flaws is much more pronounced at higher frequencies where reflections are more "specular," namely where reflections adhere more closely to straight line paths.

Dish Efficiency

Dish efficiency is a measure of the percentage of intercepted signal actually captured by the reflector/feed system. While a perfect dish system would be 100% efficient, typical efficiencies range from lows of 40% for inferior products to 70% or even higher for excellent quality products. Offset fed parabolic dishes can have efficiencies in excess of 80% because there are no structures between the target and the reflector surface to impede the capture of incoming radiation.

Efficiency is determined by a number of factors including:

- the accuracy of the reflective surface
- losses that occur when microwaves are not perfectly reflected but are absorbed by its surface
- aperture blockage caused by absorption and reflective losses from the feed, LNB and its supports that are located in the path of the incoming signal
- losses between the feed and low noise amplifier (LNB)
- "spillover" which is examined below in the section on feeds. An offset fed dish does not have efficiency losses due to aperture blockage.

Gains ranging from the theoretical maximum of 100% to as low as 50% efficiency are listed in Table 5-2. When reading these gain figures remember that a 3 dB difference means a doubling of gain; 1 dB means a 26% difference; and 0.1 dB means a 2.3% difference. The important message is that small decibel changes translate into large performance variations.

Calculated gains for 50% efficient reflectors are not half of those having 100% efficiencies because the numbers in Table 6-2 are expressed in decibels, not the raw signal concentration factors. For example, a perfectly efficient 5 foot dish concentrates signals by a factor of 4,150 while a 50% efficient dish would have half this gain, equal to 2,075. However, the higher gain translates into 36.18 dB and the lower into 33.17 dB, a difference of 3.01 dB as would be expected. A 3dB difference equals a 50% reduction in power.

TABLE 5-2. DISH GAIN in DECIBELS at 3.95 MHz

Dish Diameter (feet)	Dish Efficiency 100%	80%	70%	60%	50%
5.0	36.18	35.21	34.63	33.96	33.17
6.0	37.76	36.79	36.21	35.54	34.75
7.0	39.10	38.13	37.55	36.88	36.09
7.5	39.70	38.73	38.15	37.48	36.69
8.0	40.26	39.29	38.71	38.04	37.25
8.5	40.79	39.82	39.24	38.57	37.78
9.0	41.28	39.34	39.73	39.06	38.27
9.5	41.75	40.78	40.20	39.53	38.74
10.0	42.20	41.23	40.65	39.98	39.19
10.5	42.62	41.65	41.07	40.40	39.61
11.0	43.03	42.06	41.48	40.81	40.02
12.0	43.78	42.81	42.23	41.56	40.77

Beamwidth and Side Lobe Patterns

The field of vision of a dish system, what a dish actually "sees," is characterized by its beamwidth and side lobe pattern. This pattern can be measured by rotating a microwave source across the field of view of a test dish. The power of detected microwaves plotted against its angle towards the source then defines its beamwidth and side lobes (see Figure 5-11 and 5-12). This pattern is, in essence, a "fingerprint" of the quality and performance of the dish/feed system.

Beamwidth is a measure of how narrow a region of space can be targeted. This is a very important factor considering that satellites separated by 2° or less and situated at a distance in excess of 22,300 miles appear to be very closely spaced together. A plot of the "vision" of a dish shows that most of the received power is contained in the main lobe. The beamwidth is defined as the width of this main lobe between "half power" points where detected signal power has dropped by 50% or 3 dB (see Table 5-3).

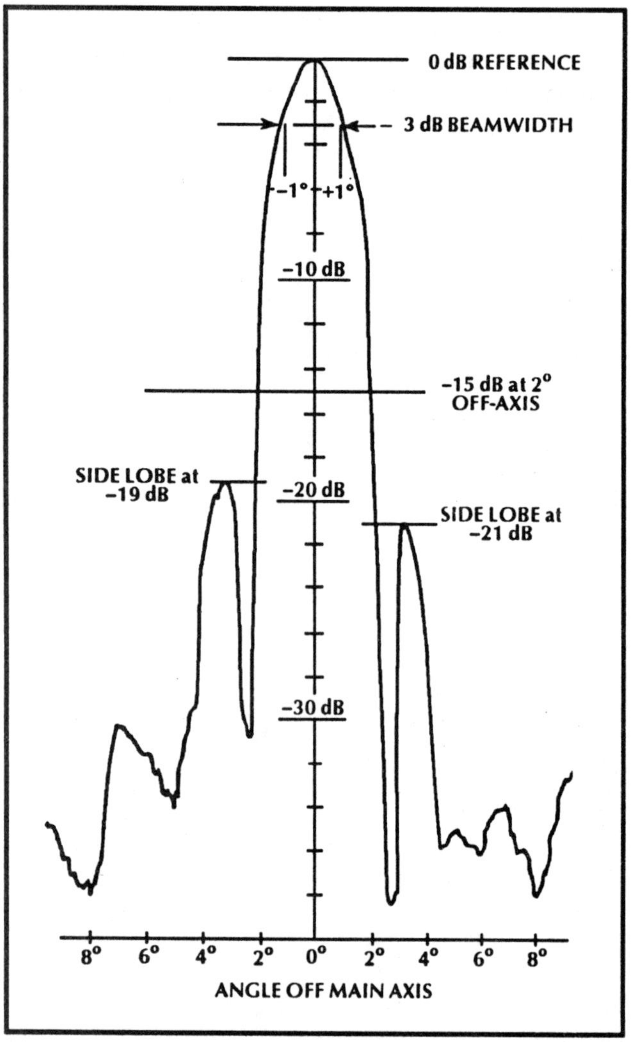

Figure 5-11. Dish Lobe Patterns. *This graph shows what power levels, expressed in relative terms, a 7.5 foot (2.3 meter) dish detects from its boresight or main axis to its periphery. The taller and more narrow the central portion is relative to the side lobes, the better a dish targets and pinpoints a satellite. Here the 3 dB or half power beamwidth is 2.2° and the side lobes located at 4.5° on either side of the main lobe are reduced 25.5 dB in power. Signals arriving at 2° are reduced to 12 dB relative to the main beam at 4.2 GHz. By contrast, the 3 dB beamwidth at 12.2 GHz is 0.73° and the side lobes, that are reduced 25.5 dB in their ability to detect signals, are located at 2.1°. Signals arriving at 2° are reduced to 26 or 27 dB depending on their position in beamwidth pattern.*

The beamwidth determines how well a dish detects off-axis radiation. For example, if the beamwidth is 1.2° then at 0.6° at either side of the main dish axis signals will be captured at half the power of those on target. In this case, a second equally powerful microwave source located at 0.6° off-axis would interfere with clear reception of the primary signal.

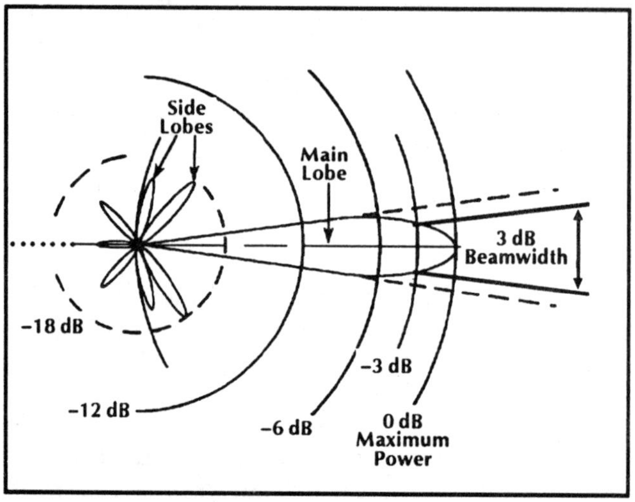

Figure 5-12. Calculated Dish Lobe Pattern. *Although most of the radiation detected for a parabolic dish is concentrated in the main lobe, this calculated pattern shows that some noise can be detected via side and back lobes in the full circle around the reflector. Beamwidth is measured between the two 3 dB or half power points.*

TABLE 5-3. CALCULATED 3 dB BEAMWIDTHS

Dish Diameter (meters)	Beamwidth (degrees) 4 GHz	12.5 GHz
0.30	17.5	5.6
0.60	8.8	2.8
0.90	5.8	1.9
1.20	4.4	1.5
1.50	3.5	1.1
2.00	2.6	0.8
2.50	2.1	0.7
3.00	1.8	0.6
4.00	1.3	0.4
5.00	1.1	0.3

The exact location and relative magnitude of the side lobes is an important determinant of performance. Side lobe patterns indicate how well a dish detects radiation arriving from off-axis directions. In order for the dish to ignore off-axis signals and noise, side lobes must be substantially reduced relative to the main lobe. For example, if a second satellite happened to be 2° removed from the target along the geostationary arc and if the first side lobe were located at 2° and reduced in power by only 5 dB, the interference between signals simultaneously detected from both satellites would be very annoying. However, if this side lobe were positioned at 3°, the second satellite would probably have no influence on the primary broadcast.

Thus the difference between an excellent and a poor quality dish can easily be seen in their lobe patterns. A wide beamwidth or a low ratio of the main lobe to the first side lobe would characterize a poor quality dish. Such a dish would be more susceptible to detecting off-axis noise and other interfering signals. Even if two dishes have identical gains their performance might be quite different. For example, if a square and a round prime-focus reflector were cut from identical larger surfaces so they had the same gains, the magnitude and placement of the side lobes would indicate why their performances might be markedly different.

Dish lobe patterns depend upon a number of factors including reflector diameter, the frequency of the incoming radiation and surface tolerances. The higher the frequency and the larger the dish aperture, the more narrow the beamwidth. Thus a 10 foot reflector has one half the beamwidth of a similar 5 foot dish. Ku-band dishes have one third the beamwidth of C-band systems because the microwave frequencies in use are 3 times higher. Surface imperfection generally tend to widen the main lobe and increase side lobes.

Side lobes stem from the inherent wave-like character of microwaves and the resulting "spreading" in the vision of a dish. This "spill-over" results when microwaves strike the edges of the reflector rim and are bent as a result of interference and diffraction, well-known effects from the discipline of optics. Microwave spillover actually allows dishes to have the rather astonishing ability to detect radiation coming from directly behind them (see Figure 5-13)!

Every dish system has a lobe pattern. The location and magnitude of side lobes as well as the size of the side lobes relative to the main lobe indicates how well power from the surrounding full circle will be detected. These theoretical patterns are altered by real world considerations such as surface tolerances and placement of the feed/LNB assembly and its support. Such considerations underlie the fact that offset fed designs, in addition to having higher efficiencies, may also have "cleaner" side lobe patterns.

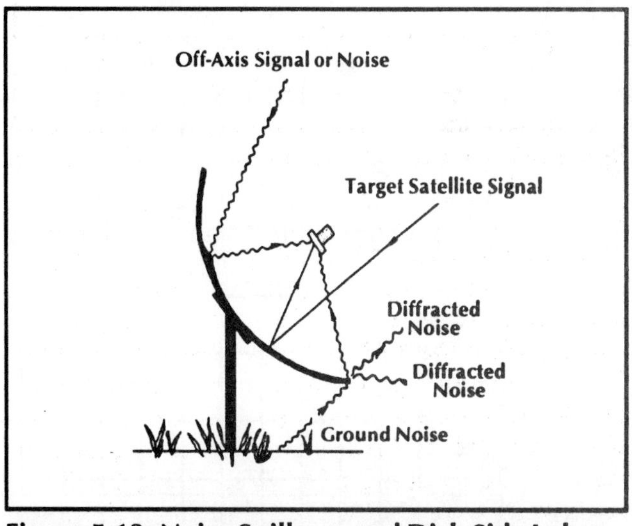

Figure 5-13. Noise Spillover and Dish Side Lobes. *A dish can actually detect microwaves and noise from behind, 180° off its main axis. Noise from the warm ground is diffracted by the reflector edges and is scattered in all directions. Some of this noise manages to enter the feed. As reflector surface inaccuracies increase, higher side lobes are formed. As a consequence, the dish is more capable of detecting unwanted signals from the ground or from off-axis communication satellites.*

Dish Noise

Dish noise temperature is a measure of how much noise the reflector detects from the surrounding environment. Some of this noise enters via the dish side lobes; a smaller amount, predominantly that from outer space, enters via the main lobe. Noise also comes from both man-made sources such as fluorescent lights which emit microwave radiation as well as from natural sources such as the surrounding terrain. In general, a satellite reception system detects mainly the natural, higher frequency sources of noise.

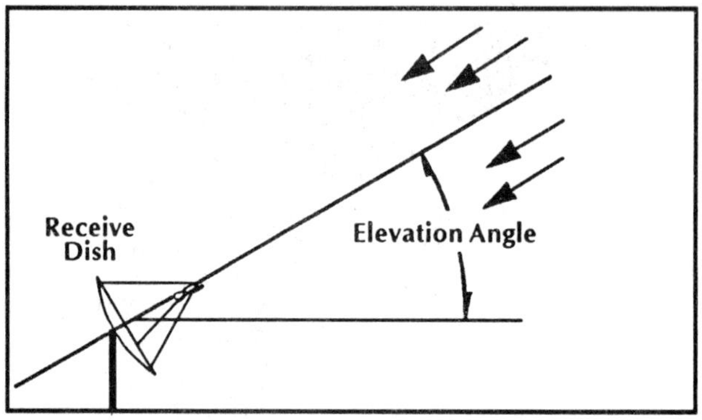

Figure 5-14. Dish Elevation Angle and Noise. *As a dish is pointed at greater elevation angles, it "sees" less of the warm ground and thus less noise.*

Since the warm ground emits microwave radiation, noise temperature increases as a dish is pointed at progressively lower elevation angles in regions further from the equator or towards satellites nearer to the horizon. Larger dishes detect less noise because they have smaller side lobes. Also, offset fed designs which can be inclined steeply towards the horizon benefit both by not collecting excessive snow loads and by having their feeds pointed farther away from the "noisy" ground (see Figure 5-14).

Typical values for dish noise versus elevation angle are listed in Figure 5-15. Noise temperatures as well as gains and half power beamwidths have all been experimentally measured on standard models. Dish noise temperature decreases with reflector diameter because the beamwidth decreases and side lobes are closer to the dish boresight. It also decreases with elevation angle as the reflector is pointed further away from the "hot" ground.

f/D – Dish Depth

The depth of a prime-focus dish, quantified by its focal length to diameter ratio, f/D, has a direct impact upon its performance (see Figure 5-16). In general, everything else being equal, deeper dishes having smaller f/D ratios will have smaller side lobes because the feed and LNB are closer to the reflective surface and thus are better screened from the surrounding environment. A similar result can be achieved by attaching a non-reflective shroud around the dish rim, similar to the tactic of using blinders on a horse so that it sees in only one direction, straight ahead. Installing a shroud results in different dish performance than lowering the f/D ratio because side lobe patterns of both these configurations differ.

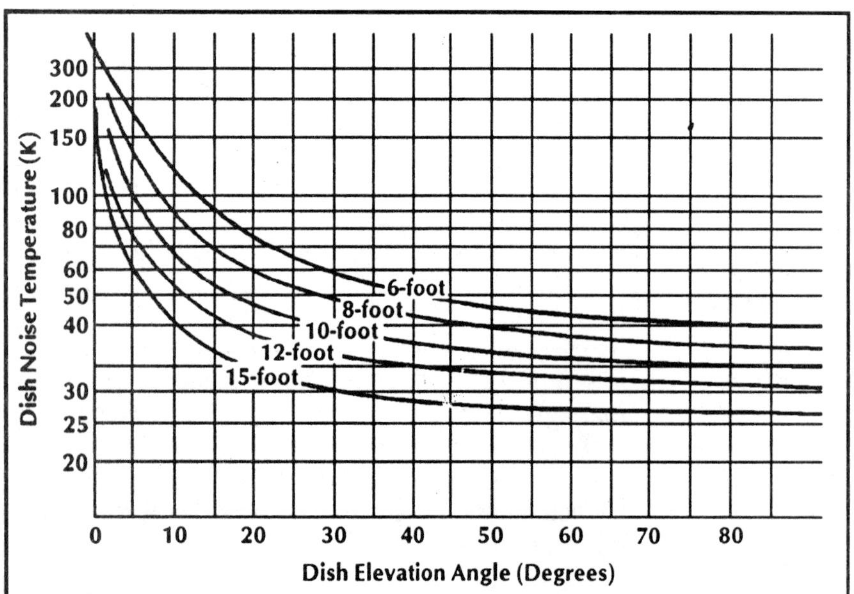

Figure 5-15. Variation of Dish Noise Temperature with Diameter and Elevation.

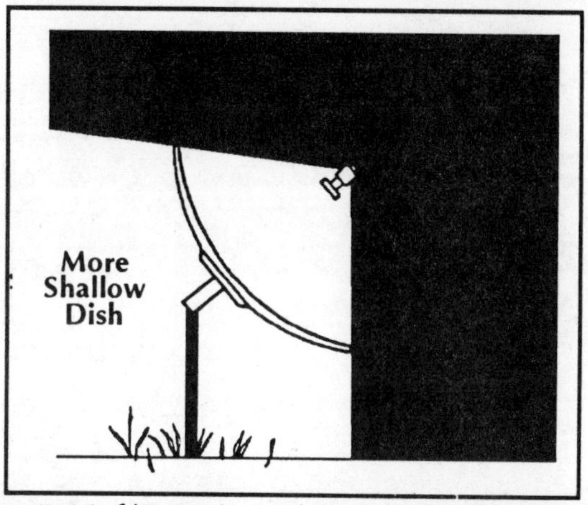

Figure 5-16. f/D Ratio and Dish Field of View. *Dishes with a lower f/D ratio have a more narrow field of view and are therefore potentially less susceptible to noise and interference. When correctly designed, these "deep dishes" have lower side lobes. However, in general, a deep dish performs more poorly in a dual-band offset fed configuration than in a shallow one.*

The f/D is used to classify dishes as either deep, average or shallow. In general, an f/D of less than 0.3 defines a deep dish; one in excess of 0.45 defines a shallow dish (see Figure 5-17). Deeper dishes are generally less susceptible to environmental noise and usually have lower side lobes and lower noise temperatures than more shallow designs. Note that offset fed antennas must generally be rather shallow to have acceptable performance.

Dish Construction

Dishes can be made from a variety of materials and by numerous manufacturing processes as long as basic requirements are fulfilled. The reflective surface must conform to its designed shape and have good performance over the product lifetime. In addition, dishes intended for residential use must be simple to assemble, aesthetically pleasing, reasonably priced, easy to ship and durable. Some of these latter criteria are not as important for commercial applications.

The details of design, materials selection and manufacturing impact on dish durability. A system may function well when new but, in time, after being subjected to environmental stresses, its performance may seriously degrade. Allowances must be made for contraction and expansion of different materials with changes in temperature, for wind loading, for rain, snow, hail and corrosion, for the effects of intense sunlight and for many other specialized environmental factors. Remember, the dish is the eyes of a satellite reception system and the most critical component needed for quality performance.

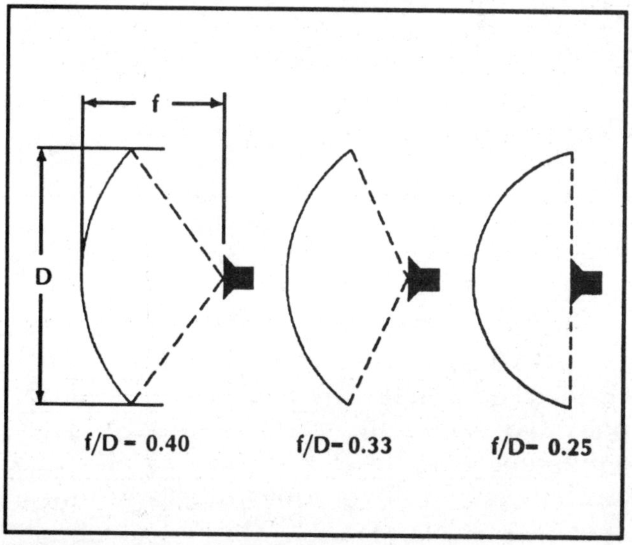

Figure 5-17. Focal Length to Dish Diameter. *A lower f/D means that the feed assembly is located more closely to a given sized reflective surface. At a 0.25 f/D, the feed lines up with the dish rim so that adequately illuminating the entire reflective surface becomes difficult.*

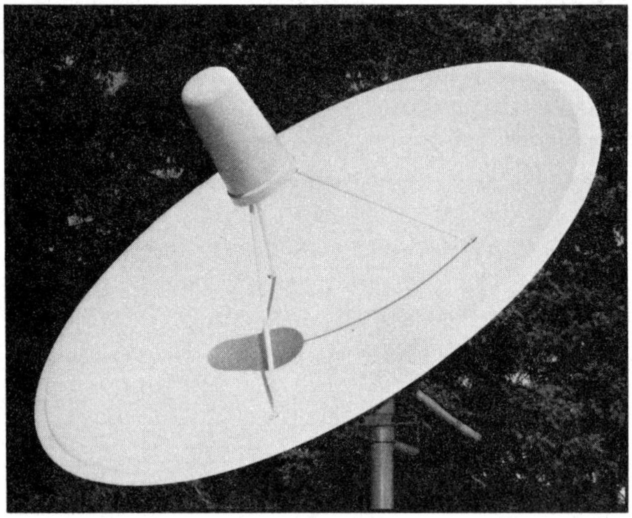

Figure 5-18. Hydroformed Dish. *In the hydroforming process a blank aluminum or stainless steel sheet is placed onto a mold and pressed into a parabolic shape using water to aid in creating a smooth and accurate reflective surfacing. (Courtesy of Paraclipse Manufacturing, inc.)*

Four major classes of dishes are manufactured today: spun aluminum or steel, stamped or drawn steel, fiberglass and expanded or perforated metal or mesh.

Spinning or Hydroforming

Spun or hydroformed aluminum or steel dishes can have very accurate surfaces (see Figure 5-18 and 5-19). They are manufactured by mounting aluminum or steel, typically of 0.080 inch thickness, onto a mold and spinning the whole assembly while pressing it into shape with rollers. When done correctly, this technique results in a highly accurate reflective surface that is well hardened during manufacturing. However, the spinning process can leave very shallow ridges created by the passage of the forming tool over the reflector surface. It is important that these ridges be substantially less than 1 mm otherwise losses due to surface irregularities can exceed 30%.

Figure 5-19. 1.8 Meter Spun Aluminum Dish. *The manufacturer of this dish used a spinning technique to mold an accurate reflective surface that can achieve high gain and efficiency. (Courtesy of Andrew Antenna Company)*

Spinning has the advantage that several different sized dishes can be manufactured from one mold; each diameter dish so produced has a different f/D ratio. But since these reflectors are manufactured in one piece, the larger ones, over 6 feet in diameter, may be difficult to ship using conventional small-package services.

Some spun dishes are perforated by punching the reflector with a regular pattern of small holes after spinning is complete. Signal losses through these holes are negligible if their diameters are less than approximately one tenth of the radiation wavelength, i.e. less than 7.5 mm for C-band signals. If the openings are too large, dish efficiency will be extremely low as microwaves "leak" through. The opposite side of this performance coin is that noise of the same frequency originating from the warm ground will pass through the reflective surface when the openings are too large and reach the feed/LNB thus increasing the amount of noise power detected.

Stamping

Stamped or drawn steel dishes are manufactured by stamping or pressing flat sheet metal into shape. The required tooling can be quite expensive because massive, thousand ton presses are needed. In addition, each mold is capable of producing reflectors of only one size. Nevertheless, reflectors with accurate surfaces can be produced if the die is true and manufacturing techniques are excellent (see Figure 5-20).

Figure 5-20. Winegard Perforated Dish. *This perforated dish can be manufactured either by stamping, hydroforming or spinning. The hole diameter should be less than 3/4 inch for adequate performance at C-band. The QuadStar 3 meter dish is constructed with 1 mm thick (0.040 inch) perforated aluminum with a hole diameter of just under 2 mm. It has an f/D of 0.278, a half power beamwidth of 1.6° and gain of 49.6 dBi at 12.7 GHz. (Courtesy of Winegard Company)*

Fiberglass

The challenge in manufacturing a fiberglass dish is to embed wire screen, aluminized mats or other reflective materials in the fiberglass resin so that the metallic substructure retains an accurate parabolic shape.

Fiberglass dishes are generally manufactured by one of three methods. In the hand layering or laminating process, a fiberglass resin called gelcoat is smeared onto fiberglass cloth laid in a mold. The gelcoat often has pigment to give the surface color and always has inhibitors to protect it against ultraviolet damage from intense sunlight. Metal is then flame sprayed onto the dish. This results in a strong, lightweight but relatively expensive unit.

In sheet-molding, polyester resin, calcium carbonate and chopped fiberglass strands are mixed together and are then pressed onto a metal screen. The resulting product is lightweight but is usually brittle and more easily damaged than other varieties of fiberglass dishes.

In the thermal injection molding process, the mat and screen are set in a closed mold and the resin is injected under pressure. Dishes manufactured by this process are strong, impact-resistant and usually have good aesthetic appearance.

Wire Mesh or Expanded Metal

The fourth type of dish is constructed from wire mesh or expanded metal. These metal petals are held together and attached to an extruded rib structure by a series of clips or by insertion into channels in the ribs (see Figures 5-21 and 5-22).

Mesh dishes have been very popular in C-band TVRO systems because of their aesthetic appeal, low visibility, light weight and lower overall wind resistance but they can be flimsy if not correctly designed. Some varieties can also be time consuming to assemble. If desired, perforated metal can be used in place of wire mesh or expanded metal petals in such a dish design.

Figure 5-21. KTI Mesh Dish. *This 6 foot pole-mounted expanded mesh dish can be shipped in four relatively small boxes. It has an f/D ratio of 0.4, a gain of 35.7 dBi at 4.2 GHz and 44.6 dBi at 12.2 GHz and weighs 60 pounds. (Courtesy of KTI)*

Wind Loading

In regions of the world where wind gusts often exceed 80 miles/hour, wind loading of satellite receive dishes is an important design consideration, especially for larger reflectors (see Table 5-4). In the worst cases, the reflector may be twisted or pulled from its mount causing structural damage or even injury to persons. Wind forces can also cause even the most securely mounted dishes to waver off target with the resultant deterioration in reception.

Forces exerted by winds can be powerful. For example, a solid 8 foot dish can experience as much as 900 pounds of force under the impact of a 75 mile/hour wind. Even a relatively small 6 foot reflector experiences as much as a 400 pound force at this same wind velocity.

In most cases, mesh and perforated dishes are subjected to lower wind loading forces than are solid reflectors. This is especially so when the wind blows directly into the reflective surface. However, when the wind blows parallel to the lip of a dish, forces are comparable to those on a solid surface. The maximum wind loading on both solid and mesh varieties occurs in just this situation, when the wind "spoons" along the reflective surface just like it does when it blows across a properly-oriented sail.

TABLE 5-4. DISH WIND LOADING (Pounds of Torque)

Dish Diameter	Wind Speed (mph) 25	50	75	100
6	40	170	400	700
8	100	400	900	1600
10	200	800	1800	3200
12	350	1400	3000	5500
14	550	2200	5000	9000
16	800	3300	7000	13000

Figure 5-22. Superior 10' Mesh Dish. *This prime focus wire mesh dish is designed to have an f/D ratio of 0.380 and gain of 40.7 dBi at C-band frequencies. (Courtesy of Superior Antenna Manufacturing, Inc.)*

DISH MOUNTS

The purpose of a mount is to accurately and securely aim a dish towards a communication satellite. While many smaller dishes permanently aim towards just one communication satellite, more flexible mounts must also be capable of targeting any other spacecraft in the geostationary arc.

Using a mount that provides stability and pointing accuracy is critical in designing and installing a reception system, especially for large dishes that have narrow beamwidths and thus target a very small portion of the sky. For example, a 10 foot reflector with a 1.7° half-power beamwidth that targets just 1/15,000th of the visible sky from any point on earth. Larger dishes have proportionately more restricted visions. A wind blowing across the sail-like surface of a dish that results in a 0.01° deviation causes the center of its vision to scan 40 miles at the geosynchronous arc leading to a decrease in signal strength and broadcast quality (see Table 5-5 and Figure 5-23).

Mounts should be rigid, strong and firmly attached to both the dish and its underlying supporting structures. Stable mounts are particularly important when larger dishes with smaller beamwidths are installed. Reflectors as large as 16 feet in diameter are not uncommon in portions of the Middle East, Africa or Latin America. These require very stable mounts that can be accurately aimed.

TABLE 5-5. SATELLITE SPACING

Angular Spacing	Separation between Satellites (miles)
0.5	594
1.0	1,190
2.0	2,378
3.0	3,567
4.0	4,757

The design of the dish-to-mount attachment is very important. Since the mount must be securely bolted onto a reflector, any mismatch can cause stresses and possibly warping. This would result in increased side lobes and impaired reception. Fiberglass or a thin-walled metal can bend when placed under such stress. The points of attachment must be properly located and sufficient in number to evenly distribute the weight under both static and wind or snow loaded conditions, and to prevent dish warping under these loads. Collars which secure the mount to the support pole must also be very solid in order to prevent twisting away from the correct north/south orientation. Pivot points which permit tracking movements must incorporate ball-bearings, bushings or other well-designed mechanical joints in order to prevent excessive wear over time.

Figure 5-23. A Rigid Mount. *The ten mount arms on this Prodelin fiberglass dish are solidly attached to the reflector in twenty places with two 13 mm (1/2 inch) bolts per arm. This rigid design counteracts the tendency of the reflector surface to sag when the antenna is aimed at low look angles. Any flexing would result in a loss of gain and an increase in side lobes. (Photo Courtesy of Brent Gale)*

Types of Mounts

There are two principal classes of mounts: those that track the arc with two movements or degrees of freedom like the azimuth-elevation (az-el) mounts and those that require only one axis of motion like the polar mounts. Either of these can be fixed in position, if desired.

Az-El Mounts

An az-el mount is simple to understand and install but more difficult to operate than a polar mount if used to track from satellite to satellite. Any communication satellite is located by first moving the mount to the correct azimuth angle, which is along a plane parallel to the surface of the earth, and then by rotating up to the required elevation angle (see Figures 5-24 and 5-25).

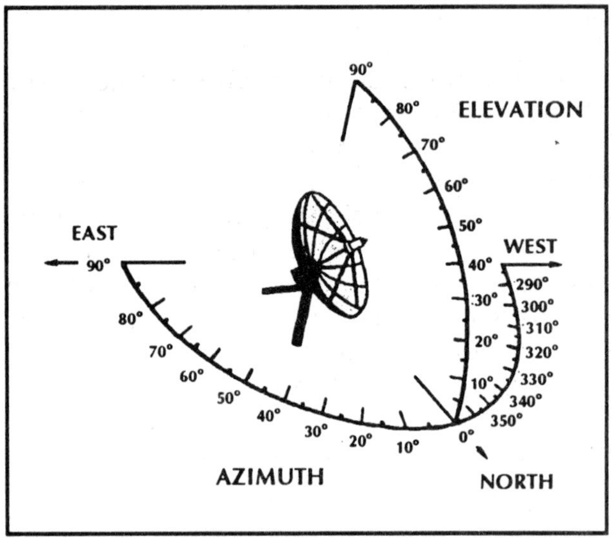

Figure 5-24. Az-El Mount Geometry. *When rotated in its azimuth, a dish moves in a horizontal plane. Adjusting the elevation angle raises the dish above the horizon.*

Linearly polarized satellite signals are oriented relative to the equatorial plane. The feed on an az-el mounted dish is also oriented with the local horizontal plane. Therefore, the reception of linearly polarized signals from satellites to the east or west of the site requires some skew of the feed adjustment for optimal reception. The geometry underlying this statement is examined below in the section on feeds.

Two motors are required to remotely move an az-el mounted dish. A third small motor or mechanism must be employed to adjust skew when linearly polarized signals are being received.

Figure 5-25. The Gibralter Az-El Mount. *This rigid, motorized az-el mount is powered by a 36 Vdc motor and is supported by a 40.5 cm (16 inch) steel pipe. Stamped aluminum dishes of 3, 3.6, 4.2 and 5 meters are available from the same company. (Courtesy of DH Antenna)*

Az-el mounts have advantages in some applications. For example, az-el mounts that support C-band dishes are commonly used in motor homes and recreational vehicles because they have the ability to hunt in two directions. They can therefore compensate for the random orientation of vehicles that are parked on a variety of terrains. There is no tracking error so each satellite can be perfectly targeted. Az-el mounts that are stabilized in both axes with gyroscopes may also be used on ocean-going vessels. The bi-directional movement is also useful in tracking satellites in unstable, inclined orbits.

Az-el mounts were historically used in conjunction with commercial C-band systems in out-of-footprint locations where larger dishes were required. In such cases, az-el mounts that supported massive dishes having relatively narrow beamwidths were mechanically more stable, stronger and simpler than polar mounts. The fine tuning of a polar mount's north/south orientation required during installation can be cumbersome when installing large diameter dishes. Today the mechanical designs have improved to the point that most large C-band tracking systems can incorporate polar mounts. Today most small fixed-dishes are mounted on az-el supports.

Polar Mounts

A dish supported by a polar mount is targeted towards any communication satellite by rotating around a properly aligned inclined axis. This one movement sweeps an elliptical curve of intersection with the equatorial plane which matches the circular Clarke Belt of geostationary satellites to within a few thousanths of a degree. This offers an accuracy that is less than the stability inherent in most mounting devices and better than the stationkeeping of most geostationary satellites.

Satellite receive dishes detect signals from satellites orbiting in the Clarke Belt, a distance that is a small fraction of the near-infinity distance at which astronomers "look". Thus polar mounts that support dishes must be modified from those that aim at stars. Stargazing polar mounts were originally developed by astronomers who realized many years ago that it would be easier to keep a telescope sighted on a celestial body if it could be swiveled around a single axis to exactly counteract the earth's rotational motion. The polar axis, or hour angle axis as it was originally called, lies parallel with a line passing through the north and south geographic poles. To achieve this orientation, the polar axis is set equal to the latitude of the installation site. For example, in Denver, Colorado, which is located at a 40°N latitude, the polar axis angle is set equal to 40° as measured from the horizon. At the equator where the latitude is 0°, the polar axis angle equals 0° because the arc of satellites is tracked by rotating along a circle directly overhead. At both these and all other locations, the polar axis must then be oriented exactly along a north/south line.

Polar mounts supporting satellite receive dishes are modified so their polar axis of rotation is tilted away from parallel to the Earth's axis of rotation (see Figures 5-26 and 5-27). The tilt is known as the declination angle. Like the astronomer's true polar mount, the mount axis is aligned in a true north-south direction.

Early attempts to adapt polar mounting to satellite reception often did not appreciate the need for modification, i.e. axial tilt. Instructions and tables directed that the axis should be inclined to exactly match the angular magnitude of the site's latitude and listed declination values that were slightly too high. The three values exactly total 90.0 degrees for any latitude. While the apex elevation angles are mathematically calculated, determining the declination angle is a matter of error minimization. Those listed in Chapter 8 are derived on the basis of minimizing "worst case" error among visible sat-

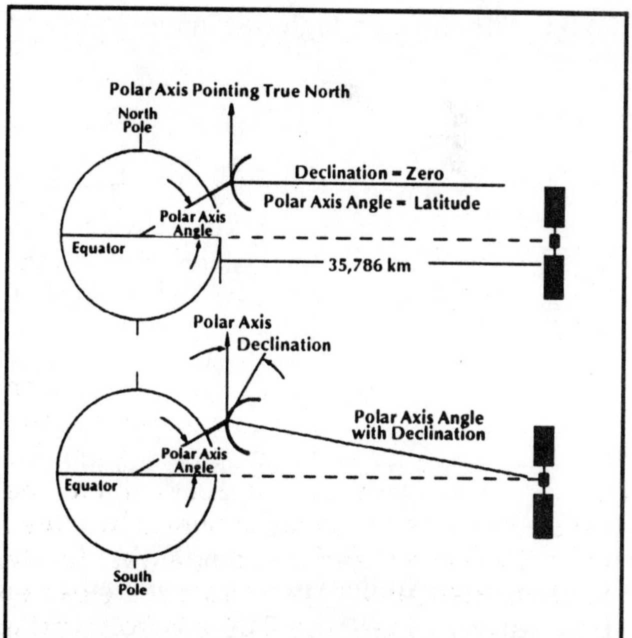

Figure 5-26. Polar Mount Alignment. *The polar axis angle, set equal to the latitude, points a dish along a plane parallel to one through the equator. Setting the declination offset angle lowers the field of view to the geosynchronous arc of satellites. Declination would not be necessary if the satellites were at a near infinite distance, as are the stars.*

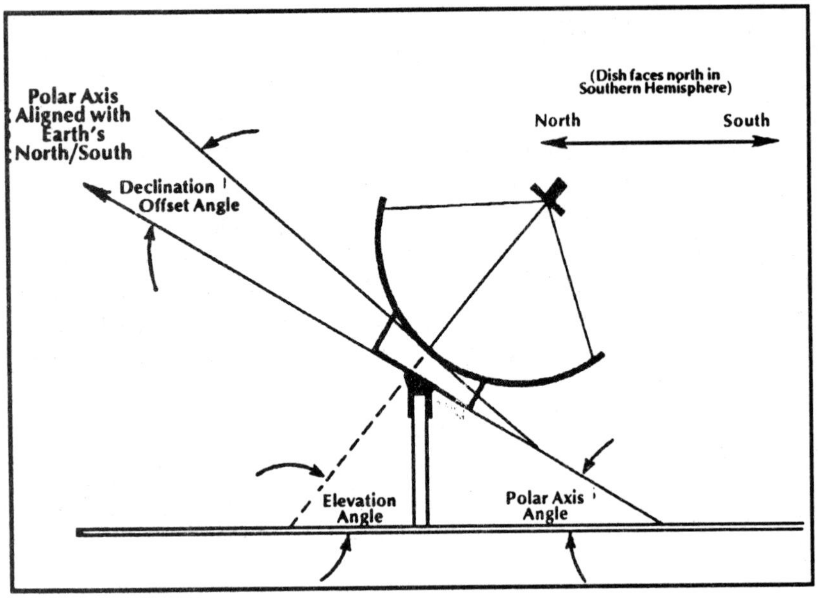

Figure 5-27. Geometry of a Polar Mount. *This illustration of a polar mount shows all the details of setting true north/south alignment, elevation and declination offset angles.*

ellites in the Clarke arc. The incorrectly high declination angles were derived by subtracting the latitude and apex angles from 90 degrees. While tilt angles are very small compared to latitude angles in most populated areas, they make a relatively large difference on declination angles that are much smaller in value. Much time and effort has been expended in the field in fine tuning a high declination angle to the optimal value.

The declination offset adjustment results in a dish whose "view" tracks a conical surface, with the site at the peak of the cone. The axial tilt away from parallel to the earth's axis of rotation, causes the conical intersection with the equatorial plane to become an ellipse that closely follows the Clarke Belt, itself a circle about the center of the earth. Without axial tilt, the conical intersection with the equatorial plane would be circular and not match the Clarke Belt since these two shapes would not be centered at the same point.

The Geometry Underlying the "Modified" Polar Mount

The aiming accuracy of the polar mount as originally conceived by astronomers can be improved from 0.1 to less than 0.01° even at the end-of-arc satellites. As described above, the "modified" polar mount geometry is achieved by fine tuning the polar axis and declination offset angles. Once this is accomplished the 0.01° pointing error is less than the tolerances of the best available az-el mount mechanisms.

The reason the polar mount tracking error is reduced can be explained as follows. Assume that the angles have been correctly adjusted in a "true" polar mount. The polar axis is then tilted slightly forward towards the arc of satellites. Recall that normally this axis points along a true north/south line. The declination offset angle is then reduced by an equal amount. A satellite due south from a receive site in the northern hemisphere (or due north from a receive site in the southern hemisphere) would be unaffected by this net zero change. However, when the dish is rotated towards either due east or due west the slight decrease in declination causes it to point higher than it would on a true polar mount. In effect, the most northerly satellite can be targeted perfectly, while the dish is also more accurately aimed at the easterly or westerly spacecraft. Tilting the polar axis effectively changes the sweep from a circle to an ellipse in the equatorial plane.

Only one motor is required to rotate a dish around a polar axis to track all the geostationary satellites within view. In addition, no correction motor is required for reception of linearly polarized broadcasts since the feed is moved around the polar axis which is, for all practical purposes, perpendicular to the equatorial plane.

ACTUATORS

Actuators provide the mechanical drive to remotely scan a dish across the arc of satellites. In the early 1980s there were a limited number of satellites in the orbit over North America so most dishes were either fixed on one target or hand-cranked between spacecraft. Most dishes for receiving C-band television broadcasts at cable TV headends in North America are still aimed at just one satellite. However, as more satellites began transmitting television programming, dishes had to be moved from one spacecraft to another in order to receive each satellite. Manually moving a 10 or 12 foot dish in cold and wet or even in good weather was not an attractive proposition.

Today, many satellite TV installations feature either linear or horizon-to-horizon actuators. Linear actuators employ a motor-driven, worm-gear assembly that attaches between a support on the dish and another on the mount. Horizon-to-horizon designs have their entire gear mechanism encased at the fulcrum of the dish and have the ability to track satellites across the whole visible sky, between both horizons.

Monitoring Dish Position

Actuators used in the early more primitive days of the C-band home satellite industry had very simple customer interfaces. An east or west button or switch would make or break a dc voltage sent to a motor that would move the gears and thus the dish. Position was maintained by the internal clutch assembly of the actuator. When some customers damaged their actuators and occasionally their dish by inadvertently driving motors past mechanical limits (see Figure 5-28), programmable east and west electronic limits were introduced. Even so, this problem still occurs at times. Even those "modern" actuators having programmable limits sometimes "forget" their east/west boundaries and can subsequently damage actuator jacks or reflectors, most often during power surges or "brown-outs."

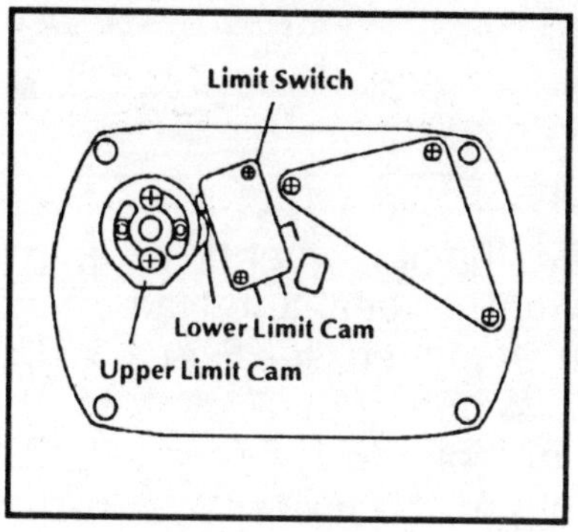

Figure 5-28. Mechanical End Limit Mechanism.

More advanced positioners are part of a closed loop control system (see Figure 5-29). The indoor controller supplies power to the motor in the positioner, while the feedback element in the positioner relays a signal back to the controller. The character of the signal depends on the position of the dish.

A satellite is selected from the front panel of the indoor controller. The position of a particular satellite is assigned a value, either a number or voltage depending on the feedback system in use. Power is applied to the positioner until the the signal returning from the positioner's feedback element matches the value stored in the indoor controller.

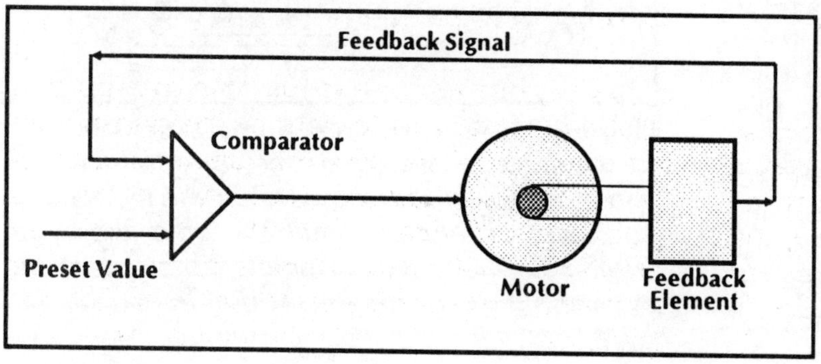

Figure 5-29. Closed-Loop Feedback in Dish Positioning. *A closed-loop feedback system used to position a dish consists of a actuator, a controller and a feedback sensor. The sensor can be a potentiometer or a reed, optical or Hall effect sensor. A preset voltage is compared with the output of the feedback element. The actuator motor is powered until the feedback value equals or exceeds the reference value.*

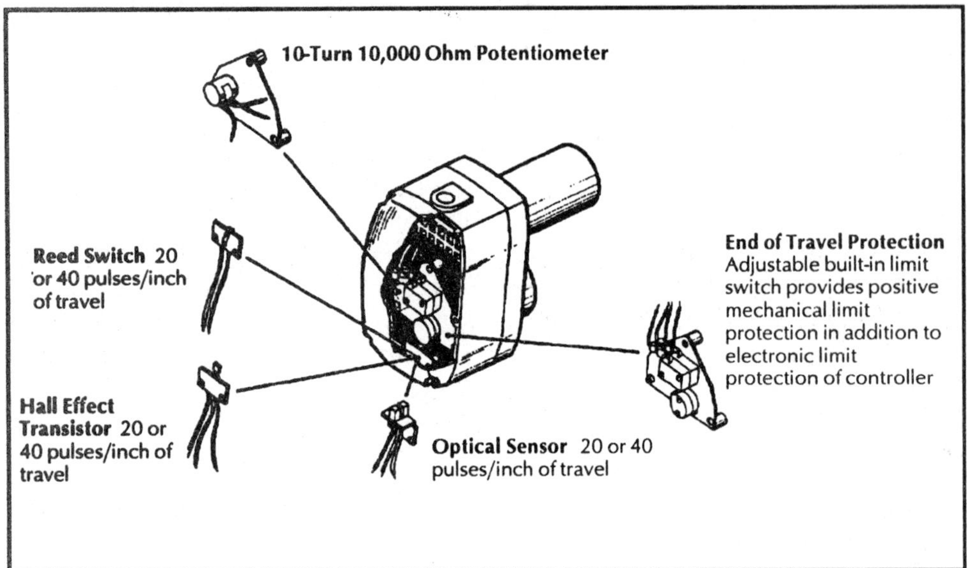

Figure 5-30. Actuator Sensor Positions. *Four types of sensors have been used to monitor dish position. The mechanical end of travel protection illustrated here is an important design feature. (Courtesy of Pro-Brand International)*

A variety of feedback elements including resistive, reed switch, optical and Hall effect sensors have been employed (see Figure 5-30). In resistive feedback, one of the earliest methods used, a ten turn voltage-dividing potentiometer, is attached to the motor by a set of gears. The voltage at the potentiometer slider is used as a feedback signal to the controller (see Figure 5-31).

A reed switch generates a pulse when moved in the vicinity of a magnet, a pulse rather than an analog voltage. One end of the actuator's limit is defined as count zero and the other limit as the maximum count. The magnet is attached to the motor by a set of gears so when power is applied to it, the magnet rotates. As a result, the reed switch closes once every rotation. The pulses thus created are relayed to the indoor control circuitry via a small diameter wire where a continuous count is maintained. The position of each satellite is defined as a specific number of pulses from zero. When a satellite is selected at the front panel, a pulse count value is loaded in from the controller's memory. The controller determines if the pulse count of a different satellite is greater or less than that of the present target and then applies the necessary positive or negative voltage to the motor until a sufficient amount of pulses has been added to or subtracted from the original count. If the motor is finely geared, this motion can be accurate to within 10 mm or less. This method is subject to electrical interference so it requires the use of shielded cable on the feedback run to the indoor controller. While ordinary cable can be used, shielded cable reduces the chance of false pulses changing the dish position.

Optical and Hall effect feedback mechanisms function similarly and generate pulses sent indoors via a shielded cable. An optical counting element uses a light emitting diode and phototransistor which is interrupted by the motion of the motor to generate a feedback pulse. A Hall effect sensor is a semiconductor that is sensitive to magnetic fields. Current flow varies as a rotating magnet moves across its surface.

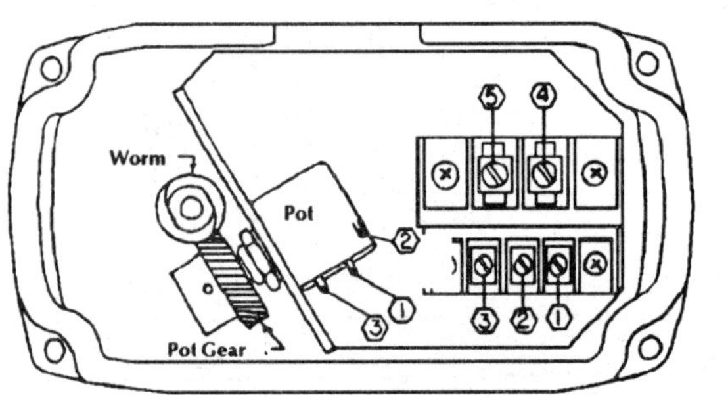

Figure 5-31. Actuator Potentiometer Sensor. *In this case, a worm gear drives the potentiometer gear and thus varies the resistance relayed to the actuator controller. Terminals 1, 2 and 3, to which the sensor wires attach, connect to points labeled with the corresponding numbers on the pot. Terminals 4 and 5 are connection points for the motor.*

Mechanical Construction

Linear Actuators

Linear actuators have a telescoping arm that moves within a fixed external tube (see Figures 5-32, 5-33 and 5-34). One end of the actuator is attached to an arm on the rear of the dish surface and the other to the supporting mount. A worm-gear which drives the internal arm can employ either acme or ball screw designs.

The acme jack is constructed from a threaded shaft that moves in a threaded collar. It is simple and lower in cost than the ball screw mechanism which is similar but has ball bearings which replace the threaded collar that moves in the tube threads. This mechanism has a much lower frictional loading than the former so that a motor of a given power transfers more force in moving a dish. This increased force and the smoother ball bearing movement reduces the chance of the shaft seizing up during cold weather or after long periods of non-use.

As a general rule, an actuator arm should never have to push or pull a dish through an angle of less than 30° as measured between the arm and the back of the reflective surface. The larger this angle, which depends upon the design of the mount and the length of the actuator arm, the less the lateral force on the arm during aiming. Proper installation reduces the chance that the internal rod could be bent and damaged. This type of actuator is limited in its motion to about a 100° sweep across the geosynchronous arc.

Figure 5-32. Internal Mechanism of a Linear Actuator. *The linear actuator arm bolts onto the mount at one end and onto the dish at the other. A dc motor extends or retracts the inner arm to track any preprogrammed satellite as selected by the indoors controller. (Courtesy of Prosat)*

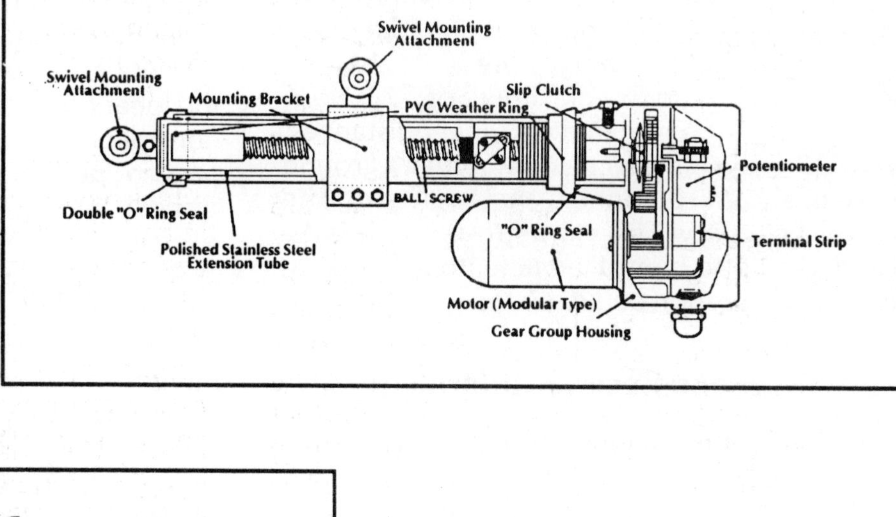

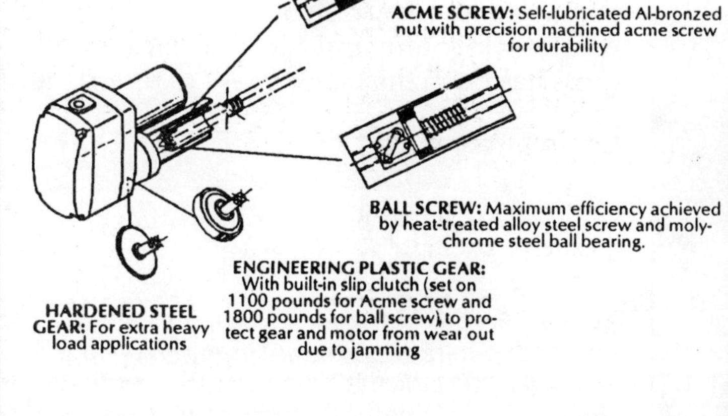

Figure 5-33. Acme and Ball Linear Jack Tubes. *An acme screwjack is constructed from a threaded screw shaft which moves in a threaded collar. This system can move up to 360 kilograms (800 pounds). Ball screwjacks are similar except ball bearings replace the threaded collar which grips the threaded shaft. This system can generate up to 700 kilograms (over 1500 pounds) of force. (Courtesy of Pro Brand)*

Figure 5-34. A Linear Actuator. *This linear actuator is mounted to an 8 foot spun aluminum dish. The points of attachment incorporate flexible joints to avoid potential binding.*

Horizon-to-Horizon Actuators

Horizon-to-horizon actuators are attached to the base of a dish. They must be solidly constructed to generate the torque necessary to rotate the reflector and to provide rigidity under wind or snow loads. Such actuators are designed to aim a reflector to near both horizons, unlike linear actuators that have a more limited range (see Figure 5-35).

Electrical Connections

A cable having four or five conductors connects the actuator motor and indoors equipment (see Figure 5-36). Two heavier gauge wires carry the 36 Vdc power required to drive the motor. While 14 gauge wire is typically adequate for runs up to a maximum of 50 meters, 12 gauge wire is required for longer distances. Two 20 gauge conductors and often a ground wire are used to connect the counter to the control circuitry. This extra ground wire is essential in pulse controlled actuators to ensure faultless operation of the microprocessor.

User and Receiver Interfaces

Controllers that operate and direct the mechanical equipment at the dish are one of two basic types. The simplest manual variety featured two buttons for east and west movements. The position of any satellite is identified by the count on a digital read-out. Most current actuator controllers have programmable memories that allow an operator to easily locate any satellite at the push of a button. Dish position is displayed by either a numerical counter or a digital read-out listing satellite code or name. Many types, which are integral components of satellite receivers, are capable of memorizing the locations of over 70 satellites. Both types of controllers should have east and west limits to prevent possible damage to motors, arms and dishes.

Improving Actuator Designs

Experiences in the C-band home satellite industry have provided invaluable guidance in designing and manufacturing actuators for Ku-band dishes. Three general conclusions can be drawn. First, time has proven that the most trouble-free sensor is a reed switch. Second, the use of mechanical limits is an important backup to any other type of end-of-arc limit to prevent damage caused by dish over-travel. Third, poor weatherproofing and the resultant water ingress has been the major cause of drive failure.

Figure 5-35. Ajax Horizon-to-Horizon Mount. *The Ajax horizon-to-horizon mount uses a chain to drive a central gear. The dish is firmly attached to the hemispherical gear structure.*

Figure 5-36. Electrical Interface Between Controller and Actuator Motor. *Five wires connect a control box with the actuator motor. Two of these send a voltage of typically 36 Vdc to the motor over a relatively heavy #14 gauge or larger cable. A three-conductor, #20 gauge shielded cable relays the counting pulses or appropriate voltage from the sensor to the indoors controller. In this case, a reed switch sensor is incorporated into the motor.*

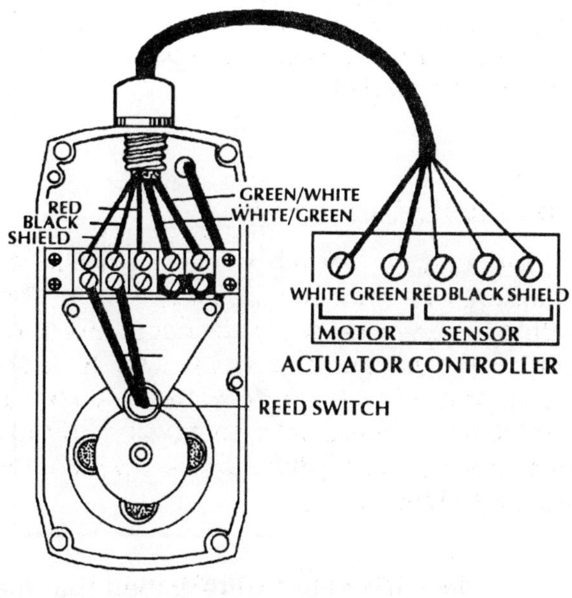

FEEDS and POLARIZERS

The feed has the important function of collecting microwaves reflected from the dish surface while ignoring noise and other unwanted signals coming from off-axis directions. This must be accomplished with minimal signal losses and without adding a significant amount of noise. A poorly designed feed can add as much as 20K of noise to a satellite receiving system. Feeds must also select the required polarity to properly detect a broadcast while rejecting unwanted signals.

Illumination Patterns and f/D Ratios

The term feed was originally derived from uplink antenna jargon. An uplink feedhorn, which in the primitive days of satellite communications actually looked like an inverted horn, "fed" microwaves onto and "illuminated" an antenna surface below. Before the advent of powerful computers, it was easier to design a receiving system by considering it as an uplink working in reverse. Consequently, the terms feed and illumination have persisted as descriptors of the microwave collection device on a receive dish.

Thus a feed is said to "illuminate" a dish surface even though it actually receives reflected microwaves. A perfectly illuminated feed would collect radiation coming from nowhere but the dish surface; it would reject microwaves from all other sources. However, in practice, prime-focus feeds illuminate dish central regions most strongly and are progressively less able to detect radiation at increasingly greater off-axis angles. Similarly, feeds employed with an offset fed reflector are designed to be most capable of collecting microwave energy from the reflector's center. The precise illumination pattern in all cases is an important design criterion. On one hand, a feed that illuminated just the central portions of a dish would introduce little environmental noise into the system but would miss some of the signal originating from the reflector edges and hence result in lowered system performance. On the other hand, a feed that over-illuminated a dish would take advantage of all the available gain but would introduce too much ground noise (see Figure 5-37). Over-illuminating a dish could be a serious problem because the ground on a typical cool summer day emits noise at a "hot" 290K.

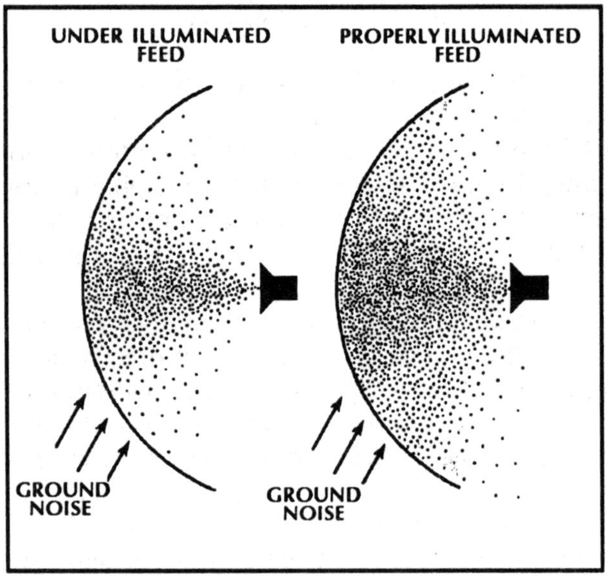

Figure 5-37. Feed Illumination Patterns. *This drawing demonstrates how a reflector can be both under- and over-illuminated by not selecting the correct feed for a particular dish f/D ratio. A poor match results in a lowering of signal-to-noise ratio. Under-illumination causes a loss of gain, while over-illumination causes detection of excessive noise from beyond the dish rim.*

Feed designers have determined that the optimal "edge taper" that maximizes signal collection and minimizes noise occurs when the received power at the edges of a prime focus dish is between 10 and 12 dB lower than that seen at its center (see Figure 5-38). Since the relative distance of a feed from the reflector surface is determined by the f/D ratio, one designed for use with deep dishes over-illuminates shallow dishes and vice versa. Nevertheless, most prime-focus feeds can be used over a range of f/Ds and still yield near optimal performance.

In many cases, feeds mated to offset fed reflectors provide a simple circular illumination pattern. Since these feeds view the reflector from an angle this circular pattern is projected as an elongated ellipse, the shape for most offset dishes. This shape is generally chosen by dish manufacturers for simplicity's sake. If other reflector shapes were used, for example one where the horizontal were larger than the vertical dimension, the feed would have to be specifically designed since the illumination pattern would not be a circular projection.

Dish efficiency can be altered by simply changing the illumination pattern of the feed. If no illumination taper were used, i.e. 0 dB edge taper, the entire surface of the dish would be equally illuminated and efficiency would be near 100%. Although the gain would be increased the additional ground noise detected would result in a lower carrier-to-noise ratio, the final determinant of system performance. In contrast, using an unduly high illumination taper of, for example, 25 dB, would result in substantially lower dish efficiency as well as detected noise. The end result would still be the same, a reduced carrier-to-noise ratio.

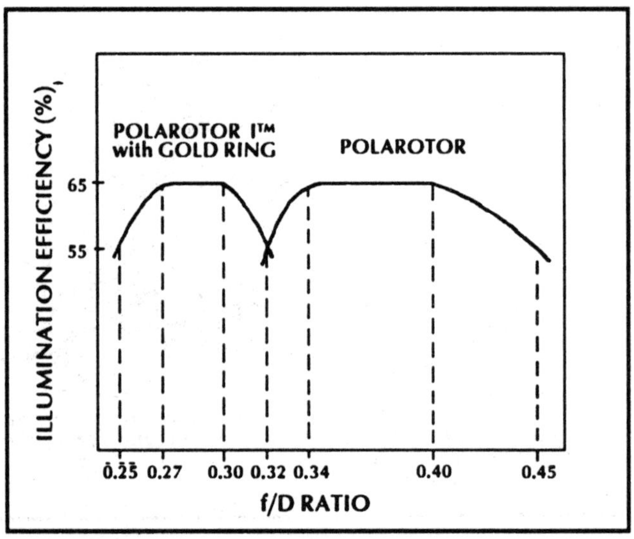

Figure 5-38. Feed Illumination Taper. *This shows the illumination taper of a prime focus feed. The design objective is to maximally illuminate the center of the reflector and to allow detected power to taper off to approximately 15 dB at its outer edge.*

Flat Plate Antennas and Feed Illumination

Active flat plate antennas that do not require feeds have the advantage of being 100% illuminated. All the incoming signal falling on a flat plate is usable. This contrasts with a conventional dish which does not maximally use the signal reflected from its outer perimeter. While the feed illumination taper strategy decreases both the amount of ground noise as well as the signal detected but still can effectively increase the overall C/N ratio, some of the incident signal is still lost. A flat plate antenna is also not oriented towards a portion of the surrounding ground and therefore is not nearly as susceptible to detecting ground noise.

The Feed/LNB Probe

The true antenna in a satellite reception system is the probe located either in the low noise amplifier or at the mouth of the feed. This finely tuned device of precisely the correct dimensions and position serves to convert the microwave energy within the feed/LNB to an electrical signal.

Varieties of Feeds

Over the years a variety of feeds has been developed for use in satellite receive systems. While the scalar feed is the most widely used today, the conical feed which looks like an inverted cone has now become commonplace with the growth in numbers of offset fed dishes.

A scalar feed consists of a section of circular waveguide with a set of concentric "scalar" rings (see Figure 5-39). The circular waveguide, the front-end, is dimensioned to detect both senses of linear polarization. These rings reduce the amount of microwave energy reflected back to the dish. Typically about 1.5% of the energy entering a scalar feed, equivalent to a loss of 0.08 dB, is lost due to reflections. Scalar feeds for Ku-band operation are 1/3 the size of those for C-band reception, because the wavelength of Ku-band radiation is 1/3 that of C-band signals.

The f/D operating range of a fixed ring scalar feed generally falls within 0.33 to 0.45. In some brands of scalar feeds, the position of the rings can be adjusted to extend this range from 0.28 to as high as 0.5 to match the f/D ratio of a dish. In general, the additional expense of an adjustable ring version is justifiable. An improvement of as much as 0.5 dB in carrier-to-noise ratio can be gained by matching a feed to a dish.

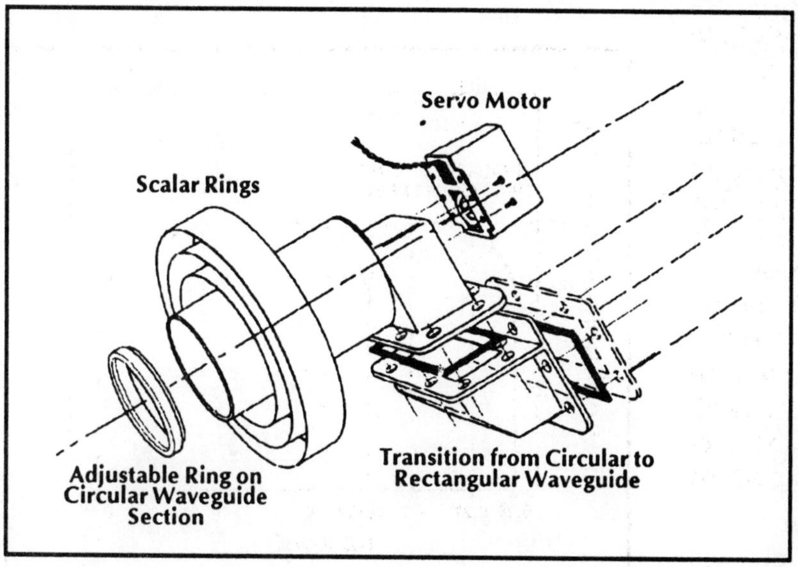

Figure 5-39. Feed Assembly. *This scalar feed is fitted with a servo motor that selects polarity by rotating a small probe in the throat of the unit.*

Feed Configurations Prime Focus and Offset

Two feed configurations are in common use today in home satellite TV receive systems: the prime focus and the offset feed.

In a prime focus system the feed is installed at the focal point of the reflector. A prime focus feed is generally employed on dishes having an f/D ranging from 0.33 to at most 0.6. Dishes with f/Ds in the 0.25 to 0.28 range often use another type of feed, a Cassegrain. Properly illuminating dishes with extremely high or low f/Ds with a scalar feed located at the prime focus can vary from difficult to downright impossible. Note that the focal point of a scalar feed is located just slightly inside its mouth.

Many smaller dishes reflect signals into a feed that is offset from its boresight. This reduction in aperture blockage can be important, especially in smaller dishes. Offset feeds are usually conically matched to a particular dish reflector shape.

Voltage Standing Wave Ratio (VSWR)

Feeds serve to collect and channel the electric and magnetic fields of electromagnetic waves into an LNB. Reception is optimized when both these fields are detected with equal efficiency. Once the incoming microwaves have been captured, they are channeled along the feed throat, a carefully designed waveguide. Both the "front-end" and the waveguide portion of a feed must be manufactured to precise dimensions to allow best reception of satellite broadcasts. Engineers have defined a quantity called the voltage standing wave ratio, the VSWR, which measures how much of the incoming radiation is reflected back towards the dish and lost (see Figure 5-40). The closer the VSWR approaches a 1:1 ratio, the more efficient is the feed. A perfect feed which captured all the incoming signals would have a VSWR ratio equal to 1:1. In general, VSWR ratios that fall below 1.5:1 are acceptable but a high quality feed should have a ratio less than 1.3:1 or better. Thus the higher the VSWR, the higher the mismatch loss.

Polarity Selection

Feeds have the important task of discriminating among various polarization formats.

Linear Polarization

The earliest method to select between horizontal and vertical C-band signals was to simply rotate the entire feed and LNA assembly. This method was cumbersome and was easily subject to mechanical failure often due problems associated with freezing. Today three less costly and more efficient polarity selection techniques are employed: mechanical, ferrite or magnetic, and pin diode devices. While the mechanical method is most commonly used in North America, the most familiar in Europe is magnetic or PIN selection. Similar techniques are now used in Africa and the Middle East.

Mechanical Rotors

Polarity can be selected by rotating a small, lightweight metal, hook-shaped probe which is housed in the circular waveguide in the throat of a feed between horizontal and vertical polarity orientations (see Figures 5-41 and 5-42). The stem of this hook is attached to a servo motor that accurately controls its position. The probe section of the hook is housed in a piece of circular waveguide allowing it to detect signals of either linear polarization.

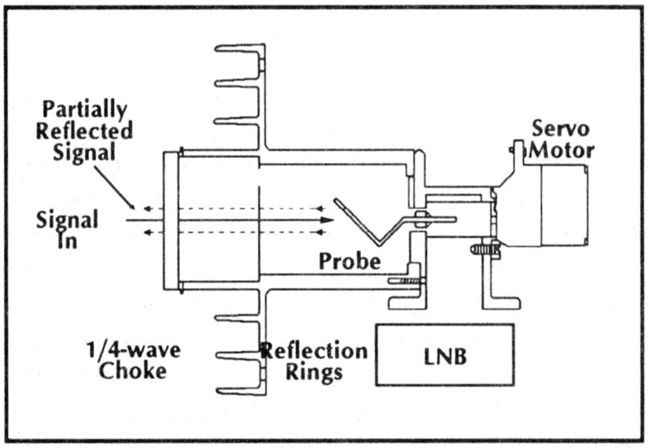

Figure 5-40. Voltage Standing Wave Ratio. *The voltage standing wave ratio, VSWR, represents how much of the incoming signal is reflected back towards the dish and lost from the feed. The closer the VSWR approaches 1 to 1, the more efficient the feed.*

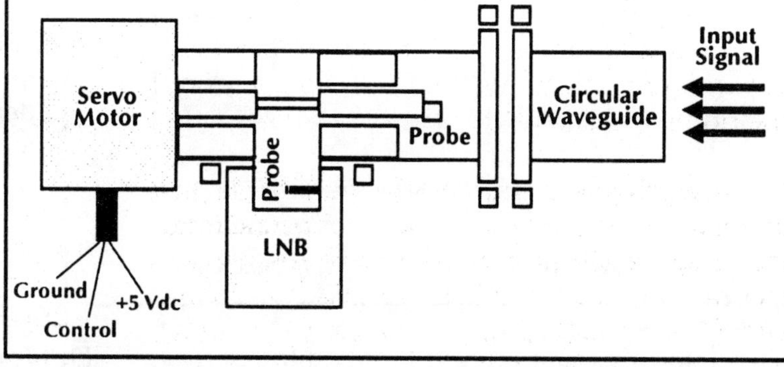

Figure 5-41. Schematic of a Mechanical Polarizer. *A mechanical polarizer rotates a probe between horizontal and vertical positions to detect horizontally and vertically polarized signals. This microwave energy is then re-radiated into the LNB where it is again detected by a resonant probe. Both senses of polarization can enter the circular portion of the waveguide at the entrance to the polarizer.*

The straight stem of the hook is encased in a plastic dielectric material. This section passes through a piece of rectangular waveguide that is closed at one end. The signal from the probe section of the hook which radiated into this waveguide retains the same polarity in the rectangular waveguide. Therefore moving the position of the probe through 90 degrees allows reception of either polarity.

Either a servo or a dc motor is used in mechanical polarizers. The servo motor can position the probe over a 180° range of motion and permits fine tuning at either polarity setting. The dc motor moves the probe through a full 360°, albeit somewhat more slowly. Most satellite receivers available today have built-in controls for interfacing with one or both types of mechanical polarizers.

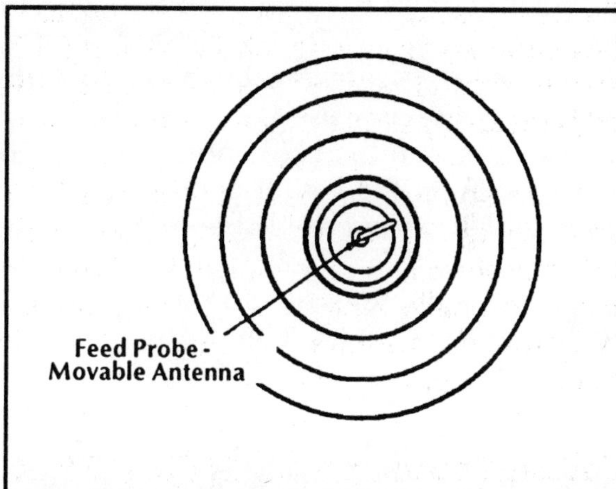

Figure 5-42. Polarity Selection. *The metal probe within most feeds that use servomotors rotates between vertical and horizontal positions in order to detect both linear signal polarities.*

Ferrite or Magnetic Devices

Ferrite devices are solid-state and have no moving parts. Polarity is changed via the interaction of the incoming microwaves with a magnetic field. The magnetic selector consists of a ferrite rod held in the center of a circular waveguide by a plastic dielectric support. The current passing through a large coil wound around the outside of the waveguide produces a magnetic field whose orientation depends upon the direction of the flow of charges. The polarity of the detected signal, in turn, depends upon the orientation of the magnetic field which rotates the plane of polarity virtually instantaneously. Magnetic devices have lower insertion losses than mechanical or PIN diode polarizers, typically about 0.1 to 0.3 dB.

Pin Diode Polarity Switching

PIN diode switching is essentially electronic switching between two probes which are mounted in circular waveguide. One probe detects horizontally polarized and the other vertically polarized signals (see Figure 5-43). Since only two fixed directions of polarity can be selected, based on the orientation of two probes, it does not permit fine tuning between horizontal and vertical polarities, namely skew control.

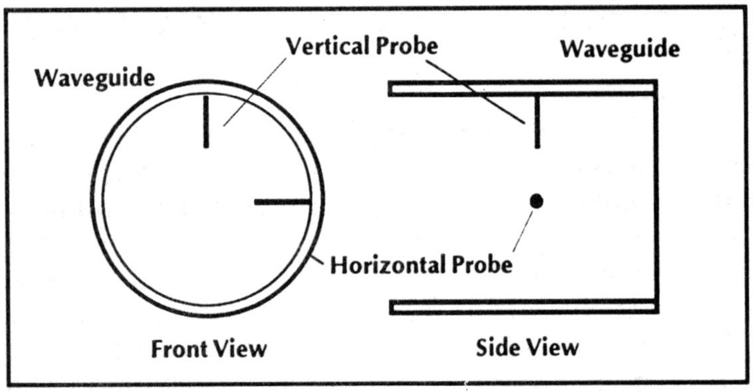

Figure 5-43. PIN Diode Polarizer Schematic. *This polarizer selects polarization via two probes, one vertical and one horizontal, contained in a waveguide. Signals from the probes are each amplified and fed to two PIN diode attenuators. One sense of polarity is then attenuated and one is selected for detection.*

Changing the supply voltage on the center conductor of the cable feeding the LNB switches between the two probes. Two versions are employed: voltage polarity switching and voltage level switching. In voltage polarity switching, not widely used, the polarity of the supply voltage on the LNB cable is changed. For one polarity, the center conductor would be $+V_{CC}$ and the shield would be 0 Volts. For the opposite polarity, the shield would be $+V_S$ and the center conductor would be 0 Volts. In this case the input socket must be isolated from the LNB casing.

Voltage level switching is one of the most common forms of polarity switching in use. A voltage drop of typically 5 Vdc is employed. One polarity is received at 17 or 18 Vdc and the opposite polarity at 13 Vdc. Which sense is received is merely a question of which PIN diode is biased "on" and which is biased "off."

The dc resistance of the coaxial cable must be considered when using voltage switched polarity. For example, if dc resistance to current flow is 2.2 ohms/100 meters, the voltage drop that occurs when a typical value of 0.2 amps of current flow through 100 meters of coax is 0.2 x 2.2 = 0.44 Vdc. Therefore over a 100 meter run, levels of 13 and 17 Vdc would drop to 16.56 and 12.56 Vdc. However, if 200 meters of poor quality coax with a resistance of perhaps 5 ohms/100 meters were used, the voltage drop would be 0.2 x 10, equal to 2 Vdc, a substantial decrease that might affect polarity switching performance.

Detection of Circularly Polarized Signals

Circularly polarized signals differ from linear signals because the vibrations of electric and magnetic fields do not lie in one plane but follow a helical or screw-like path as they travel through space. In earlier days of domestic satellite TV feeds designed specifically for circular polarization were rarely seen. However, such feeds are now relatively common in Africa, the Middle East and Europe for systems that, for example, receive transmissions from Intelsat 601.

Circularly polarized signals have one principal advantage over linear formats. Reception of linearly polarized signals requires use of a skew adjustment because the polarity can vary continuously between horizontal and vertical. The direction of the plane of polarization can vary substantially between satellites. For example, the American GSTAR A1 spacecraft had its plane skewed nearly 26° relative to most other domestic Ku-band satellites. In contrast, circular polarized signals have either left or right orientations. There is nothing in between and thus no need for skew adjustment when moving from satellite to satellite.

Feeds designed for reception of circularly polarized signals can have relatively simple designs. Two internal probes at right angles to each other detect the incoming signal. The output of one is delayed one quarter wavelength relative to the other and the signals are added to detect one sense of polarization (LHCP or RHCP); reversing the delay detects the other polarity sense. A PIN diode switch controlled by a small dc voltage can be used to select polarity. Such a simple device easily fits into the throat of an LNB.

The most common form of circular polarity feed encountered in the field is a scalar feed that has been modified to receive circularly polarized signals by inserting a dielectric slab, simply a rectangular piece of teflon, into the circular waveguide at an angle of 45° to the internal probe. It acts by "delaying" the circularly polarized signal thereby translating it into a linearly polarized signal. While, standard scalar feeds cannot distinguish between LHCP and RHCP transmissions, this simple retrofit permits detection and discrimination of both circular polarity senses. A conventional feed with skew adjustment is then capable of detecting and discriminating between vertical and horizontal polarities as well as between LHCP and RHCP signals. In this case, the skew of the probe must be adjusted when receiving circular polarized signals so the teflon slab lies between the two linear positions.

A scalar feed designed to receive linearly polarized transmissions, when modified for circular, attenuates circularly polarized signals by 2 to 3 dB, cutting the received power nearly in half. It also causes about 1 dB loss in linearly polarized signals and reduces linear polarization discrimination to as low as 15 dB, well below the acceptable 25 to 30 dB range. This reduction in the ability to discriminate between horizontal and vertical polarized signals is often overlooked and can become a problem if the reflector surface is not "true" but warped or if the feed has not been well installed.

Orthomode Transducers (OMTs) and Dual LNB/Feeds

Orthogonal mode transducers (OMTs), also known as orthomode transducers, can simultaneously detect both vertically and horizontally polarized signals. Such devices are commonly found in commercial C-band installations such as satellite master antenna television (SMATV) and cable television headends because they allow all satellite transponders on one satellite to be viewed at one time.

Two orthomode feed designs are available. One is based upon the property of waveguides that prevents the passage of microwaves having wavelengths longer than twice their internal dimension. For example, if a signal of 1.0 cm wavelength entered a waveguide having a 2 by 0.3 cm cross-section, waves polarized in the direction of the 2 cm dimension would pass through while those polarized at right angles to this direction would be rapidly attenuated. A dual polarity feed uses two waveguides of the required dimensions joined at right angles to each other. One arm transmits horizontal polarity; the other vertical polarity. Two LNBs are used to independently amplify each of these signals.

Another type of feed designed to simultaneously detect both polarities, a dual LNB/feed, has no waveguide section but instead has two independent probes and two LNBs built into one housing. It is designed so each probe/LNB detects signals of only one polarity.

A dual-band feed is another type of orthomode feed that can simultaneously receive signals on both C- and Ku-bands (see Figure 5-44).

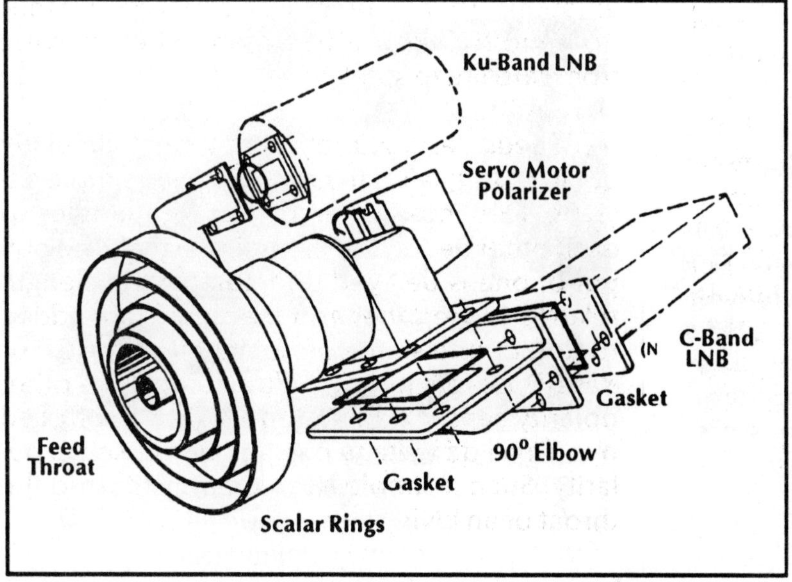

Figure 5-44. Schematic of a Dual-Band C/Ku Feed.

Feed/Customer Interfaces

Most systems control linear polarity selection in two ways. First, receivers can be programmed for an automatic polarity selection whenever channels are switched. Some more sophisticated units also fine tune channel center frequency and optimize polarity alignment by employing feedback circuits which maximize signal strength to adjust these parameters. All those brands which are not microprocessor driven can be manually switched between polarity formats. Second, whenever a mechanical feed is used, a skew adjustment allows for fine tuning of the final probe position.

Many satellite receivers provide connections for mechanical polarizers. Three 20 gauge or larger wires are necessary to control polarity selection and fine tuning. Other types provide the required voltage pulses via the coax to switch PIN diode polarizers.

LOW NOISE BLOCK DOWNCONVERTERS (LNBs)

The low noise block downconverter, known as an LNB, has the important function of detecting the signal relayed via the dish and feed and converting it to an electrical current, amplifying this extremely weak signal by 50 to 60 dB (a factor of 100,000 to 1,000,000) or more and downconverting or lowering its frequency. The downconverted signal is subsequently relayed along cable to an indoor satellite receiver. An LNB is the first "active" or electronic component in the series of steps by which a satellite signal is processed. In concert with the dish it is the most important component in determining how well a TVRO functions (see Figure 5-45).

All LNBs are not created equal. Parameters such as input and output frequency ranges, noise temperature and phase noise must be understood before selecting the right LNB for the job at hand.

The Evolution of Modern LNBs

An LNA amplifies the satellite signal, while an LNB downconverts (lower the signal frequency) as well as amplifies the incoming signal. It is instructive to examine how LNAs evolved in order to more clearly understand the function and operation of LNBs. A more thorough explanation of downconversion is presented in the next section.

The typical power of signals reaching the input of an LNB via a 3 meter C-band dish is less than 10^{-14} watt/m^2. Therefore, this device must contribute very little noise power or signals would be drowned out in the roar of amplifier internal thermal noise. This feat is made possible by recent advances in transistor technology. Without such progress, satellite broadcasting would not exist as we know it today.

Amplifiers used in the early days of radio astronomy were based on standard parametric transistor circuits that were immersed in baths of very low temperature liquid nitrogen or helium. This technique dramatically reduced noise levels by slowing down molecular mo-

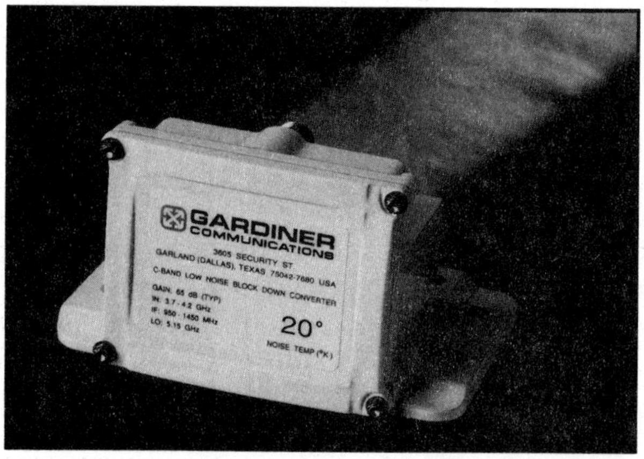

Figure 5-45. Gardiner C-Band LNB. *This C-band LNB has a 20K noise temperature, a 65 dB gain and a local oscillator frequency of 5.15 GHz. (Courtesy of Gardiner Communications)*

tion, the source of background noise. The development of the gallium arsenide field effect transistor, known as a GaAsFET, made the modern "low noise" amplifier a reality. These special transistors make an amplifier operate as if it were cooled to near absolute zero where all molecular motion ceases. GaAsFETs, a miracle of modern engineering, are at the core of consumer satellite reception systems. Since GaAsFET transistors are inherently less noisy than conventional transistors, they are used as the "front end" of an LNB.

Today there are two types of low noise transistors available to low noise amplifier designers and manufacturers. These are HEMTs and MESFETs, acronyms for high electron mobility transistor and metal semiconductor field effect transistor. MESFETs have been used since the early 1970s while HEMTs have been commercially available only since 1987. The basic difference between the two devices is that the HEMT performs better at higher frequencies, namely it has better noise figure and gain than the MESFET.

Sources of Noise

An understanding of electronic noise is critical in a study of satellite communications because the very low power signals transmitted are just a step above the noise generated within electronic systems and from the surrounding environment. Noise is caused by molecular and atomic motions, tiny electrical currents that create electromagnetic waves. Some of this radiation is in the same frequency band as satellite transmissions.

There are two main types of noise in electronic systems: thermal noise and noise associated with the operation of transistors. Thermal noise arises from the random physical motion of conduction electrons and holes that occurs in all materials at temperatures above absolute zero. It depends both upon the ambient temperature inside electronic equipment and upon the bandwidth of the signal being processed. As signal bandwidth increases, more noise is present and can be detected. So decreasing the bandwidth of signals entering an amplifier decreases the noise and interference associated with a broadcast.

Interference is also considered a form of noise even though it usually originates from some man-made communication device used for a purpose other than satellite broadcasting. When a dish has high side lobes or a wide beamwidth, interference may originate from a satellite adjacent to the one being targeted. Higher quality LNAs and LNBs are less susceptible to random internal noise but, as a result, are more capable of seeing organized interfering signals from either non-targeted satellites or other man-made sources.

Noise generated within transistors is classified as either shot or partition noise. Shot noise is caused by the quantized nature of current flowing across the PN junction in transistors. Like thermal noise, it also depends upon the signal bandwidth. Partition noise, similar to shot noise in being caused by quantized jumps in current is caused by variations in current between the emitter and collector in a transistor.

Noise Temperature and Noise Figure

The noise characteristic of an LNB or any other electronic device for that matter, is described either in terms of noise temperature or noise figure. These yardsticks are interchangeable. The noise figure is a measure of the degree by which an LNB degrades or decreases the satellite signal as it passes through this device. Noise temperature, the other measure of noise is expressed in degrees Kelvin. This scale is based on the fact that at a temperature known as absolute zero, 0K (equal to minus 273.16 °C or minus 459.72 °F), molecular motion ceases and consequently noise disappears. There no temperature colder than absolute zero.

C-band LNBs are typically rated according to noise temperature. Noise temperature is interchangeable with a parameter known as noise figure (see Table 5-6). The conversion between noise figure and temperature is outlined in Appendix B.

TABLE 5-6. EQUIVALENT NOISE FIGURES and TEMPERATURES

Noise Temperature (°K)	Noise Figure (dB)
20	0.290
25	0.359
30	0.428
40	0.561
50	0.691
60	0.819
65	0.881
70	0.942
75	1.002
80	1.061
85	1.120
90	1.177

How does substituting an LNB having a lower noise temperature affect system performance? A simple method to calculate this change is demonstrated in the following example. If a 40K LNB were replaced by one having a noise temperature of 20K in a system that has a dish and other components that add 40K of noise, the C/N ratio would improve by the relative decrease in system noise temperature, everything else remaining the same:

$$(40 + 40)/(20 + 40) = 1.33$$

This factor of 1.33 decrease in noise temperature or 33% performance improvement equals a 1.25 dB (10 log 1.33) increase in carrier-to- noise ratio.

Judging LNB Performance

Noise temperature or figure is the main parameter used to judge LNB performance and quality, assuming that its gain is sufficient, the correct frequency range has been selected and that other important design issues such as waterproofing and lightning protection have been adequately addressed. Digital-ready LNBs must also have relatively "clean" frequency stability and low phase noise characteristics.

Noise Temperature

An LNB is the "front end" of a satellite TV reception system. The noise it adds to the incoming signal sets the noise "floor" and plays a large part in determining picture quality. Reasonably-priced, high-performance LNBs are now available. Modern C-band LNBs have noise temperatures as low as 15K.

It is fascinating to examine how rapidly LNA/LNB technology has progressed in recent years. Commercial GaAsFET low noise amplifiers became available in the late 1970s. At that time, respectable noise temperatures were 220K while these units had price tags in excess of US $3,500. These LNAs were used primarily with dishes that were five meters or larger. However, the rapid growth in the home satellite industry changed everything. LNA performance was gradually improved and prices began to decline rapidly. In mid-1985, an 85K LNA purchased directly from the manufacturer sold for under $100. During this time, when Ku-band GaAsFETs were introduced, C-band noise temperatures as low as 40K became common. With the development of "HEMT" devices, $50 LNBs have noise temperatures approaching 20K.

The improved noise performance of LNAs and LNBs has brought benefits to many end-users living in difficult reception areas. For example, in regions where daily temperatures regularly soar to above 38°C, products are now available having enough noise figure margin to overcome the associated performance degradation. In weak footprint regions of the world, ultra-low noise LNBs can greatly enhance reception of low power satellite signals. In particular, in those areas that both receive weak downlinked power and are subject to a hot climate, ultra-low noise performance is a must.

The noise contributed by an LNB can vary from the rated value that is usually measured at room temperature. An LNB operating at high noon in a desert with a container temperature of 120°F, generates substantially more noise than it would at 50°F. Most manufacturers shield their products from such temperature extremes with a rubber or enamel paint.

LNB noise temperatures vary across the design frequency band. All manufacturers specify the maximum average noise temperature or figure over the design frequency range while some provide these measurements at three or more frequencies on an attached plate or product literature. The overall noise temperature or figure rating should reflect the highest measurement over the range in question.

Operating Frequency Range

Each LNB is designed to accept a specified frequency range. Bandpass filters serve to sharply attenuate signals that fall outside this range to protect the receive system from out-of-band interference. For example, an American 3.7 to 4.2 GHz C-band LNB cannot manage the extended 3.625 to 4.2 GHz extended C-band range that may be used in Africa and other parts of the world.

Phase Noise

One of the main functions of an LNB is to "downconvert" or lower the frequency range of the input signal. In the process, an output signal with some frequency "jitter" is unavoidably created. While current levels of phase noise in standard LNBs have not been a concern for analog broadcasts, digital-ready LNBs must adhere to more stringent phase stability requirements so that a QPSK signal can be properly demodulated in the satellite receiver.

What is phase noise? In a perfect LNB a pure sine wave signal of, for example, 4 GHz might result in a output sine wave signal of exactly 1.15 GHz after downconversion. However, local oscillators in LNBs tend to randomly drift in frequency. Thus the signal might end up bouncing about in a range of maybe 10 MHz centered on this 1,150 MHz frequency. As a result, the signal thus becomes fuzzy around its edges. When a QPSK-modulated carrier instead of a pure sine wave signal is fed into the same LNB, it also becomes less perfectly defined. If this spreading of frequency or phase noise is too pronounced, the demodulator could have trouble detecting when a change in phase occurs, thus leading to a misinterpretation in the value of a bit. High phase noise results in an increase in bit error rate.

Phase noise is measured by how fast the power of the signal drops off about its center frequency. Thus if the signal power at 1 kHz away from the center frequency has fallen by 60 dB, the phase noise is given as -60 dBc/Hz. The term dBc refers to decibels relative to the power of the center frequency.

A digital signal with a higher bit rate is spread about a wider bandwidth than a signal with less information. The wider this bandwidth relative to the width of the phase noise, the more immune the system is to phase noise errors. Thus an MPEG-2 digital transmission occupying a full 27 MHz bandwidth and carrying 40 Mbits/second for a total of perhaps 5 TV programs, is less susceptible to phase noise than a digital signal of 3.9 Mbits/seconds and carrying just a single video program.

Gain

LNB gains today usually range from 60 to 70 dB. Each GaAsFET and subsequent parametric amplifier stage in an LNB typically contributes between 10 and 12 dB of gain.

LNBs are designed to have enough gain to overcome cable losses as well as the effects of noise contributed within the block downconversion stage. The motivation to produce LNBs with at least 50 dB gains stemmed from a requirement of earlier TVROs. A typical 30 to 40 meter run of RG-213 coax had approximately a 13 dB loss when relaying signals at 4 GHz. N-connectors then contributed another 2 dB of loss. Today F-connectors and lower loss cable are used to relay lower frequency signals.

In modern TVROs, LNB gain must exceed a minimum value. Past a certain point adding extra gain does little to improve system performance. Tests have shown that this point occurs somewhere between 50 and 60 dB, but 50 dB should be considered a minimum gain, especially when using an under-sized dish or unduly long cable runs.

LNB gain varies with ambient temperature and signal frequency as does noise figure. As temperature increases gain decreases, typically by about 0.6 dB for each 10°C temperature rise. This variation is of little significance except for operation in extremely hostile environments such as outer space. Small variations in gain also occur across the frequency range. A perfectly "flat" LNB would amplify every frequency equally. In general, the variations are small enough to be negligible. Large deviations from linearity which could cause an LNB to become unstable and oscillate are possible at extremely low ambient temperatures below -40°C.

A perfect LNB would amplify signals only in the designed bandwidth and reject those with higher or lower frequencies. In practice, gain drops off rapidly but measurably at frequencies outside this band. Out-of-band rejection is most important when there is a possibility that other communicators are using similar or adjacent frequency ranges which could interfere with clear reception. While this can be a serious concern for C-band satellite systems, it has not generally been a major factor in Ku-band TVROs.

VSWR

LNBs, like feeds, are rated according to voltage standing wave ratio, the VSWR. This ratio, a measure of how efficiently the input signal is collected, is found by dividing the power of the input signal by that of the signal actually entering the LNB. When no signal is reflected back towards the feed in a perfect device, the ratio is 1 to 1. VSWRs should be less than 1.3 to 1 in good quality products.

LNB Design and Operation

All C- and Ku-band LNB designs are similar in shape because the waveguide section must have the appropriate dimensions to channel C- and Ku-band microwaves (see Figure 4-46). The input flange and waveguide of a C-band LNB, known technically as a WR-229 choke with waveguide, has dimensions of 2.29 by 1.145 inches. Occasionally, fine tuning controls are built into the waveguide portion of an LNB to allow slight changes in its internal dimensions to improve VSWR. These permit the precisely manufactured LNB to be even further tweaked to maximum performance on sophisticated equipment. Such tuning minimizes the amount of signal lost in entering the LNB.

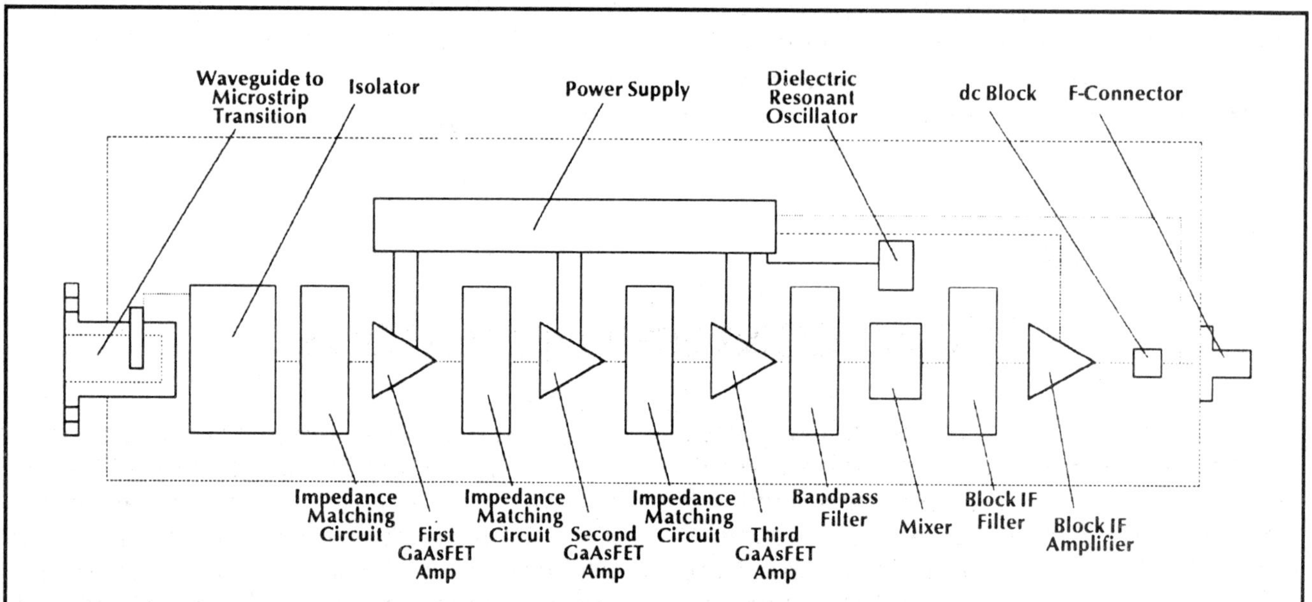

Figure 4-46. LNB Schematic. *A low noise block downconverter combines the functions of an LNA and block downconverter in one case. The LNB has a single female F-connector output which feeds directly over coax to a satellite receiver. Each stage contributes noise and gain. The GaAsFET or MOSFET stages which are responsible for the low noise temperature characteristics, contribute approximately 12 dB of gain; the two or more bipolar stages contribute a total of 30 dB additional gain.*

Every LNB has an internal probe which is the real microwave antenna. This small metal antenna receives microwaves and converts them into electrical currents. A DC shorted probe prevents high voltages which are caused by nearby lightning strikes from frying the internal components. Of course, a direct strike would destroy any LNB. The probe is set at precisely the correct position to maximize signal reception and should never be tampered with even if it appears to be slightly bent.

The electronic components of an LNB are enclosed in a hermetically sealed box. Water vapor has a very destructive effect on the operation of electronic components and is carefully avoided by such design.

The amplification section of an LNB consists of several GaAsFET transistor stages (usually two or three) in a cascaded arrangement followed by several conventional amplification stages. A voltage regulator is also included in the circuit. LNBs usually draw from 80 to 150 milliamps of current and operate at 15 to 24 Vdc (see Figure 2-53).

Electrical Connections

The electrical connections to an LNB are simple. LNBs have a F-connector output that is joined to the satellite receiver via a single run of coaxial cable. Power and signal are relayed via the same cable.

Mechanical Packages

Techniques for packaging LNBs have evolved rapidly. Early designs were large and bulky; some devices weighed nearly 3 kilograms. However, pressures to lower price and improve performance have led to the creation of smaller, more cost-efficient designs. Today smaller, lighter LNBs are available. This is especially important when using orthomode, hybrid C/Ku or linear/circular C-band feeds that can be bulky. In these cases, guy wires to support and center the feed/LNB assembly can be used.

The electronic components of an LNB must be enclosed in a hermetically sealed box. Water vapor has a very destructive effect on the operation of printed circuit boards and other electronic devices and is avoided by watertight encasing.

The Future of LNBs

Modern C-band LNBs are approaching their theoretical minimum noise temperature limits, even with the use of HEMT devices. 15K can now be regularly attained. While thermoelectric or cryogenic cooling can also improve performance beyond 15K, this technology is cost-prohibitive in the home TVRO market.

DOWNCONVERSION METHODS

Pioneer TVROs relayed the high frequency C-band microwave signal that had been amplified by an LNA to an indoor receiver where it was downconverted or lowered in frequency. Relaying such high frequencies required the use of expensive coaxial cable that was characterized by high signal losses. This design constraint mandated that that cabling distances had to be minimized. So dishes were mounted as closely as possible to the satellite receiver and, in some cases, expensive line amplifiers were used to boost the signal. Furthermore, installing such high frequency cable was difficult since fitting connectors required expertise and the coax had to have a maximum turning radius and so could not be sharply bent. All in all, early TVROs had severe limitations, were difficult to install and were expensive.

The radical design improvement was to separate the satellite receiver into two components: a downconverter at the dish and the remaining electronics. Then just a short length of cable rated for higher frequencies could be used to relay signals between the LNA and downconverter. This less expensive, lower frequency cable could be used to transmit signals more than 100 meters without the use of line amplifiers.

The final logical step was to include the downconverter within the LNA. This device thus evolved into the low noise block downconverter, the LNB.

Downconversion in Satellite Receivers

Three distinct schemes have been used in satellite receive systems: single, dual and block downconversion (see Figure 5-47). The design of downconverters is intimately related to the method used to select individual channels on satellite broadcasts. Each channel typically occupies a bandwidth ranging from 18 to 28 MHz. Any channel is selected or tuned by translating this band of frequencies down to one with the same bandwidth but centered at the final intermediate frequency (IF). Most C-band receivers use a 70 MHz IF. Thus, for example, if a channel having a center frequency of 3900 MHz in a 3886 to 3914 MHz band had been selected, the downconverter would lower this to one spanning 56 to 82 MHz and centered on 70 MHz. Both these bands are 28 MHz wide and contain exactly the same audio, video or data information. Similarly, if a channel centered on 4140 MHz were selected, it would also be downconverted to one centered on 70 MHz. All other channels could likewise be downconverted and tuned into one by one.

Downconversion is accomplished by a process called heterodyning or mixing. To illustrate, if a 1000 MHz signal is mixed with one of 900 MHz, the resulting sum and difference frequencies of 1900 and 100 MHz are created. If a filter is used to remove the higher frequencies, the effect has been to lower the 1000 MHz signal to one having a frequency of 100 MHz. If "intelligence" had been superimposed or modulated onto this 1000 MHz signal, the final IF of 100 MHz would still contain the original information.

The tuner in a satellite receiver selects channels by sending a voltage to a voltage tuned oscillator, a VTO, that produces the desired mixing frequency. In pioneer single or dual downconversion systems the variable-frequency VTO was located in a separate outdoor unit, the "downconverter." In block downconversion systems the variable oscillator is a component of the satellite receiver and the oscillator contained within the LNB is set to a fixed frequency.

Single Downconversion

Single downconverters lower the chosen satellite channel to the final 70 MHz IF in one step. Rotating or pressing the tuner control knob or button selects a voltage to be relayed to the VTO in the downconverter at the dish. A mixing frequency 70 MHz lower or higher than the center frequency of the desired satellite channel is produced so that, after mixing, the resultant difference frequency is 70 MHz. For example, if a channel having a center frequency of 3900 MHz is selected, the mixing frequency is set at 3830 MHz and the difference and sum frequencies of 70 and 7730 MHz are produced. The higher frequency is filtered out, leaving the desired 70 MHz IF. Similarly, if a channel having a center frequency of 4040 MHz is chosen, the VTO generates a mixing frequency of 3970 MHz.

Downconverters can use either high-or low-side injection. The previous two are examples of low-side injection. In the example above, if the channel at 3900 MHz were selected using high-side injection, 3970 MHz would have been chosen to mix with the 3900 MHz center frequency. When low-side injection is employed, the highest frequency in the signal block appears as the highest frequency in the downconverted block. However, with high-side injection, the highest frequency in the signal block appears as the lowest frequency in the downconverted block and vice versa. This inversion is the reason why video inversion switches used to be incorporated on the back panel of most C-band satellite receivers.

Figure 5-47. Downconversion Methods. *Single, dual and block downconversion methods all accomplish the same end result, converting the extremely high microwave signal to a lower frequency IF and selecting one of the many channels to be viewed. The intermediate frequency (IF) is often 70 MHz while the RF signal falls in the 950 to 1750 MHz range. Here VTO is a voltage tuned oscillator, LO is a fixed local oscillator and DSO is a fixed dielectrically stabilized oscillator*

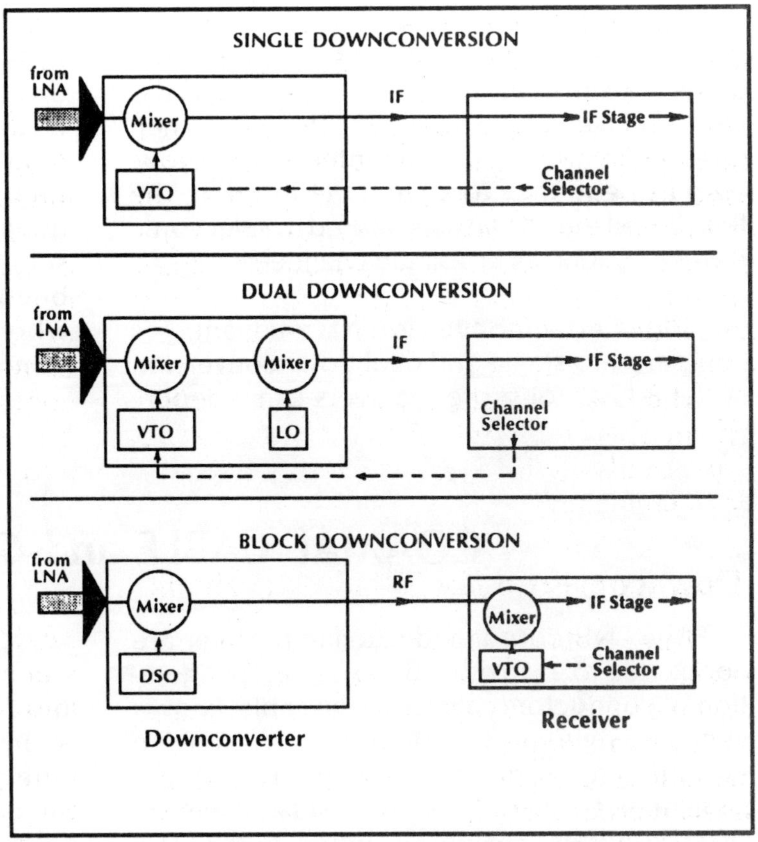

Single downconversion systems have a potential problem because the oscillator frequency is 70 MHz above or below the desired channel center frequency. This signal falls within the C-band range and can leak back into the system to cause interference on nearby channels. Only the highest frequency channel using high-side injection and the lowest frequency channel using low-side injection are immune to this difficulty. Thus single downconversion C-band satellite receivers use circuits called image rejection mixers to counteract this effect.

Dual Downconversion

Dual downconverters attain the final IF in two stages. Both the first and second mixing occurs outdoors in the downconverter. Often 810 MHz is used as an intermediate frequency. For example, if a channel having a 3900 MHz center frequency is chosen, the VTO mixes it with a signal of 3090 MHz, equal to 3900 less 810 MHz. Then the intermediate 810 MHz is mixed with a local oscillator signal of 740 MHz to downconvert the signal to the final 70 MHz IF to be transmitted to the receiver.

Dual downconversion systems use an oscillator frequency well outside the C-band and there is no concern with interference and the need for image rejection circuitry and ferrite isolators is eliminated.

Block Downconversion

Block downconverters in LNBs use a fixed frequency local oscillator to lower the entire satellite band to an intermediate range. Until the development of dielectric resonant oscillators (DROs, also known as DSOs, dielectric stabilized oscillators) this was not practical because the frequency produced was not sufficiently stable. DROs are stable to within 10 MHz, and can be stable to nearly 1 MHz, over the whole satellite band. A high-side injection local oscillator running at a fixed 5.150 GHz is typically used in C-band systems. Thus, for example, the 3.7 to 4.2 GHz C-band signal is lowered in a block to a 950 to 1450 MHz signal; both the original and downconverted signals have the same 500 MHz bandwidth.

The entire satellite bandwidth that contains the same intelligence as the higher frequency band that had originally entered the LNB is relayed to the satellite receiver. It is then mixed with a signal from an internal VTO, lowered in frequency to a band centered on the final IF and then bandpass filtered to select only those frequencies in a single channel.

Block downconversion have advantages compared to single and dual downconversion systems. Two or more receivers can independently select channels because each one using its internal mixer and VTO receives the entire downconverted block which contains all the channels on any given satellite. Both single and dual downconversion systems relay only one channel indoors at any time. Block downconversion satellite receivers also are less subject to drifting away from a target channel because channel selection occurs indoors where electronic components are protected from large temperature and humidity swings.

COAXIAL CABLE and CONNECTORS

The LNB output and satellite receiver are connected to each other by a special configuration of conductors called coaxial cable, known as coax. Single wires of copper or aluminum are adequate for conducting electricity at lower frequencies encountered in most familiar electrical devices. However, when higher frequency microwaves are relayed, single strands of metal behave like antennas and can radiate most of the power away before it reaches its destination. With exceptionally high frequency signals that are characteristic of satellite broadcasts, specially designed cable must be used to prevent almost complete loss or attenuation of the transmitted signal.

Coax is composed of two concentric conductors separated by an insulating material called a dielectric (see Figure 5-48). This whole assembly is sheathed in a non-conducting jacket for protection against the elements. The signal travels along the central wire and the external cylindrical conductor is grounded. Using this configuration greatly reduces radiative losses at high frequencies. It is interesting to note that at higher frequencies the signal travels progressively closer to the surface of the center conductor. If downconversion did not occur in the LNB, extremely expensive coax having a large diameter inner conductor would be required to transmit the high frequency signal since the "skin effect" would be so pronounced.

Coaxial cables are usually sweep tested by the manufacturer to ensure that there are no breaks or discontinuities. In this test signals spanning a range of frequencies are fed into one end of the cable and the output is measured at the far end.

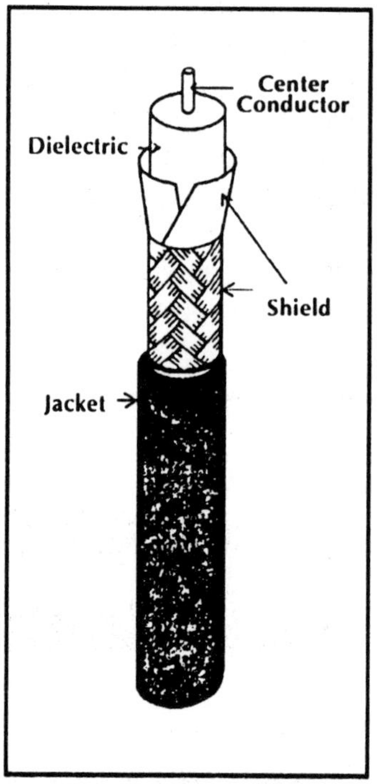

Figure 5-48. Coaxial Cable. *Coax cable is composed of two concentric conductors separated by an insulator called the dielectric. Signals are transmitted on the central conductor while the outer shield is designed to contain the signal and eliminate ingress interference.*

Types of Coax

A wide range of coaxial cables is available. These fall into three broad categories depending upon the construction of the dielectric sheathing material. These are hardline, conventional coax, and foam or air dielectric coax. Regular coax has one or two pliable metal grounded layers wrapped around a plastic dielectric. Dielectrics in common use include polypropylene, a hard, translucent substance and polyethylene foam, a soft, white material. The outer conductor, which is often a braided copper or aluminum sheath, is occasionally doubled up or replaced with a solid casing to further lower losses and lessen the chance of interference from leakage of unwanted signals into the coax. The degree of cable braiding is rated by what percentage of the dielectric surface is shielded. 67% shielding is a typical value. Completely shielded cables are recommended where high levels of ingress interference from local communicators, whose signals have frequencies similar to those relayed from the LNB to the satellite receiver, may occur.

Foam or air dielectric coax uses foam or compressed air as the dielectric material; it generally has lower loss and is more expensive than ordinary coax. Hardline is similar in construction to either of the previous two types except that it has even lower losses because it has a more rigid, metal sheath and a minimal amount of high quality dielectric (see Figure 5-49). Hardline is usually used in the main trunk lines of cable TV or other similar communication networks such as large SMATV installations.

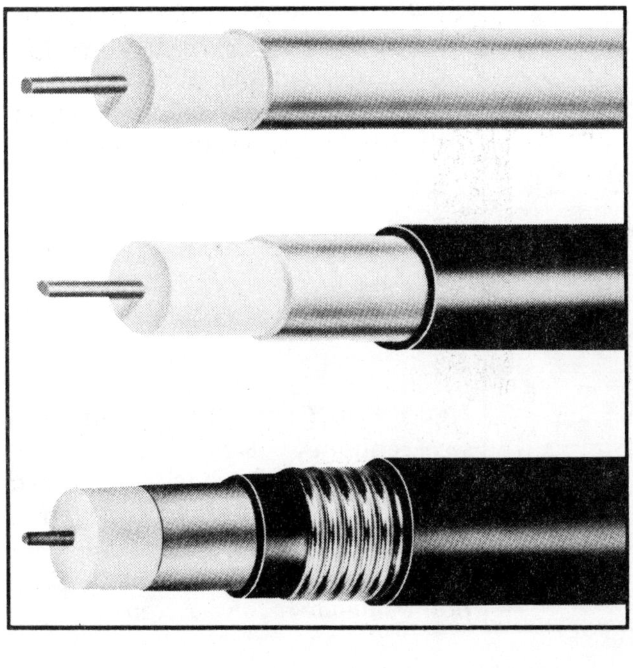

Figure 5-49. Hard-line Coaxial Cables. *These three types of coaxial cable designed to transmit high frequency signals are protected from interference to increasing degrees. The middle one has a polyethylene jacket. The lower hard-line coax has two metal jackets, one a corrugated chrome plated steel armor that prevents sharp cable bends. This or any other deformation could cause impedance mismatches and signal attenuation. (Courtesy of M/A COM, Comm/Scope Marketing)*

Coax has two principal uses in satellite reception systems: to relay the LNB output to one or more satellite receivers and to distribute the receiver output(s) to one or more televisions. In general, a cable similar in characteristics to RG-6 is recommended for the former use. The type of cable used for distribution systems depends upon distances involved and design details.

Judging Cable Performance and Use

Coaxial cable is rated according to performance criteria which include characteristic impedance, loss of signal power with distance, and material composition. These factors determine the proper uses of various brands.

Characteristic Impedance

Every conductor has a specific resistance to current flow which causes power losses. In addition, the interaction between electrical charges on the inner and outer conductors, known technically as capacitance and inductance, causes delays and losses. These factors together determine a value called the characteristic impedance. Cables used in satellite reception and distribution systems are almost always rated at 75 ohms, although a few systems still employ some runs of 50 ohm cable. Higher frequency coax, such as RG-214, has a 50 ohms characteristic impedance.

Knowing the characteristic impedance is crucial in designing electronic systems. Every electronic device also has a characteristic impedance. If the cable impedance does not match that of the device it feeds, substantial reflective power losses, known as return losses, can occur. This mismatch would be similar to sending water from a larger diameter into a smaller diameter pipe and experiencing the resistance to flow that is heard as knocking. Similarly, if the output impedance of an amplifier is not matched to the coaxial line impedance, reflections and the ensuing power losses will occur.

This idea is similar to the concept underlying VSWR, used to characterize internal losses due to signal reflection for both feeds and LNBs. If the impedances are matched between the dish, within the feed components, between the feed and LNB and further downstream to the satellite receiver, losses will be controlled. Any weak point in this chain can seriously degrade performance.

When cables are bent too sharply, incorrectly crimped, crushed or otherwise changed in internal dimensions, a local change in characteristic impedance is created. This causes reflections and subsequently signal losses above and beyond those normally encountered.

Signal Losses

Coax is rated according to decibels of signal loss per distance the signal travels. Losses increase with frequency for any given type of cable. Therefore, for example, RG-6 which has acceptable losses at 1 GHz is unusable at 4 GHz. It has 8.7 dB loss per 100 feet at 1450 MHz. Therefore, a cable run of 150 feet results in a loss of 13.1 dB, a drop of a factor 20 in signal power. It is clear that cable runs should be as short as possible and that the correctly rated cable should be installed. Figure 5-50 graphs loss characteristic of more familiar coax types.

Minimizing the length of cable runs has a number of advantages. The main ones are lower cost, improved signal-to-noise ratio and, for ground mounted dishes, less trenching. Values for signal losses over distance at 100 MHz, 1450 MHz and 4 GHz for cable encountered in cable and satellite systems are listed in Table 5-7. The higher the signal frequency, the greater the losses (as also shown in Tables 5-8 and 5-9 and Figure 5-50).

Attenuation is greater at higher signal frequencies. Errors made when installing cables and connectors can cause more deterioration in performance when higher frequency signals are used. Thus downconverting signals within the LNB has two principal advantages: less money is spent on cable and problems caused by potential installation errors are minimized. Losses in each variety of cable can also be dependent upon details such as the diameter of the central conductor and the type of braiding material.

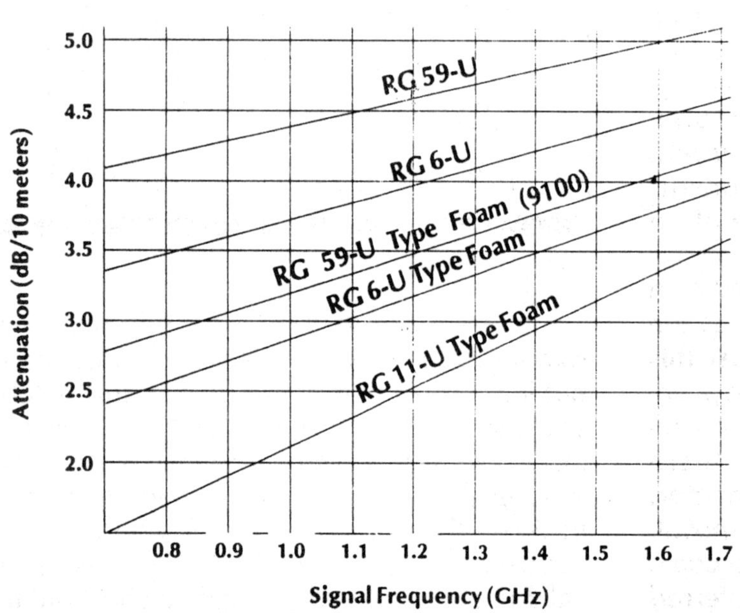

Figure 5-50. Cable Attenuation versus Frequency. *Signal losses in coaxial cable versus frequency are graphed here. The attenuation is measured in dB per 10 meters for the range of frequencies typically encountered in home satellite receive systems..*

In most satellite systems the coax run from the dish to the indoors electronics carries signals in the 950 to 1450 MHz range (or the 950 to 1750 MHz range in Europe). RG-6 or equivalent coax should be installed. The slightly smaller diameter RG-59 should be used only when distributing lower frequency modulated signals from the output of the satellite receiver.

Installation errors can cause attenuation in excess of rated values. If cables are bent too severely, an impedance mismatch at the sharp bend can occur which will increase reflective losses. To illustrate, RG-8A has a foam dielectric and keeping the center conductor aligned directly down the middle of the grounded shield is difficult if a wide turning radius is not used, particularly when the coax gets hot. A suggested minimum cable turning radius is 5 cable diameters.

dc Voltage Losses

dc voltages are often relayed via coaxial cable to switch polarity in LNBs or to activate voltage-switched relays. As discussed previously in the section on voltage switched LNBs, it is important to select coax with reasonable dc resistive losses. Acceptable values fall in the 2 ohms per 100 meters range.

Coax Connectors

The majority of LNBs on the market are attached to coaxial cable by F-connectors, an abbreviation for feedthrough connector (see Figure 5-51). The F-connector has a 75 ohm impedance, a measure of resistance to current flow, and is generally mated with components such as LNBs with output impedance of 75 ohms. This connector which is commonly referred to in Europe as the F584 connector is an American creation that uses the center of the coaxial cable as the connector pin. Because this wire is easily bent it can potentially cause problems when either the LNB or the receiver is repeatedly disconnected. Nevertheless, there are two distinct advantages to an F-connector: its ease of construction and its ease of removal. It is extremely easy to fit F-connectors to coax, perhaps the reason why this system is preferred by some installers.

TABLE 5-7. CHARACTERISTICS OF COMMONLY USED COAX

Cable Type	Signal Loss (db/100 feet) 100 MHz	1.45 GHz	4 GHz	Impedance (ohms)
RG-59	3.40	11	N/A	75
RG-6A	2.70	8.7	N/A	75
RG-11	2.30	7.0	N/A	75
RG-8A	1.90	23.0		50
RG-213	1.90	21.5		50
RG-214	2.30	21.5		50
9913	N/A	11.0		50
9914	N/A	13.0		50

TABLE 5-8. CALCULATED MAXIMUM CABLE

Frequency (MHz)	Attenuation (dB/100 feet) Cable Type RG-59	RG-6	RG-7	RG-11
5	0.52	0.42	0.33	0.26
55	1.76	1.40	1.11	0.89
83	2.16	1.73	1.37	1.11
187	3.28	2.62	2.08	1.69
211	3.49	2.79	2.22	1.80
250	3.81	3.05	2.42	1.97
300	4.18	3.35	2.67	2.18
350	4.53	3.64	2.89	2.36
400	4.86	3.90	3.10	2.54
450	5.17	4.15	3.31	2.71
500	5.46	4.39	3.50	2.87
550	5.74	4.61	3.68	3.02
600	6.01	4.83	3.86	3.17
650	6.26	5.04	4.03	3.31
700	6.51	5.25	4.19	3.45
750	6.75	5.44	4.35	3.58
800	6.99	5.63	4.50	3.71
850	7.22	5.82	4.66	3.84
900	7.44	6.00	4.80	3.96
950	7.65	6.18	4.95	4.08
1000	7.87	6.35	5.09	4.20
1450	9.59	7.77	6.25	5.18

Power losses can also be substantial where connectors join cables. If these connectors are not installed correctly, impedance mismatches and losses occur. It is important to examine the inside of each connector before mating. This ensures that the center pin or conductor is not broken off and that it is extended far enough out from the connector to make secure electrical contact but not far enough to short out to a chassis and damage circuitry.

Figure 5-51. Coax Connectors & Adapters. *F-connectors are commonly used with RG-59, RG-6 or RG-11 coax. Numerous adapters are available. For example, RCA male-to-female F-connectors connect an RCA plug to a coax line. F-81 barrels are sometimes used to splice two F-connector lines together.*

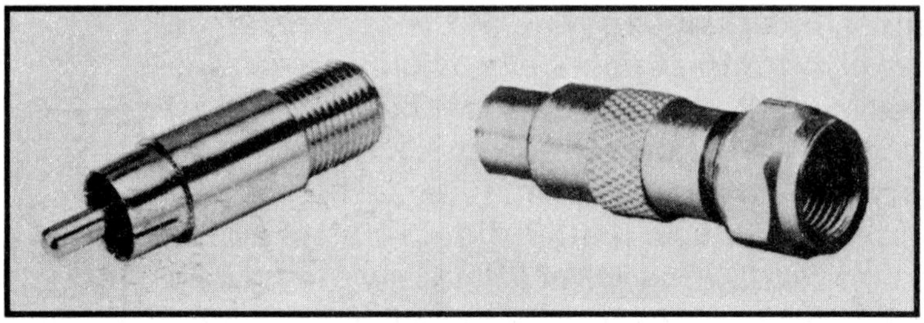

Water and Aging

Coax can easily be destroyed by intrusion of water, especially salt water. Also, any leaks at connectors or along the cable body can short out a signal to ground and possibly damage an LNB or receiver.

Underground moisture, which inevitably comes into contact with buried cables, can cause rapid corrosion of metal components. Tubular aluminum outer conductors have been almost completely destroyed within 90 days. Even small pinholes in outer jackets can allow this to occur. Poor installation or cable handling techniques or even the presence of rodents may cause such damage.

Most direct burial cables on the market today are sufficient to do the job. But time will tell whether or not conduit should be used to protect them. Commercial installations employ, when necessary, more expensive coax having a sticky, flooding compound under the jacket which protects against water intrusion. For maximum reliability against rodents, a steel tape armor with over-jacketing or rigid conduit such as gray electrical PVC is suggested.

Summary

In summary, coax must be rated to carry the frequencies in use and connectors must be installed securely and be watertight. Coaxial cable should also be carefully selected so that impedances are properly matched and so that the frequency carrying ability is adequate. The distances between the LNB and satellite receiver should be minimized. Recommended cables for various length runs are listed in Table 5-9, assuming no amplification is used.

Multi-Run Cables

Satellite TV cables are generally available in a multi-strand form as a "one run" or "dual run" ribbon (also known as direct burial cable). Single run ribbons incorporate one coaxial cable, a 3-wire polarity servo motor control run and a 5-wire actuator power/control cable all bound together. Dual run cables have two coaxial lines (see Figure 5- 52 and 5-53).

Ribbon cables are much easier to install than individual wires. Tangles are eliminated and any concern about one or more of the individual cables being too short is eliminated. It is wise practice to use dual run cables if an up-

TABLE 5-9. RECOMMENDED COAXIAL CABLE SIZES

Maximum Usable Frequency with no Amplification	Cable Lengths (Meters) 25	50	100
70 MHz	RG-59	RG-59	RG-6
950 MHz	RG-6	RG-6	RG-11
1,750 MHz	RG-6	RG-6	RG-11

grade to a dual-band system might be a future possibility. In some cases, installing an additional 3 or 4 conductors might be considered if electronic switches, known as coaxial relays, are later to be used in dual-band systems. The individual wires in multi-run cables that relay power to drive servo or actuator motors and to transmit pulses to count actuator position have improved during the years. In the earlier days of satellite TV, solid wires (Romex) were used. However, the multi-stranded wires now incorporated into ribbon cables are generally less likely to break and have lower resistances per unit length. Multi-strand wires are also better able to contain the signal within the confines of their insulating jacket than solid core conductors. As a result of these improvements, longer runs are possible.

Polarizer control wires should be contained in a grounded sheath and be of the minimum gauge as outlined in Table 5-10. In dual-band installations where the lines to two polarizers are wired in parallel, the recommended cable diameter should be increased by two gauge numbers (e.g. 20 to 18 gauge).

The actuator line consists of two #14 gauge 36 Vdc motor wires and three #20 gauge sensor lines including one for a ground connection. These are usually adequate for runs up to about 200 feet. For longer distances of up to 400 feet, #12 gauge can be used to power the motor.

TABLE 5-10. WIRE SIZES RECOMMENDED for POLARIZERS

Maximum Cable Length (meters)	Wire Gauge (shielded)
25	20
50	18
50 to 100	16

TABLE 5-11. RECOMMENDED WIRE SIZES for ACTUATOR CABLES

Maximum Cable Length (meters)	Wire Gauge Motor	Wire Gauge Shielded Sensor
25	16	20
50	14	20
100	12	20

Cable runs to the actuator and LNB should be minimized (see Table 5- 11). When runs exceed 100 meters, heavier gauge lines must be used to prevent problems such as servo hunting or incorrect actuator counting. (These phenomena are explored later when discussing troubleshooting in Chapter 11).

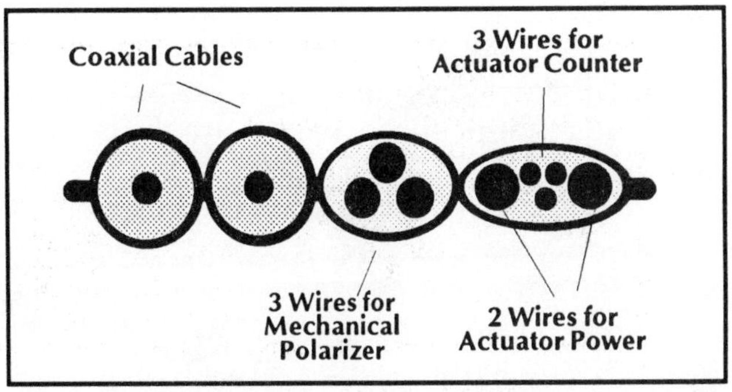

Figure 5-52. Multi-Run Cable. *A standard multi-run cable consists of two coaxial cables, a 3-wire cable for control of mechanical polarizers and a 5-wire cable for actuator control. Not all cables are used in all situations.*

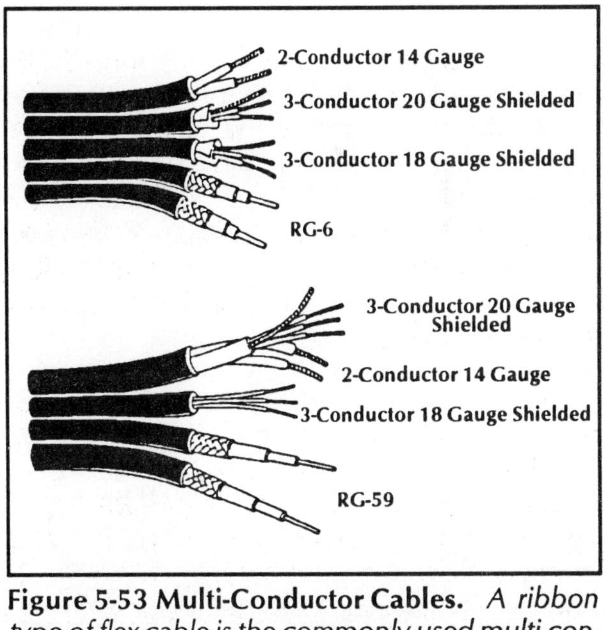

Figure 5-53 Multi-Conductor Cables. *A ribbon type of flex cable is the commonly used multi-conductor cable in satellite installations. These have all the necessary coaxial cables and voltage carrying conductors to control and receive the signals to and from the satellite dish. This cable uses dual RG-6 with 67% braid coverage with 20 gauge center conductor with 3-conductor 18-AWG, 3-conductor 20-AWG and 2-conductor 14-AWG wires. All cables are stranded and shielded with a ground wire. (Courtesy of M/A COM)*

SATELLITE RECEIVERS

The function of a satellite receiver is to select a channel for viewing or listening from the available block of information received from the LNB and then to process this signal into a form acceptable by a television, TV monitor or stereo. Gone are the days when receivers were clumsy devices whose weight was almost a measure of cost. Modern receivers, which are light and attractively packaged, are the control stations for TVROs (see Figure 5-54).

Receivers designed to process digital video broadcasts are in many respects similar to the earlier analog components. Both types are explored in this section. In general an analog satellite receiver has to perform the following functions:

- provide power to the LNB
- control polarity switching at the feed
- provide power and control the actuator
- downconvert and select one channel from those available on a satellite
- bandpass filter and amplify the signal
- demodulate and extract the original information
- amplitude modulate the signal for viewing on a conventional television set

An integrated receiver decoder (IRD) feeds an unclamped baseband signal into a decoder contained within the same case. This is discussed in more detail below.

A digital IRD functions like its analog counterpart up to the point at which a single transponder band is selected. However it also must perform additional steps. It extracts and decompresses the information from the demodulated data stream, decodes the audio, video and data signals and then converts the audio and video into analog forms. Like an analog receiver, it outputs both the raw or baseband audio and video signals at rear panel ports and provides an AM modulated video signal for a conventional TV set.

Analog Receiver Design

The operation of an analog satellite receiver is outlined before looking more closely into the function of a digital IRD since both perform many of the same functions. This device consists of a power supply, downconverter/tuner, final IF stage, discriminator, video and audio processors and a built-in modulator (see Figure 5-55). Commercial-grade receivers are similar except that they feed the output raw video and audio signal into a stand-alone, commercial modulator to improve signal quality.

Power Supply

A power supply used in satellite receivers is a standard electronic component, similar to those found in FM radios and other sound equipment. It consists of an off-on switch, a line-voltage fuse, a transformer, a number of rectifiers and smoothing capacitors as well as integrated circuit regulators. Occasionally, dc fuses in the 300 milliamp range are included on the LNB supply line.

The last stage in a power supply is a voltage regulator that provides "clean" power necessary for proper receiver operation. Voltage regulators are rated at 24, 18, 12 and 5 Vdc. Almost every receiver uses a 12 Vdc regulator because most transistor and integrated circuits are powered by this voltage. Regulators with other voltage ratings are used for a variety of circuit components. LNBs draw about 100 to 200 milliamps at 12 to 28 Vdc while the polarizer draws 500 milliamps at 5 or 6 Vdc. These components often have their own built-in regulators. While receivers draw about 1 amp of current, actuators draw up to 6 amps regulated at plus or minus 18 to 36 Vdc.

Downconverter/Tuner

The purpose of the downconverter/tuner stage is to select one channel from the block of frequencies relayed from the LNB and to lower its frequency to the final IF. The core of the tuner is a voltage tuned oscillator (VTO) that mixes its output with the LNB-to-receiver input signal. A signal frequency centered on the IF is then passed on to the receiver's final IF stage.

The industry standard intermediate frequency of 70 MHz is an arbitrary choice. Other IF frequencies such as 134, 400, 479 or 612 MHz are occasionally used. The 70 MHz standard arose because a circuit called a phase lock loop (PLL) detector that was required for conventional TV receivers, was readily available at a reasonable cost when satellite receivers were first developed. These electronic components had originally been used in the telephone microwave industry and operated at a maximum frequency of 35 MHz. The 70 MHz satellite IF was therefore divided by two so that off-the-shelf, lower cost devices could be used. Today, off-the-shelf PLL detectors operate up to as high as 612 MHz.

Tuner – Selecting Channels

The design of the tuning receiver section reflects its intended use. For example, in cases where a receiver is used exclusively to decipher transmissions from just one transponder on one spacecraft, a fixed VTO frequency can be employed. Commercial C-band systems are occasionally fixed-channel receivers. Other brands are termed semi-agile whereby channels can be selected by interchanging highly stable, fixed-frequency crystal oscillators. Home satellite receivers are completely agile.

Receivers use either rotary, preset or detent, and synthesized tuning to select channels. Rotary tuned receivers, rarely produced today, use a front panel-mounted potentiometer configured as a voltage divider to transmit the required voltage to a VTO. The quality of the rotary tuning method depends upon the stability of both the potentiometer and the tuner's VTO. Most brands have the tendency drift off channel. Just as rare today, preset tuners select channels via one or more preset resistors or potentiometers. A user then chooses the required channel by means of a switch that selects the tuning voltage from the relevant preset.

Figure 5-54. Pansat XR4600 Low Threshold Analogue Satellite Receiver. *This receiver features 2 IF inputs, dual C/Ku-band compatibility over a 950 to 2050 MHz tuning range and an advertised 4 dB C/N threshold. It has on-screen graphics capability, a built-in VCR timer, 3-output SCART connectors as well as the ability to be pre-programmed to receive up to 300 channels via as many as 50 satellites. It also features two built in positioners that support automatic inclined orbit tracking. The PAL B/G and I modulator covers the channel 30 through 45 UHF range. It provides the necessary 22 kHz tone for low and high band LNB selection via 14/18 Vdc and has ferrite and mechanical polarizer controls. The audio section provides dynamic noise reduction stereo, continuous audio tuning as well as a choice among 50 and 75 microsecond or J-17 de-emphasis. (Courtesy of Panarex Electronics)*

Another form of preset tuning, also known as detent tuning, is used in the areas where C-band channels are evenly spaced and follow an alternate pattern with the odd channels being on one polarity and the even channels on the other. This regularity in frequency spacing allows a voltage divider chain to be used to set voltages in selecting these channels. The channel voltages are then chosen via a front panel switch. Two presets are incorporated in the voltage divider chain for optimizing the tuning voltages.

Synthesized tuners, employed in most modern satellite receivers, create a stable mixing frequency via either phase-locked loop (PLL) circuits or voltage-synthesis. The PLL is the most accurate approximation to the 0.005% accuracy of a single crystal oscillator. This tuner incorporates a divide-by-256 circuit to divide the VTO oscillator frequency by 256 and lower it to a frequency where it is compared to a very stable crystal oscillator. Any deviation from the correct value causes a dc correction voltage to be sent to readjust the tuner oscillator.

Quartz-locked or voltage-synthesis circuits use a microprocessor to generate a digital signal for each specific channel. This digital signal is converted to an analog voltage to drive the VTO. A more stable mixing signal is generated than with detent tuning methods. This variety locks tightly onto a selected channel.

Modern satellite receivers that are capable of automatically processing satellite signals of varying transponder center frequencies and bandwidths use microprocessor controlled synthesized tuning. These are essentially computers that remember which satellites have which center frequencies and bandwidths.

Final IF Stage

Once the signal has been downconverted and amplified it is transmitted to the final IF stage, composed of a bandpass filter, an amplifier and a limiter. The bandpass filter sets the channel bandwidth at typically 28 MHz or less by selectively eliminating any out-of-band signals. A filter as narrow as 18 MHz for those satellites relaying "half-transponder" broadcasts may be required. An amplifier then restores losses incurred during the downconversion process and strengthens the signal passed to the detector/demodulator. Finally, a limiter produces a square wave output from a sine wave input because most, if not all, demodulators operate more efficiently with such an input. If the signal entering the demodulator were not properly limited, the video would have sparklies.

Demodulator

The demodulator (detector) circuits process the FM modulated satellite TV signal and strips off the carrier to convert the wideband FM television signal back into the more familiar "raw" composite video signal and FM audio subcarrier known as the baseband signal. This has all the original audio and video information contained in a bandwidth of approximately 10 MHz. The baseband signal is used as an input to decoders but not external modulators or VCRs.

Baseband Processing

The demodulator output contains the composite video signal and the FM audio subcarriers. Baseband is the term used to describe this combined output. At this stage in the receiver, the baseband is split into its component signals. The video must be fed to the video processing circuitry and the FM audio subcarriers to the audio demodulation circuitry.

Video Filtering

A low pass filter is next used to extract the composite video signal from the baseband. Satellite broadcasts relay this signal in a variety of video formats. The American NTSC standard that uses a video bandwidth of 4.2 MHz. PAL B or G as well as SECAM are transmitted in Europe and Africa with a video bandwidth of 5.0 MHz. PAL I is the Irish and English PAL standard having a video bandwidth of 5.5 MHz.

De-Emphasis Circuit

Uplinked FM video signals are pre-emphasized and de-emphasized at the receiver. At the uplink, the video signal is passed through a network that attenuates only the lower frequencies effectively boosting the high frequency portion of the signal. At the receiver, a de-emphasis network attenuates the upper frequencies to restore the original video waveform. The objective is to improve the S/N ratio by about 2 dB. This method is effective because the noise in an FM satellite transmission increases from zero to the upper limits of the baseband signal, about 6 MHz (a triangular noise distribution). Pre- and de-emphasis reduces noise by lowering the higher power, higher frequency noise.

Pre- and de-emphasis circuits for audio signals are simple RC (resistor/capacitor) networks with typical time constants of 50 or 75 milliseconds.

Clamp Circuit

At this stage, the analog video signal has been filtered and de-emphasized but still rides on top of a "dispersal waveform", also known as the "dithering" waveform. The purpose of the clamp circuit is to remove the dispersal waveform. An annoying flicker or pulsation would be evident if this waveform were not removed. However, unnecessary clamping would cause a tearing of the whites and distortion of captions and text in the television picture.

The motivation for including an energy dispersal waveform originated in the earlier days of satellite television when satellite frequencies were shared by terrestrial microwave links, still the case with American C-band transmissions. The traveling wave tube amplifiers (TWTA) used on board pioneer satellites were not very robust. If a transponder was left vacant, the only signal transmitted would be a black or a white level, effectively a high concentration of energy at one particular frequency. This sometimes resulted in failure of the amplifier. The dispersal waveform caused the energy to be spread or dispersed over a wide range of frequencies. This tactic also reduced the potential for interference with terrestrial microwave links which operated in mainly narrowband ranges. Without the dispersal waveform, the satellite signal could have severely affected terrestrial transmissions because its energy would have been concentrated near one frequency.

American C-band systems use a 30 Hz waveform while Russian systems use a 2.5 Hz triangular waveform on some C-band transmissions. The standard European and African dispersal signal is a 25 Hertz triangular waveform.

Filtering

Following clamping, this signal is filtered to remove audio subcarriers above the upper edge of the video band. The final step is another amplification stage.

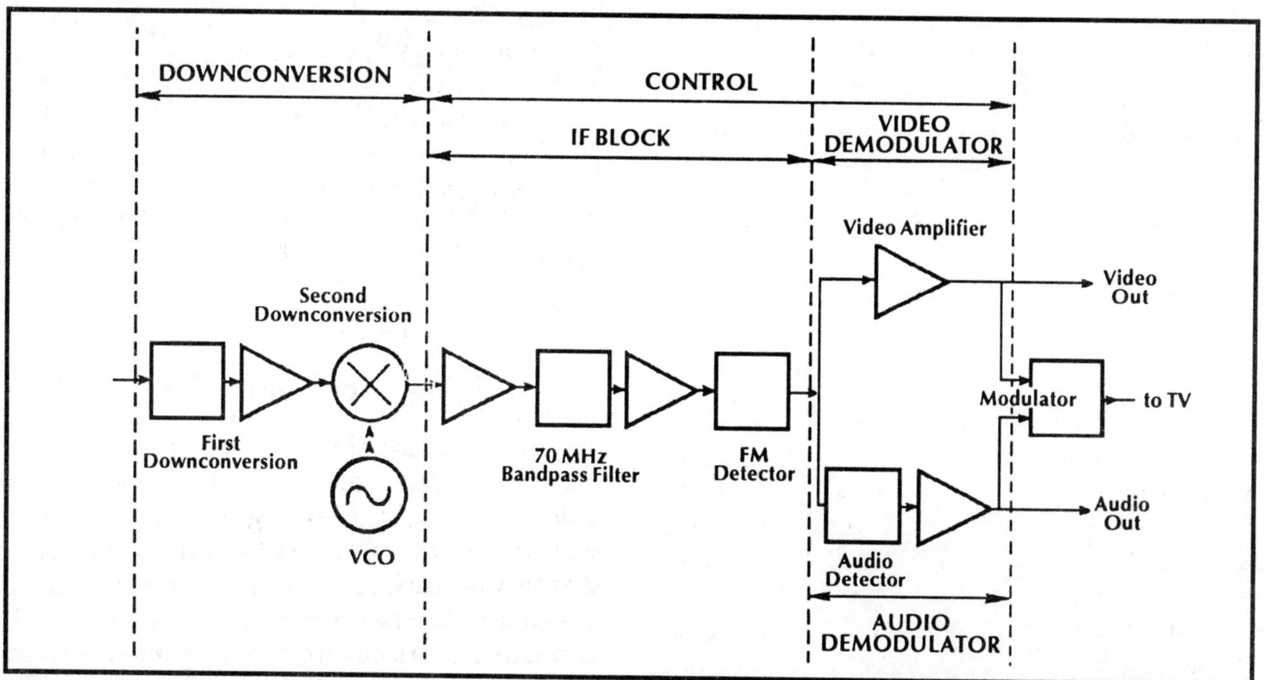

Figure 5-55. Analog Satellite Receiver Block Diagram. *An analog satellite receiver accepts the IF frequency from the block downconverter and selects one channel by downconverting one portion of the satellite bandwidth to a single IF of typically 70, 130, 134 or 612 MHz. This band centered on the IF is then amplified and fed to the detector/demodulator where the FM carrier is removed and the signal is converted to baseband. These signals are then remodulated so that AM video and FM audio can be transmitted to the TV set.*

Audio Processor

The audio processor demodulates the audio signal carried by any selected subcarrier. The output from the audio processor is the original audio information contained in a band ranging from 30 to 15,000 Hz.

Modulator

A modulator is required to "rebroadcast" the composite baseband audio and video signals onto an AM form that can be interpreted by a conventional television. Terrestrial channels, unlike FM satellite signals, are amplitude modulated. The selection of modulation frequency determines which television channel will receive the satellite programming.

North American receivers have a back-panel switch that permits selection between either VHF channels 3 or 4. Receivers in Africa and Europe have the capability to AM modulate the baseband analog signal onto most if not all UHF channels in the 21 through 69 range. The target channel is typically selected via a back-panel screw.

Digital IRD Design

In a digital IRD analog signals are not present except for those available as outputs on the back panel. All other processing is digital. However, the first step in a digital IRD is tuning of the carrier of an individual transponder, the same as occurs in an analog brand.

Next the QPSK demodulator extracts a digital signal from the incoming carrier. The heart of this process is the demultiplex or D-chip. Here the packets are identified and transmission errors are corrected. The reassembled audio and video data streams are then routed to their respective decompressors.

A microprocessor, hard-wired with read-only memory (ROM), interfaces between the audio and video decoders and the conditional access system. Encryption keys are routed from the D-chip where they are decrypted and relayed to the audio and video decoders. Tracking and billing for services are managed within a Smart Card or within the decoder, depending upon the system in use. The microprocessor also receives and decodes commands from the infrared remote control and front keyboard commands. Other digital devices in the IRD are also controlled via address and data lines.

Video Decoding

The reconstructed video bit stream is processed by an MPEG-2 DVB decompresser and decoder. Finally a digital-to-analog chip converts the bit stream into an analog NTSC, PAL or SECAM signal.

Audio Decoding

An MPEG-2 decompresser and decoder translates the bit stream into the raw stereo signals. Then another digital-to-analog converter translates this into left and right channel analog audio signals.

Energy Dispersal in Digital Broadcasts

The triangular energy dispersal waveform used in analog satellite broadcasts to spread the energy throughout the bandwidth is not present in digital transmissions. Instead this function is managed by the channel encoding system. The objective of a channel encoder is to create the most random possible bit stream from the MPEG-2 bitstream so energy is spread as evenly as possible in the channel bandwidth. This allows the channel to be utilized as efficiently as possible.

Judging Receiver Performance

The quality of a satellite receiver is ultimately judged by clarity and fidelity of the television picture and crispness of the sound it produces whether the source be digital or analog. While the viewer makes the final decision about picture quality, these subjective criteria can be well estimated from receiver video bandwidth and threshold. In digital broadcasts other factors such as the type of error correction and modulation method used may allow a smaller dish to be used. However, the ultimate limit is still bandwidth and receiver threshold, the same as for analog transmissions.

Signal-to-Noise Ratio and Picture Quality

Signal-to-Noise Ratio

The central factor in determining the quality of an analog communication signal at the output of a receiver is the ratio of the signal to noise power, the S/N ratio, usually expressed in decibels. Even when noise power is relatively high, if the S/N ratio is acceptable, a clear picture can be received. Thus, as a crude example, an S/N ratio of 40/1 can be achieved both by a signal of 400 watts and a noise power of 10 watts, as well as by a signal power of 4 watts and a noise power of 0.1 watt.

The concept underlying the use of a ratio of signal to noise power can be understood by imagining a listener in a room full of people. If everyone was speaking and the noise level was high, then someone would have to shout to be heard. Signal power would have to be higher. However, if it was so quiet that everyone could easily hear a pin drop, if the noise level was low, then a whisper at much lower audio power would suffice.

Picture Quality

Subject experiments have determined that TV picture quality is directly related to the input S/N ratio. For example, in 1959, the FCC in the United States, authorized a task group to study the problems associated with television broadcasting and reception. This group was known as the Television Allocations Study Organization, or TASO. In one experiment, a standard TV set was viewed by numerous people who were then asked to rate the picture quality as S/N was varied. The results were that for an acceptable picture a S/N of at least 35 dB (a ratio 3,162) was necessary. Other studies have confirmed these results (see Table 5-12).

Video Bandwidth - Analog Transmissions

Analog satellite video broadcasts are transmitted with bandwidths ranging from nearly 30 MHz to below 18 MHz. Although a watchable television picture can be reproduced with a bandwidth as narrow as 15 MHz, reducing bandwidth lowers picture quality. Using a limited bandwidth allows additional channels to be transmitted over a satellite circuit but also increases its susceptibility to noise.

Picture fidelity worsens as receiver bandwidth is narrowed. Details are slowly lost and color shades begin to vary. This appears as a "softening" of the video, as streaking in rapidly changing scenes and as the presence of "sparklies" in the saturated red and green colors. However, receivers that are designed to detect a more narrow bandwidth also see less noise power so that "sparklies" are avoided in all but the most saturated colors. Expressed in another way, picture clarity is traded for fidelity. Subjective studies have shown that the picture quality using a 24 to 28 MHz bandwidth is virtually indistinguishable from that with a 36 MHz bandwidth. Therefore, a bandwidth reduced from 36 MHz to approximately 28 MHz has been adapted as a de facto industry standard for analog satellite receivers. Home satellite brands often use a more narrow bandwidth to minimize detection of noise.

The judgment of what constitutes reception quality is of course based on subjective criteria. Consumer studies in which picture quality was rated according to the signal-to-noise ratio (S/N) demonstrate that, although judgments vary, analog satellite receiver threshold can be an excellent indicator (see Table 5-13).

TABLE 5-12. CCIR 5-POINT SCALE of IMPAIRMENT

S/N Ratio (dB)	Quality	Grade	Subjective Rating of Picture Quality
above 46.6	Excellent	5	Excellent quality - near perfect picture
42.3	Good	4	Good quality - some interference perceptible
38.0	Fair	3	Passable quality - interference not objectionable
33.2	Poor	2	Marginal - poor quality with annoying interference
29.3	Bad	1	Poor - just watchable picture

What is Analog Receiver Threshold?

The threshold of a satellite receiver determines how weak an input signal can be, as measured by the carrier-to-noise power ratio (C/N), before a picture is judged unacceptable. For a given C/N ratio, a particular receiver generates a given output S/N ratio.

If the input C/N ratio is plotted against output S/N ratio for an analog receiver, a straight line results over most of this range (see Figure 5-56). This linear relationship means that for a given change in input, there is a proportional change in output. To illustrate, if an input of 1 watt results in an output of 5 watts, then an input of 2 watts creates a 10 watt output. A simple example shows why this linear relation between input and output is important for picture clarity. A photograph of a scene having one area twice as bright as another would not look proper if the reproduced picture had that area only 50% brighter; that is, if the reproduction were not linear.

Threshold is defined as that point where the deviation from a linear or straight line plot equals one decibel. Near or just above threshold "sparklies" begin to appear. More exactly, threshold is the point at which the peak white of the demodulated video signal has decreased to the point that it equals the amplitude of the impulses noise spikes. This is observed as white, comet-like streaks in the television picture.

Sparklies appear at low C/N ratios because the FM discriminator is no longer able to distinguish between the signal and noise but instead creates its own white noise that is added to the video signal. This is observed as sharp peaks of noise exceeding the video signal. If the tuner center frequency is tuned up or down at low C/N ratios, this can make either white or black sparklies the predominant form of noise.

TABLE 5-13. PICTURE QUALITY and ANALOG RECEIVER THRESHOLD (Threshold = 8 dB)

Carrier-to-Noise Ratio (dB)	Picture Quality
5	Extremely noisy, tearing, audio noise
6	A little better, sparklies
7	Watchable but sparklies
Threshold	A few sparklies
9	Very good picture; sparklies on only saturated colors
10	Video tape quality
11	Cable TV quality

Bit Error Rate – Digital Transmissions

The quality of a compressed digital video broadcast depends upon a number of factors including the type of compression/decompression scheme applied, the modulation technique used as well as the bit error rate after forward error correction has been applied. Unlike an analog satellite receiver, a digital IRD must "acquire" the signal. As power and therefore BER falls below the threshold value, the picture will begin to "tile." The picture freezes and individual blocks of pixel groups appear. Then suddenly the picture is entirely lost. Above threshold, an excellent picture is received. The difference between picture quality just above threshold and well above threshold is hardly noticeable, one of the advantages of MPEG-2 DVB broadcasts.

Electrical Connections

Most satellite receivers have fairly standard rear panel input and output connections. Coax from the LNB feeds via an F-connector into the IF input. Receivers having built-in actuators have connections for two heavier gauge motor and three fine sensors wires. Units that drive a mechanical polarizer have three wires, the ground, white/pulse and 5.7 Vdc red/power wires, which are usually attached via screw or lug connections onto a barrier strip. A separate set of appropriate screw connectors for a dc motor polarizer are sometimes also provided. Receivers that drive ferrite polarizers require connection of just two wires. PIN diodes polarity selection is relayed via the coax to the LNB so no additional connections are required.

Receivers have an RF output sent via coax to the television set. The clamped and filtered baseband signal, known as the composite baseband audio and video, is typically available via two audio ports and one video port for connection to TV monitors, VCRs and external modulators in commercial-grade rebroadcast systems. The unclamped and unfiltered baseband signal is often accessible for relay to a descrambler or stereo processor. SCART connectors are extensively used in Africa and Europe and standard F-connectors are almost always used in North America.

User Interfaces and Adjustments

All satellite receivers feature some standard controls including power on-off switch, channel selection, audio subcarrier selection, polarity and skew adjustment, and usually, video fine tuning. Many other controls are found on other brands. These include switches and buttons to select from among stereo formats, channel scan mode, narrow/wideband audio filters and video bandwidth control. Some receivers come equipped with signal strength meters and, occasionally, frequency centering meters.

The trend has been towards simpler customer interfaces in non-commercial units. Many consumer brands are now available which automatically fine tune both the video and skew and have no customer interfaces for these functions.

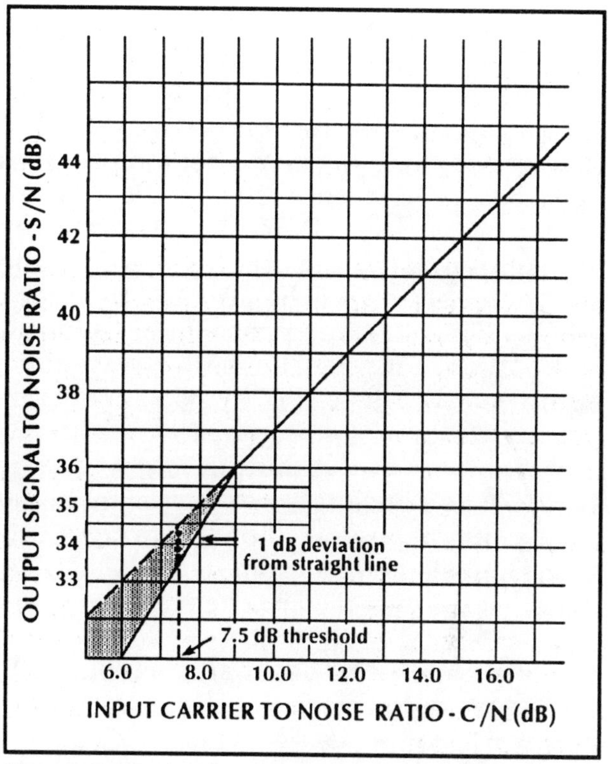

Figure 5-56. Analog Receiver Threshold. *A satellite receive system should provide a sufficient C/N ratio to exceed receiver threshold. Threshold is measured at the point where the departure from linearity is 1 dB.*

MODULATORS

A modulator is a "rebroadcast" device that processes the satellite composite baseband signal, the raw video and audio information, into an AM form that is acceptable by television sets. This device is essentially an interpreter that adds the clamped and filtered baseband signal onto an AM carrier frequency that a TV set has been designed to receive.

Televisions accept AM modulated signals. AM was chosen for terrestrial broadcasts and later for transmission over cable TV networks because relatively high signal power levels are available, as required for clear AM reception, and also because the transmission bandwidth must be maintained in a relatively narrow range to permit a maximum number of channels to be carried in the allocated frequency space. Satellite broadcasts are frequency modulated for precisely the opposite reasons. Power is limited but wide frequency bands are available.

The composite baseband signal, which contains all the video and audio information required to recreate a television picture, can be fed directly into a TV monitor without being remodulated. However, a conventional television receiver has built-in AM demodulation circuits designed for detecting terrestrial AM broadcasts. Unlike a TV monitor, it can accept only remodulated signals. In general, higher quality reception will be attained by using a monitor since these additional modulation/demodulation steps are eliminated. Note that a home video cassette (VCR) also accepts the composite baseband signal and has its own built-in modulator for output to a television. Thus, VCRs are capable of recording higher quality videotapes from the satellite composite baseband signals than from demodulated and remodulated over-the-air broadcasts.

The composite baseband signal can be modulated onto any chosen TV channel by selecting the correct carrier frequency, channels 3 or 4 in North America. African consumer satellite TV receivers have built-in modulators that feed signals into UHF channel 38, although one of channels 21 through 69 can be selected. European modulator output frequencies, also user adjustable in the UHF range, are generally set to UHF channel 36. If any of these channels are occupied by local terrestrial broadcasts, satellite signals can be modulated onto any other unused frequencies. Then all satellite programs can be selected by leaving the television set to this channel and tuning via the satellite receiver. Home satellite receivers modulate video and audio information onto channels commonly used for microcomputers, video recorders and video games. Whenever necessary, a modulator may be retuned to another channel so that it does not interfere with these other sources.

Commercial Grade Modulators

Modulators which are nearly always integrated into home satellite TV receivers are adequate for residential purposes. However, a more complex commercial installation, such as a satellite master antenna TV (SMATV) system for a large apartment development, requires that a higher quality signal be fed into its distribution network. Commercial modulators are of higher quality and are generally more expensive than those built into satellite receivers. In such applications, each receiver is permanently fixed to a given satellite transponder and has a dedicated external modulator tuned to any chosen television channel. Signals from the external modulators are then combined at this "headend" for transmission into a cable leading to the distribution network.

Channels in commercial systems are selected in a variety of ways. Each modulator in a headend can be set to a unique channel so a customer then tunes to any one of these via the television set. Or all the modulator outputs can be fed via a distribution system into a set-top converter that is modulated onto a single television channel. Then a customer would tune channels via the set-top converter, not the television tuner. Of course, in either case the result is the same.

Poorly designed home modulators can wreak havoc in the UHF band when connected into a UHF distribution system. Distortion usually appears as intermodulation, caused when the video and audio carriers interfere or "beat" with each other. In some receivers the problem results when the UHF signal is fed to the back panel connector by a thin piece of coaxial cable. Replacing or rerouting this cable can sometimes cure the problem. Some home TVRO receivers currently on the market have inexpensive, built-in diplexers, devices that allow the user to mix various channels together, that also can cause intermodulation. In these cases, it is better to use an external diplexer to combine the terrestrial and satellite television signals. More details about signal distribution systems are presented in Chapter 10.

CHAPTER 6

TERRESTRIAL INTERFERENCE

Terrestrial interference, unfondly known as TI to satellite dealers, results when a satellite TV system detects unwanted earth-based communication signals. The effects on picture quality can range from a mild case of sparklies to complete picture wipe-out. Some forms of TI are easy to avoid or cure but severe interference is difficult to deal with and can completely ruin reception of a satellite broadcast.

Like an outbreak of acne, TI cannot be ignored. It has been estimated that 10 to 20 percent of all new satellite TV installations experience some form of TI, from mild sparklies to complete picture wipe-out.

The good news is that learning to understand and combat nearly any form of TI is not as difficult as first appearances might indicate. But it is also crucial to know when eliminating terrestrial interference might be nearly impossible or prohibitively expensive. Many satellite dealers have lost hundreds or even thousands of dollars on that unforgettable installation.

The most reasonable approach to combating TI follows a two stage strategy. First, TI should be avoided by correctly choosing satellite TV equipment and by properly locating the antenna. When this strategy fails, interference can be suppressed by using a combination of filters and man-made shields.

SOURCES of TI

Any signals sharing frequencies encountered at any point in a satellite TV receiving system can potentially result in interference. This includes transmissions ranging from the microwave C-band and above to as low as the audio and video baseband range. Thus, for example, if an interfering 70 MHz signal leaks into a satellite receiver, pictures may deteriorate because processing occurs at the same frequency. Or if a nearby communicator is using a 980 MHz relay, the 950 to 1450 MHz block downconverter range might be affected.

The potential interference band can be broken down into two segments: the ingress interference band below 1 GHz; and the antenna interference band from 1 GHz to approximately 8 GHz (see Figure 6-1). The difference between these two bands is that while lower frequencies cannot enter via the antenna/feed/LNB route, microwave signals in the 1 to 8 GHz range can do so.

Frequency (GHz)		Nature of Potential Offender
0.960-1.350		Land-based air navigation systems
1.350-1.400		Armed Forces
1.400-1.427		Radio astronomy
1.427-1.435		Land-mobile: police, fire, forestry, railway
1.429-1.435		Armed Forces
1.435-1.535		Telemetry
1.535-1.543		SAT–maritime mobile
1.605-1.800		Radio location
1.660-1.670		Radio astronomy
1.660-1.700		Meteorological radiosond
1.700-1.710		Space research
1.710-1.850		Armed forces
1.990-2.110		TV Pick-up
2.110-2.180	← 1 →	Public common carrier
2.130-2.150		Fixed point-to-point (non-public)
2.150-2.180		Fixed omnidirectional
2.180-2.200		Fixed, point-to-point (non-public)
2.200-2.290		Armed forces
2.290-2.300		Space research
2.450-2.500		Radio location
2.500-2.535		Fixed SAT
2.500-2.690		Fixed point-to-point (non-public)
		Instructional TV
2.655-2.690		Fixed SAT
2.690-2.700		Radio astronomy
2.700-2.900		Armed forces
2.900-3.100		Maritime radio navigation
2.900-3.700		Radio location
3.300-3.500	← 3 →	Amateur radio
3.700-4.200	← 1 →	Common carrier (telephone)
		Earth stations
4.200-4.400		Altimeters
4.400-4.990	← 3 →	Armed forces
4.990-5.000		Meteorological /radio astronomy
5.250-5.650		Radio location (coastal radar)
5.460-5.470		Radio navigation - general
5.470-5.650		Maritime radio navigation
5.600-5.650		Meteorological ground based radar
5.650-5.925		Amateur
5.800		Industrial and scientific equipment
5.925-6.425		Common carrier and fixed SAT
6.425-6.525	← 2 →	Common carrier
6.525-6.575		Operational land and mobile
6.575-6.87		Non-public point-to-point carrier
6.625-6.875		Fixed SAT
6.875-7.125		TV pick-up
7.125-8.400		Armed forces
8.800		Airborne Doppler radar

Interface frequencies are listed in the order of occurrence on TVROs:

1. Telephone carrier spectrum co-located with TVROs.
2. Widely distributed common microwave carriers.
3. Seldom occurring frequencies close to TVRO Band.

Figure 3-1. Potential Antenna Interference Frequencies. *The most likely sources of TI come from communicators operating in or near the 3.7 to 4.2 GHz C-band range of frequencies. (Courtesy of Microwave Filter Company)*

The Ingress Interference Band

Interfering signals below about 300 MHz in frequency usually enter through poorly grounded or improperly connected equipment. For example, a local TV station might be received along with the satellite TV signal if a ground connection is "floating" and has not been properly attached. At frequencies above 300 MHz, wavelengths are short enough that signals can leak into poorly shielded equipment cases or via breaks in cables (see Figure 6-2).

This type of interference is usually easy to cure by properly grounding cables, using wall plug filters, shielding equipment interconnects and closing unnecessary gaps in metal cases. Ingress has the recognizable effect of causing similar problems on all satellite channels seen, for example, as a characteristic picture defect or heard as an audio buzz.

The Antenna Interference Band

While ingress interference usually arises when an installation is not "tight," TI can enter via the dish in even the most professionally installed system. The antenna interference band can be subdivided into two distinct regions: in-band TI centered on 3.7 to 4.2 GHz; and out-of-band TI spanning the 1 to 8 GHz range but excluding C-band frequencies.

In-Band TI

The source of in-band interference nearly always originates from land- based microwave repeater networks which share the C-band with satellite TV transmissions. Common carriers transmit voice, video and data traffic for a fee via such familiar microwave antennas which are usually situated on towers or tall buildings. AT&T, Sprint, MCI, and Allnet are among the better known common carriers in North America but many regional and local companies also use C-band relays.

Figure 6-2. One Potential Source of Ingress Interference. *This 10-foot Prodelin fiberglass antenna was installed near power lines. Although these lines have little effect on the incoming satellite signal, the 60-cycle power could be a source of ingress interference. All precautions against ingress TI including proper grounding techniques should be taken to eliminate this possibility (Courtesy of Brent Gale)*

Line-of-sight C-band networks have proliferated as telephone, data and other communication needs continue to grow. Experts estimate that new license applications and requests for network modifications are being submitted at a rate of 30 to 50 every month. In North America, the growth rate has been a minimum of 10 to 20% yearly. Today, such networks criss-cross the North American continent like a giant spider web centered on the major metropolitan areas and spanning rural regions (see Figure 6-3). In-band TI accounts for approximately 95% of all interference problems.

Each antenna on a repeater station can transmit up to six different frequencies having either vertical or horizontal polarization. In addition, each tower can be fitted with multiple antennas. In some areas, only one frequency is used; in others, all 25 possible common carrier channels may create serious problems for those receiving satellite TV broadcasts (see Table 6-1).

TABLE 6-1. COMMON CARRIER CENTER FREQUENCIES

Transponder Number	Satellite Center Frequency (MHz)	Common Carrier Center Frequency (MHz)
		3730
1	3720	
		3730
2	3740	
		3750
3	3760	
		3770
4	3780	
		3790
5	3800	
		3810
6	3820	
		3830
7	3840	
		3850
8	3860	
		3870
9	3880	
		3890
10	3900	
		3910
11	3920	
		3930
12	3940	
		3950
13	3960	
		3970
14	3980	
		3990
15	4000	
		4100
16	4020	
		4130
17	4040	
		4150
18	4060	
		4170
19	4080	
		4190
20	4100	
		4210
21	4120	
		4230
22	4140	
		4250
23	4160	
		4270
24	4180	
		4290

In the United States, the FCC wisely allocated these "Ma-Bell" transmissions different center frequencies than those used by satellite TV broadcasts. The allocated frequencies lie 10 MHz above and below the center frequency used by each satellite channel. For example, channel five is centered on 3800 MHz, so common carrier relays using frequencies of 3790 and 3810 MHz can be potential sources of interference to this channel. A total of 25 frequencies is assigned to common carriers; 23 between each channel and one each above and below the satellite range.

To complicate matters, line-of-sight relays in many countries do not conform to a standard pattern of frequencies or polarities. Transmissions may have the same or different center frequencies than those used by C-band satellite TV broadcasters.

Common carriers usually transmit signals that occupy bandwidths of 3 to 5 MHz around their center frequency. These "narrow-band" transmissions typically for telephone links can often be easily "notched out" from the signal seen by a satellite receiver. However, "wide-band" formats often used for videoconferencing links and data service networks are grad-

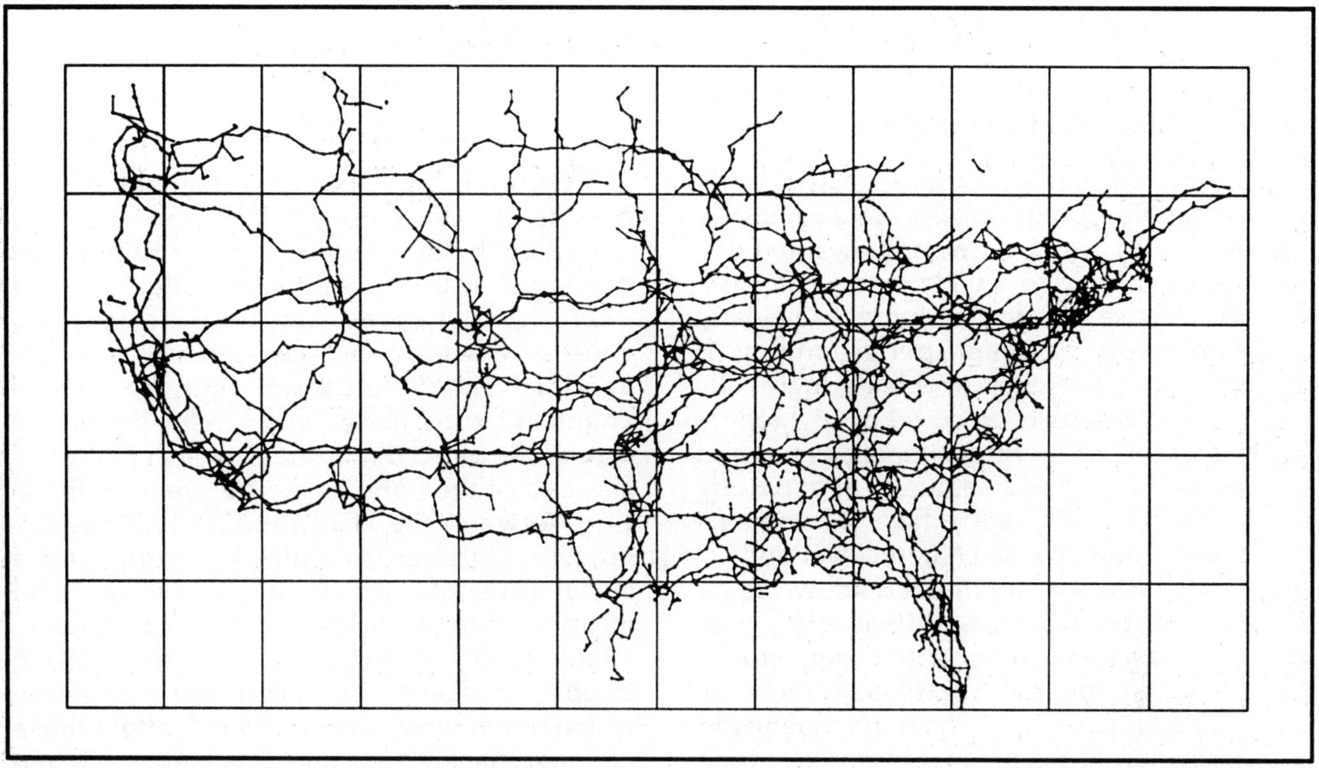

Figure 6-3. C-Band Common Carrier Network. *This map shows the locations of microwave towers and their routes. This network has more than doubled since this map was drawn. (Courtesy of Microwave Filter Company)*

ually becoming more commonplace (see Figure 6-4). Such transmissions having 5 to 30 MHz bandwidths severely interfere with satellite broadcasts and cannot be eliminated with notch filters without destroying a great deal of the picture information. More expensive and sophisticated "phase cancellation" methods are required to overcome this form of TI.

TABLE 6-2 . MAJOR SOURCES OF OUT-OF-BAND TI

1.990 - 2.110 GHz	TV studios to relay towers and betweenmetropolitan areas
2.110 - 2.180 GHz	Common carrier bands
6.425 - 6.525 GHz	for relaying data
5.925 - 6.425 GHz	"Fixed SAT" for carrying remotely originating programs to uplinks for satellite broadcasts

Out-of-Band TI

Out-of-band TI has carrier frequencies that lie in the 1 to 8 GHz band. Four major communication bands account for most out-of-band TI (see Table 6-2).

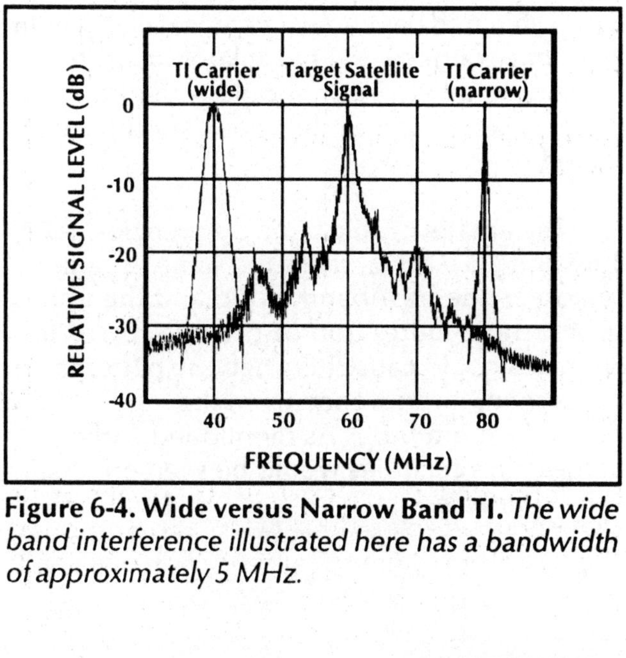

Figure 6-4. Wide versus Narrow Band TI. *The wide band interference illustrated here has a bandwidth of approximately 5 MHz.*

EFFECTS of TI on SATELLITE TV

Ingress Interference

Ingress interference characteristically affects all satellite channels with identical disturbances. It usually occurs when unwanted carriers in the 5 MHz to 1 GHz range enter the system because of poor equipment shielding, faulty connectors or improper grounding or through poorly filtered power lines. Ingress interference can usually be cured by "tightening up the system," properly grounding cables and connectors, shielding all equipment and using power line filters. The advantage of using a familiar test system, i.e. an LNB, receiver and TV set, during installation is that it is known to be working properly so time is not wasted in a wild goose chase tracking down a bad connector or some other component. Therefore if ingress interference appears later, it can be diagnosed and easily cured. Even if ingress interference does show up during a site test, it should not be a cause for undue concern.

Ingress TI can appear when the signal from VHF TV channels 2 through 6, which occupy the 54 to 88 MHz band, leaks into a poorly shielded receiver. Co-channel interference caused by off-air channels being detected on a poorly shielded modulator-to-TV cable can also cause similar symptoms, a faint second picture in the background or "venetian blinds" across the screen. In order to avoid this problem, a shielded flat two-wire cable called a twin lead that connects the modulator in a satellite receiver to a television, should never be used. Instead, a well-grounded coax should be installed.

The amateur radio band frequencies, 1.79, 3.58 and 7.15 MHz, share the same frequency region as the baseband video and the audio subcarriers. Detection of the two lower frequencies could cause herringbone patterns on the screen; interference from the latter could cause an audio buzz. As mentioned earlier, all of these irregularities would be seen on all satellite channels.

Out-of-Band TI

Out-of-band TI has characteristic effects on satellite TV reception. If its power is sufficiently high, the interference can ruin the reception of all transponders on a satellite. However, in most cases, picture deterioration caused by out-of-band TI is seen most strongly at either end of the C-band on those satellite channels closest in frequency to the interfering signal. The effect diminishes towards the opposite end of the band as the frequency separation between the TI and the transponder increases. Thus, for example, if a strong 2 GHz signal were detected by a dish and transmitted into a satellite receiver it would affect the lower channels most strongly. If this interfering signal happened to be either horizontally or vertically polarized, it would most strongly affect those channels having the same polarization. Generally, only channels having the same polarization as the TI will be affected. So a strong indicator of out-of-band TI is picture deterioration on every other channel near either edge.

The picture deterioration caused by out-of-band TI generally appears the same on all affected channels. Horizontal lines across a picture are often seen. Out-of-band TI can result when broadcasts from a local multichannel, multipoint distribution system (MMDS) are detected. MMDS systems, also known as wireless cable systems, operate at a frequency in the range of 2 GHz. Wireless cable systems broadcast signals via a local microwave transmitter to many small, "tin-can" antennas at homes of individual subscribers.

It is rare that weak to moderate out-of-band interference below 2577 MHz will disturb a satellite receiving system. The WR-229 waveguide flange on an LNB input severely attenuates signals having frequencies below 2577 MHz. However, very strong out-of-band signals can overpower the lower frequency cutoff of this waveguide and interfere with reception.

Out-of-band TI is not as rare an occurrence as ingress interference but is still quite unusual because most LNBs and LNAs have built-in bandpass filters (see Figure 6-5).

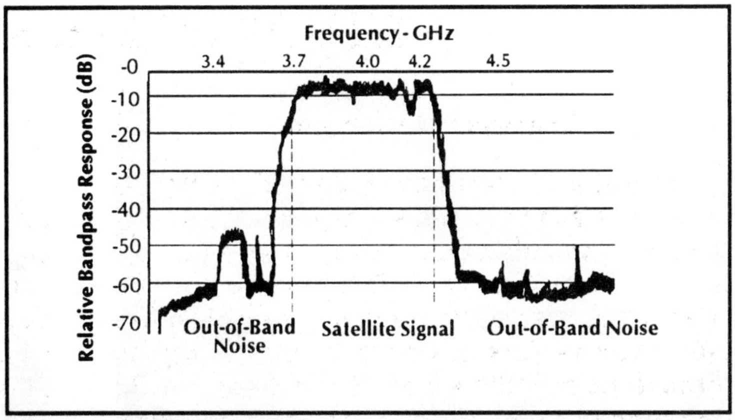

Figure 6-5. Frequency Response of an LNB with Bandpass Filter. *This illustrates how frequencies outside the satellite C-band range are rejected by a bandpass filter. (Courtesy of Microwave Filter Company)*

In-Band TI

In-band TI on analog broadcasts also has characteristic signs. Since each antenna on a microwave repeater station transmits up to six horizontally or vertically polarized carriers spaced 80 MHz apart, picture distortion is often seen on every other channel. If interference from one relay were intercepted, six satellite channels spaced two channels apart could be affected. If fewer carriers are in use, as few as one channel could be disturbed. As power levels of the interfering carriers increase, channels adjacent to these can be affected.

The interference pattern across channels caused by two sets of carriers is somewhat more complex. Generally if two transmitting antennas are used, they have opposite polarity signals or widely spaced frequencies. Twelve horizontal and twelve vertical channels can be affected in these cases.

If multiple line-of-sight transmitting antennas each having different polarities and frequencies were in use, all 24 channels could be affected. Typically, patterns of good and bad channels are seen when in-band TI is present. Even in those cases where all 24 channels were affected, the degree of interference generally varies among channels.

Both wide- and narrow-band TI have similar effects on satellite picture quality. But wideband interference carries more power and therefore has more pronounced negative effects. For telephone relays the bandwidth is determined by how many conversations are being relayed at a time. As more people come on line the bandwidth increases until the transponder is completely wiped out. This occurs most often in the late afternoon when telephone traffic is at a peak. The varying signal strength meter often seen when TI is present reflects the changing bandwidth as thousands of people make and break telephone links.

Thus, in-band TI caused primarily by voice and data telephone relays generally appears as a pattern of disturbance across channels with alternating good and bad pictures. It can vary in intensity from mild sparklies to complete white-out or black-out. At this point a signal strength meter in a receiver would usually be driven off the top end of its scale. As the TI power level increases, so does picture deterioration. Table 6-3 illustrates the effect of interference on a satellite broadcast. TI levels are expressed relative to the satellite signal received at the downconverter. To measure these relative levels in the field a spectrum analyzer must be used.

TABLE 6-3. RELATIVE TI LEVELS and SYMPTOMS - ANALOG

TI Relative to Satellite TV Signal	Symptom
Less than -18 dB	No Problems
-18 to -10 dB	None to Light Sparklies
-10 to -3 dB	Heavy Sparklies - Picture Barely Watchable
-3 to 0 dB	Lines or Random Pattern - No Picture.
Greater than 10 dB	Complete Wipe-out

−15 dB

−10 dB

−5 dB ("sparklies")

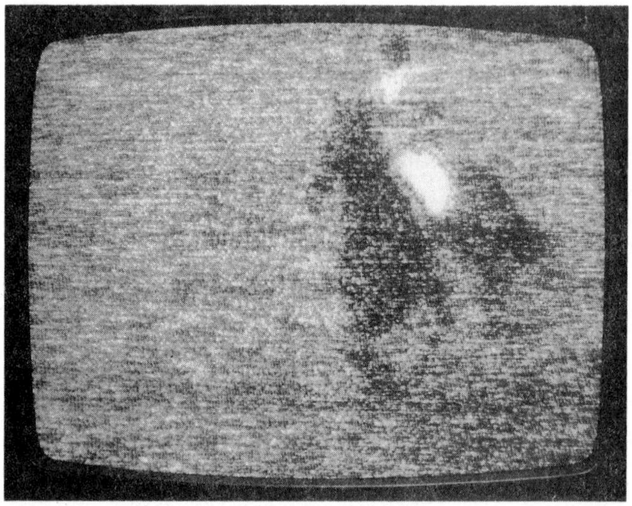

−3 dB ("flurries")

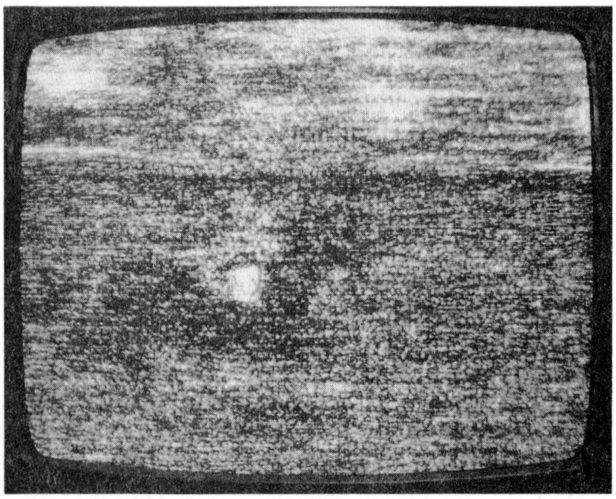

O dB ("blizzard")

+3 dB ("coarse screen")

Figure 6-6. The Effects of TI on Picture Quality. *Picture quality deteriorates as the power of an interfering carrier increases relative to that of the incoming satellite signal. The interfering carrier in all these cases is at 10 MHz below the satellite center frequency. (Courtesy of Microwave Filter Company)*

+5 dB (blank, uneven screen)

+10 dB (blank screen with fine, uniform texture)

Figure 6-6. continued...

Like out-of-band interference, in-band TI may disappear when the antenna is aimed at another satellite. Microwave interference, like satellite broadcasts, can be very directional.

As mentioned in the previous paragraph, the effect that TI has on satellite television depends upon its power relative to the satellite signal. If it is more than 18 dB below the level of the desired signal, it generally is not noticeable on a TV screen (see Figure 6-6). As TI powers rise to about -10 dB sparklies begin to appear on the screen. These increase until about -3 dB when the dots reach a "flurry" level. At 0 dB, the "blizzard" level, the picture is nearly lost. Above this power, TI begins to detune a receiver, namely, the automatic frequency control (AFC) circuit begins to track the TI instead of the satellite signal. At +3 dB the screen has no detectable picture and a coarse appearance. At +5 dB the screen is blank and uneven. Above +10 dB the screen is completely whited out and has a fine, even texture.

If the AFC circuit is disabled, usually with an external or internal receiver switch, the satellite frequency can be manually tracked. Then a watchable picture may be seen even with TI levels as high as 10 dB (see Figure 6-7). Note that if a receiver has frequency synthesized tuning it must have a very "stiff" AFC circuit which deviates less than 3 MHz from the selected frequency. If not, it may track even lower levels of TI than a receiver normally would.

The effects of in-band TI are also illustrated in Figures 6-7a and b presented on the following pages. These diagrams clearly show how in-band TI appears as patterns of "good" and "bad" channels.

+5 dB TI at −10 MHz

+7.5 dB TI at −10 MHz

+10 dB TI at −10 MHz

Figure 6-7. The Effects of TI and Shifting the Center Frequency. *If a receiver's AFC circuit is manually disabled, it no longer tracks interference. The center frequency can then be tuned away from the TI to maximize reception of a satellite signal. As a result, a TVRO can bear higher levels of TI than otherwise would be the case. (Courtesy of Microwave Filter Company)*

KEY to FIGURES 6-8a and b

- ▱ **Light Interference**
- ▨ **Medium Interference**
- ▩ **Heavy Interference**
- ■ **Wipe-out**
- ↓ **Vertical Interference**
- ⇣ **Horizontal Interference**

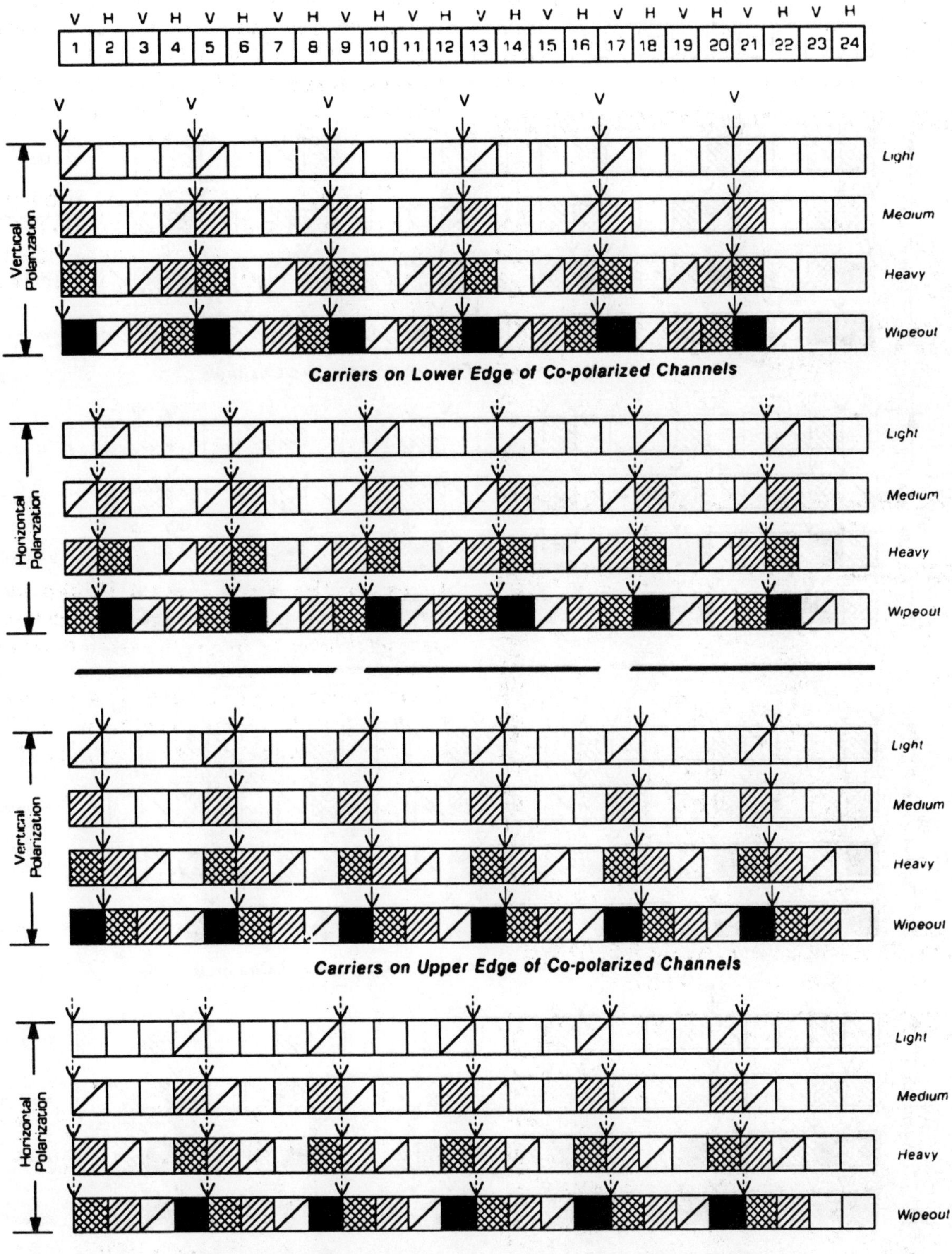

Interference patterns due to a set of six carriers (4 GHz) from one transmitting antenna.

Figure 6-7a. Patterns of Interference on Satellite Channels. *Six carriers from one earth-based transmitting antenna show a characteristic regularity in patterns of channel disturbance. The key to reading both of these figures is shown on the left. (Courtesy of Microwave Filter Company)*

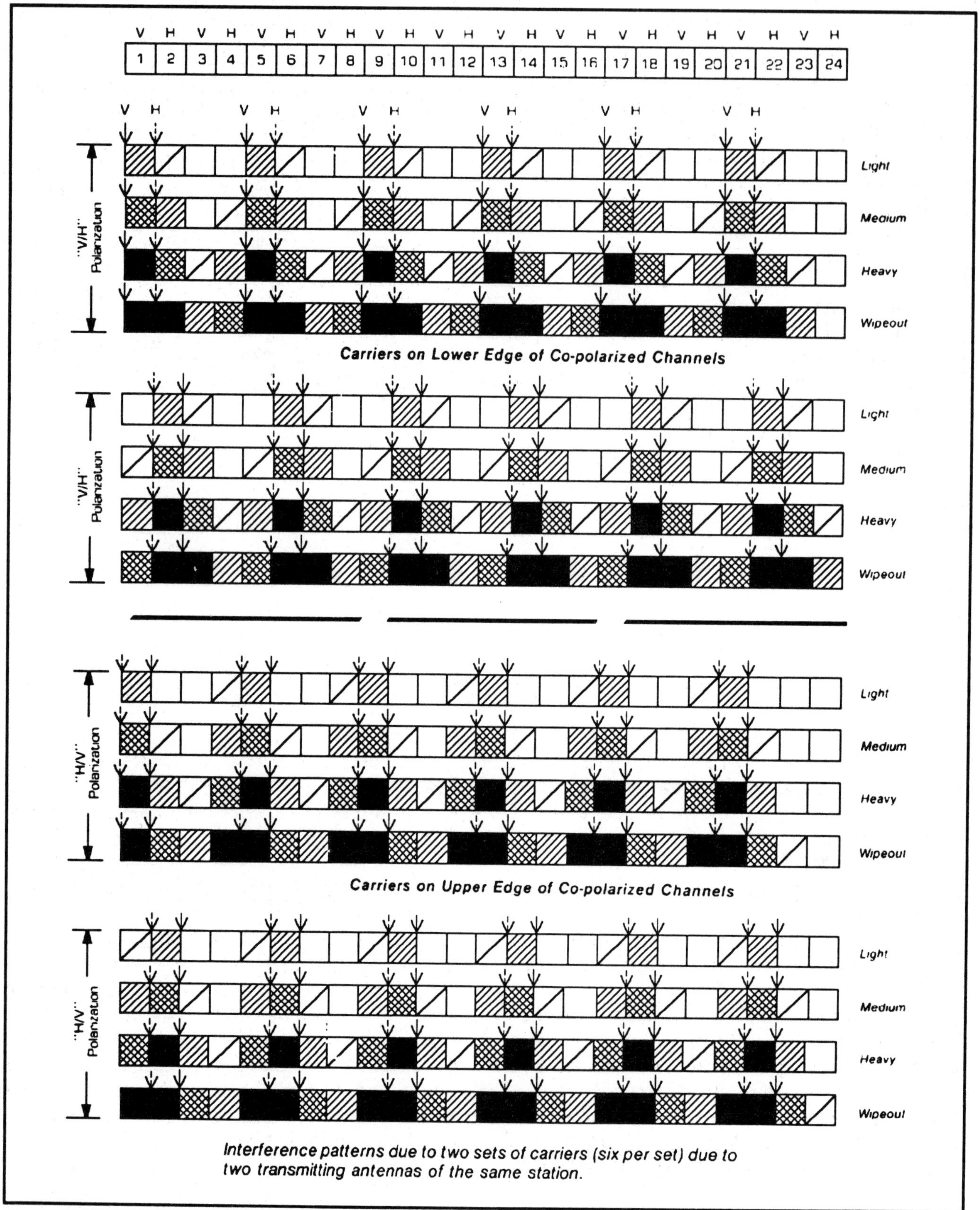

Figure 6-7b. Patterns of Interference on Satellite Channels. *Twelve carriers from two earth-based transmitting antennas show a characteristic regularity in patterns of channel disturbance. The key to reading this figure is shown on previous page. (Courtesy of Microwave Filter Company)*

In-Band TI – Digital Broadcasts

In-band TI can also have a pronounced effect on compressed digital broadcasts. The nature of this picture degradation depends upon whether frequency or time division multiplexing has been employed. In frequency division multiplexing (FDM) each television program is carried by a separate carrier; in TDM all programs on a given transponder are relayed on a combined carrier.

In an FDM transmission individual MPEG-2 DVB signals for each program are combined at the end of the compression/encryption chain. An interfering signal would thus impact upon one or more programs depending upon its bandwidth since each program has a distinct "identity". Given the nature of the digitally compressed signal, the effect would not be noticeable until a "threshold" was exceeded. When this occurred the result would be catastrophic – the picture would disappear as the receiver lost the necessary information to reconstruct the original bit stream. In an MPEG-2 compressed digital video broadcast approximately 2% of the original information is relayed, so past a threshold level no amount of error correction could provide a sufficient bitstream.

In a TDM broadcast, all programs on each transponder are interlaced in time. The threshold level of complete picture loss would likely be higher than in the case of an FDM broadcast since each program is effectively spread across the entire transponder bandwidth and the TI would affect just a portion of each signal. However, when this level is passed, all programs would be lost, not just one or a limited number as in the case of FDM.

The importance of avoiding TI in C-band MPEG DVB cannot be over-stressed, especially since there is still little information available about what TI level would cause complete loss of signal. The field is still too new for such information to have been gathered.

COMBATING TI

The best strategy in combating TI is to avoid the problem at the time of installation through proper equipment selection and dish siting. Only when this approach fails should alternative methods such as building artificial screens, installing filters or moving the antenna be attempted. It is not unusual for TI to appear years after an installation when a new, local line-of-sight transmitter becomes operational.

Ingress interference is usually cured by "tightening up" the system by properly shielding and grounding all leads and ground connections. Reasonable levels of out-of-band TI can generally be eliminated by proper placement of either microwave or RF filters. In-band interference usually calls for RF notch filters or sometimes microwave traps. The most difficult situations arise when the TI level is high relative to the satellite signal or when the carrier has a wide bandwidth ranging from 5 to 30 MHz. Then either more expensive phase cancellation methods or screening methods must be used.

TI has annoying symptoms that can often mimic those caused by component failures. However, a trained technician should be able to recognize and diagnose TI without much difficulty. One who understands and can treat cases of TI, especially in metropolitan areas having heavy data and telephone traffic, is often in the position of a doctor who has a cure for cancer. Business will be good.

Equipment Selection and Susceptibility to TI

The receive dish, the eyes of a TVRO, sees not only the targeted satellite but also noise and interference. Proper selection of the dish/feed/LNB assembly avoids the need for later "band-aid" remedies to combat poor performance. In fact, a system with inadequate gain and poor noise rejection will often act as if it were detecting TI. Furthermore, if TI is present or becomes a problem at some point in the future, the system will be much more susceptible to its effects.

The first step in avoiding TI is to choose a dish/LNB combination that can provide a C/N ratio in excess of the receiver threshold. If a clear picture cannot be received under normal conditions without any interference, then if even light levels are present quality will be very poor.

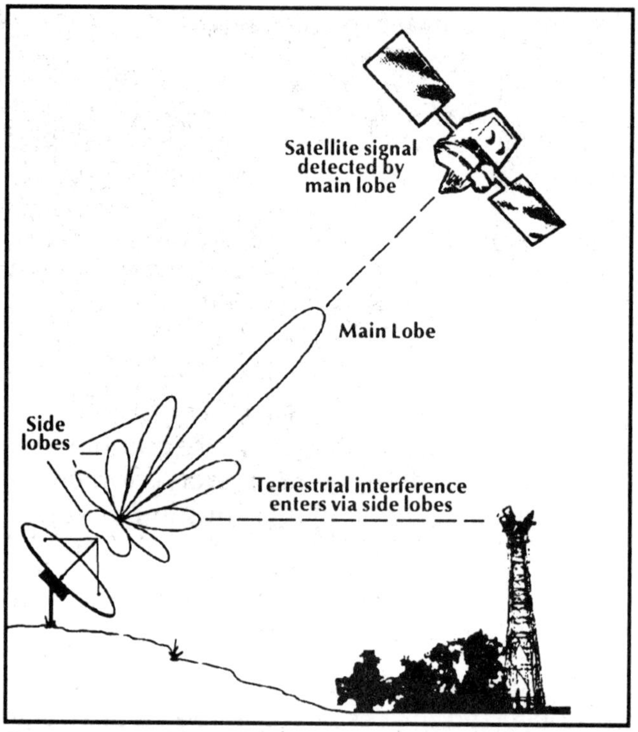

Figure 6-8. Side Lobes and TI. *A dish usually detects interfering signals via its side lobes. If these lobes are weak relative to the main lobe, TI has less effect on reception.*

Optimally, a dish having a narrow beamwidth and low side lobes should be chosen (see Figure 6-8). Interfering signals almost always come from directions off the main antenna axis and are detected by the side lobes. If the first side lobe is 30 dB rather than 20 dB below the main lobe, it will provide 10 dB more protection against TI. This may mean the difference between a watchable and a wiped-out picture. Dishes having higher quality reflective surfaces perform better since side lobes are substantially higher for dishes having rougher surfaces or slight warping. One simple but perhaps costly solution for light to moderate levels of TI would therefore be to install a larger dish having a better surface accuracy. This selection would increase the main lobe and decrease side lobes. Another less costly strategy would be to use a deeper dish which generally has lower side lobes.

It is important to understand that TI is detected by side lobes and very rarely enters via the main lobe. This explains why TI might be seen when aiming at one satellite and may disappear when targeting an adjacent one. Moving the dish even a few degrees could cause the directional source of interference to fall between two side lobes and therefore be detected at much lower strengths.

These considerations suggest that very weak levels of TI may be encountered during a site survey if a relatively small dish is used. It would have a lower gain so the carrier to noise power would be decreased relative to the power from the interfering signal. A small site check dish can therefore be an effective diagnostic tool.

A feed must be chosen to be compatible with the dish f/D ratio for the same reason. A feed which over-illuminates the reflector would cause an increase in detected side lobe power. On the other hand, one that under-illuminates while having significantly lower side lobes would not take advantage of the possible dish gain (see Figure 6-9).

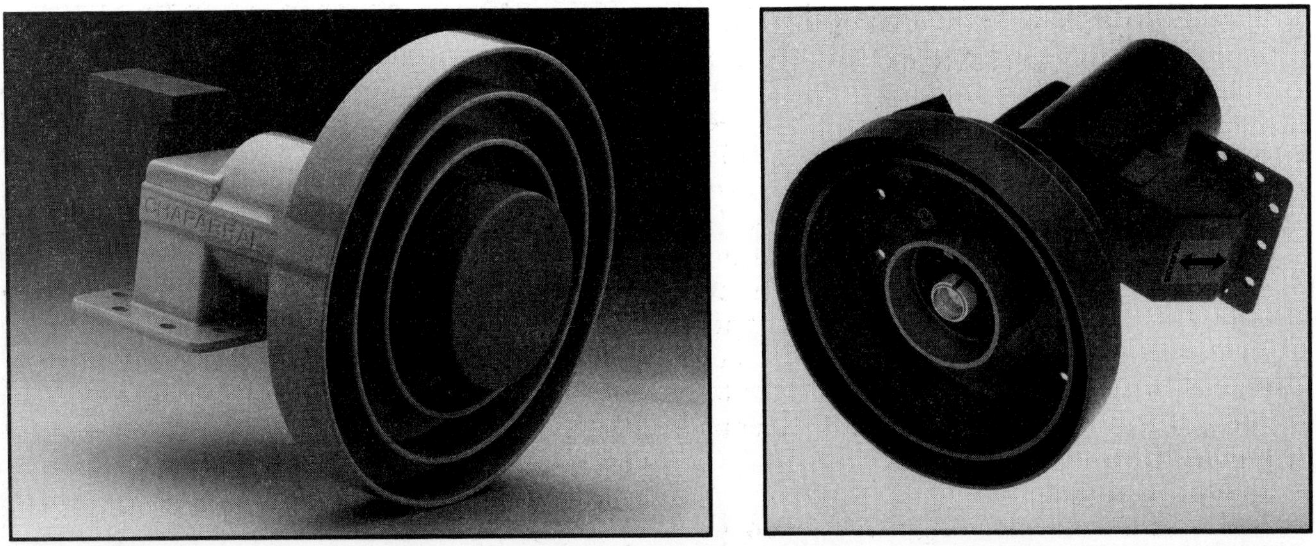

Figure 6-9. TI and Polarizer Design. *The front end of the CalAmp "Centerline" feed (photo at right) is designed for improved rejection of terrestrial interference. Chaparral's Polarotor I features the standard configuration of scalar rings. (Photos Courtesy of California Amplifier and Chaparral Communications)*

An LNB with an acceptably low noise temperature should be chosen. Most LNB manufacturers publish performance curves that graph how rapidly gain falls off outside the 3.7 to 4.2 GHz band. An LNB having a good quality, built-in bandpass filter affords protection against out-of-band interfering signals. Technicians should also realize that when using an LNB, it is impossible to insert either a microwave bandpass or notch filter between the low noise amplifier and downconverter. The alternatives in the case of severe microwave interference are either use of effective screening techniques or placement of a more expensive trap between the feed and LNB.

Medium power out-of-band TI can usually be eliminated by inserting a microwave bandpass filter between the low noise amplifier and downconverter sections when a separate block downconverter has been installed. The bandpass designation means that the filter allows only those frequencies in a given band to be transmitted while rejecting all others. Note that modern LNBs are factory sealed and cannot be opened without causing potentially severe moisture problems. However, effective bandpass filters are incorporated into nearly all LNBs.

In rare cases where the interfering signal is sufficiently powerful it can overdrive a low noise amplifier. This condition is known as driving a low noise amplifier into compression. Then a more expensive microwave bandpass filter between the feed and the LNA or LNB input would be required to eliminate TI.

Downconverters found in older systems are susceptible to TI. A single conversion unit can be a source of interference and should be mounted behind the dish. The oscillator mixing frequencies created for channel selection lie in the C-band range. If the downconverter were mounted directly onto the LNA, a slight leakage of signal could result in these frequencies being re-radiated to the reflective surface only to be detected and re-amplified as TI.

When selecting a satellite receiver, it is useful to have an easily accessible AFC defeat switch. In cases of light to moderately heavy interference, defeating the AFC circuit and manually tuning away from the TI onto the satellite signal can restore a picture. Similarly, a receiver with an adjustable bandwidth control can be a useful diagnostic tool during a site check or installation. If a reasonable picture can be restored by defeating the AFC or by narrowing the bandwidth, chances are good that a low-cost notch filter would overcome the problem.

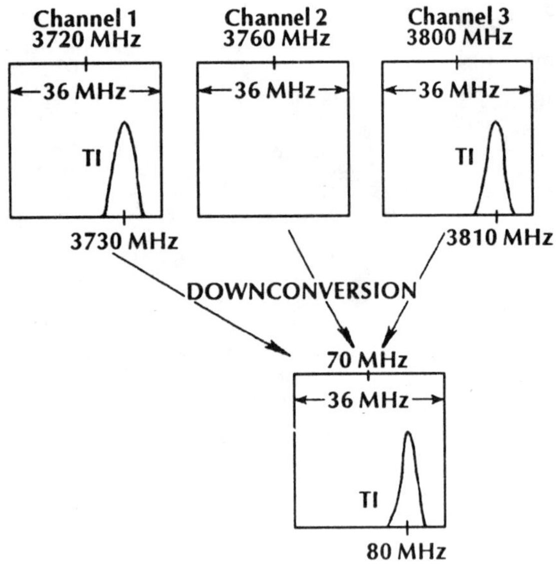

Figure 6-10. Downconverting Interfering Carriers. *In a satellite receiver, both the satellite and interfering signals are downconverted on each channel to the final IF band of frequencies. The interference can lie 10 MHz above or below the satellite IF center frequency. This drawing illustrates how TI carriers at 10 MHz above the satellite center frequency on both channels 1 and 3 are downconverted to the same 80 MHz location at the final IF. Other TI carriers on any other transponder would also be downconverted to the same frequency.*

Combating In-Band TI

Microwave notch filters should be used only when necessary to combat in-band TI because they eliminate some of the satellite signal and therefore degrade picture quality. Carefully choosing equipment and siting the antenna can often reduce or eliminate interference. However the use of filters is generally preferable to relocating an existing installation or to building expensive and often unattractive screens.

The type of filter used to combat in-band TI depends upon the severity of the interference. The least expensive notch filters operate in the IF range. For low to moderate levels below -3 dB, IF notch filters, which remove a narrow band of signal centered on either 60 or 80 MHz, can restore pictures to near normal quality. In cases where non-standard C-band and/or line-of-sight center frequencies are used, notch filters that have variable frequency notches must be chosen. Notches can also be adequate for levels from -3 to about 0 dB relative to the satellite signal. At 0 dB the picture is still usually visible. But IF filters are rarely effective beyond this level, since the receiver AFC begins to follow the interference. However, disabling the AFC and manually re-tuning the receiver can extend the useful range of traps and watchable pictures can result. Such traps may be set for the wrong frequency range for half transponder broadcasts such as those that originate from some Intelsat communication satellites.

When using receivers having an IF different than 70 MHz, such as the DX-700 which operates at 134 MHz IF, specially made notch filters operating at 124 and 144 MHz must be used. A wide selection of 60 and 80 MHz filters and traps for more uncommon IFs such as 124 MHz or 134 MHz IF are available.

In order to understand the operation of notch filters, remember that a satellite receiver lowers the center frequency of each channel to the IF frequency, usually 70 MHz (see Figure 6-10). Since TI signals are centered at either 10 MHz above or below the satellite center frequency, they are also downconverted to either 10 MHz above or below the IF frequency, usually 60 or 80 MHz. Therefore a filter that notches out 60 or 80 MHz can remove the interfering signal (see Figures 6-11 and 6-12). Just two notch filters can remove any interfering signal on any channels.

At TI levels above the -3 to 0 dB range, IF notch filters are not sufficient and microwave filters or traps must be used. One trap is required for each interfering frequency. At levels above +10 to +15 dB a special type of microwave filter, a resonator trap, must be installed between the LNA and downconverter. The complexity and cost of finding a technical solution for TI increases as TI power grows. At levels above +35 dB, very costly specially-made traps must be placed between the feed and LNA to allow the LNA to function. Note that when an LNB is used traps cannot be placed between the low noise amplifier and the internal downconverter. So if TI is severe, an LNB system may have to be replaced with an LNA and single or dual downconversion receiver. In some cases, the sensible alternative is simply not to install a system.

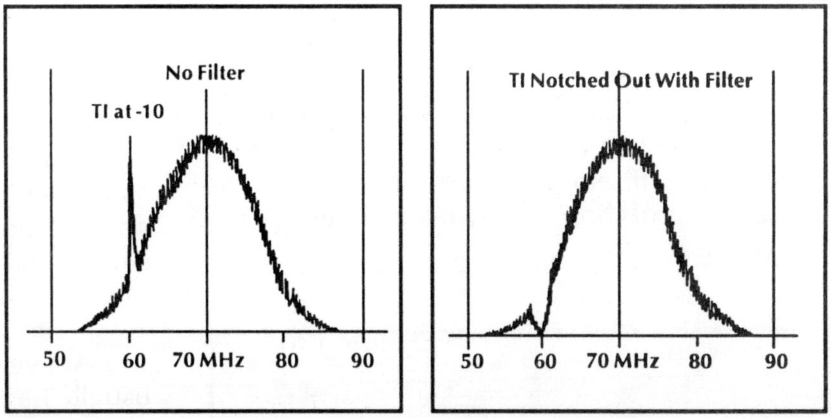

Figure 6-11. Notching Out TI. *These drawings illustrate how an interfering carrier at 60 MHz, 10 MHz below the center frequency, affects the waveform of a satellite signal (left) and how it can be "notched out" by a notch filter (right). In practice, any filter also removes a portion of the desired satellite signal when it eliminates interference.*

Filtering wide band interference is not possible with conventional notch filters which remove narrow bands of signal and interference. If the interfering signal covers the full satellite broadcast bandwidth, then notching out the TI also removes the TV signal. The solution to this problem aside from moving the dish or using artifical screens is called phase cancellation. This method is simple on paper. A sample of the TI which is taken with a second feed and amplifier is subtracted from the satellite signal plus TI by shifting its phase 180 degrees and adding the signals together. This leaves just the satellite signal and eliminates the TI. Unfortunately, this method requires relatively expensive, sensitive equipment and is not one used by the average dealer on the average installation.

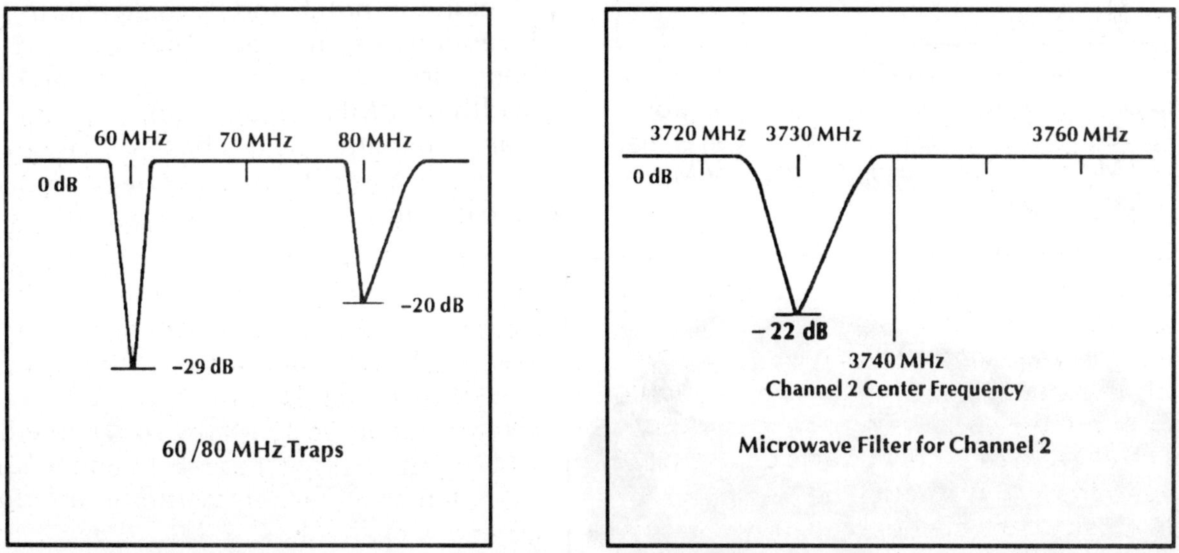

Figure 6-12. Notch Filter Characteristics. *A notch filter is characterized by the depth of the notch and its bandwidth. The left drawing shows the effects of 60 MHz and 80 MHz notch filters installed in series. The lower frequency filter has a notch depth of 29 dB; the upper of 20 dB. The right illustration displays the effects of a notch filter that operates at microwave frequencies, in this case channel 2 which is centered on 3740 MHz. This filter notches out a wider bandwidth of signals than an IF filter.*

Selection of RF and Microwave Filters

Filters can be grouped into two basic categories: notch filters or traps, and bandpass filters. "Threshold extension" filters are a subclass of bandpass filters designed for use in the IF range. Most of these are available in either active or passive forms.

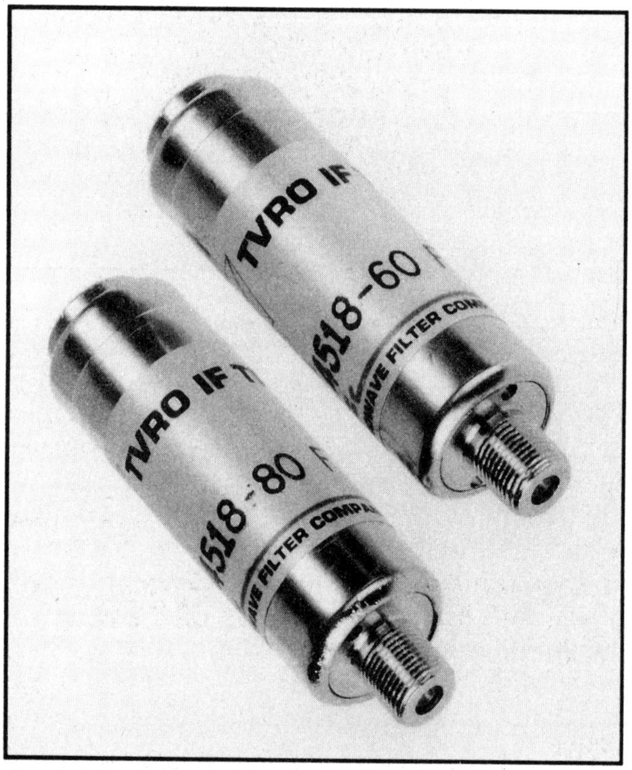

Figure 6-13. Passive Notch Filters. *The MFC 4518-60 and 80F notch filters can be used separately or in series to reduce interfering carriers in a receiver that has a 70 MHz IF loop-through. (Courtesy of Microwave Filter Company)*

Passive filters, relatively inexpensive and uncomplicated (see Figures 6-13 and 6-14), are constructed from standard electrical circuit elements and do not need to be "plugged in." Since these use no power, they provide no gain as the signal passes through. In fact, they have an "insertion loss," ranging from nearly zero to substantial decreases in signal power depending upon the type used.

Active devices need power to function and usually have an integrated circuit and either a ceramic filter or a surface acoustic wave (SAW) filter to limit the range of rejected frequencies. They can be smaller than passive units and amplify as well as filter the processed signal. Their disadvantages include a need for external power and higher cost (see Figure 6-15).

All notch filters are rated by the amount of TI reduction, the depth of the notch. For example, a filter which was rated for -40 dB would reduce an 8 dB interfering signal to -34 dB.

Notch filters, designed for use in the IF range, eliminate a portion of the frequency spectrum during the process of wiping out interfering carriers. The notches, which are centered on or close to either 60 or 80 MHz, are characterized by the relative suppression of power in this notch and by the width or range of frequencies affected. Obviously, if a telephone carrier centered on 60 MHz and having a width of 3 MHz is attacked, the ideal filter would eliminate only this band of frequencies. Eliminating additional frequencies would cause undue loss of the desired satellite signal.

Figure 6-14. Dual Switchable TI Filter. *This filter combines a 60 and 80 MHz TI notch in one compact housing. It provides effective suppression of mild to moderate levels of TI. (Courtesy of Pico Products, Inc.)*

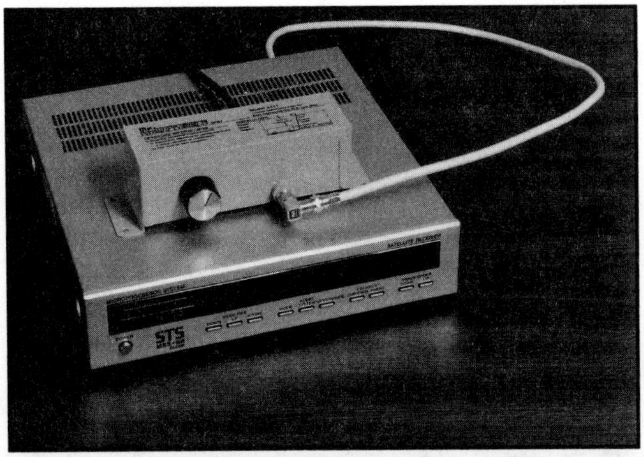

Figure 6-15. Model 5234 Block Notch Filter. *This filter inserts directly into the IF line between an LNB and receiver. The single-turn knob tunes a narrow 15 dB notch through the entire standard block downconverter band of 950 to 1450 MHz. (Courtesy of Microwave Filter Company)*

Some notch filters are equipped with adjustments on one or all of depth, width and center frequency. Some can also be switched in and out of action so that when not required they do not degrade picture quality by unnecessarily removing a portion of the spectrum. For example, the Microwave Filter Company 3217LS-60 and 80 passive notch filters separately or in series can reduce TI carriers by about 53 dB. They are field-tuned by two screwdriver adjustments. These filters, like most other notch filters available, can easily be inserted in the IF line to the receiver by using an F-connector input and output. MFC model 4616 dual 60/80 notch filter is also field tunable, reduces carriers by about 29 dB and can be switched into and out of action without being disconnected (see Figure 6-16).

Threshold extension filters are similar to those used by some manufacturers to lower the apparent receiver threshold. They restrict the bandwidth of the signal entering a receiver and eliminate noise, some satellite signal and hopefully, in the process, the interfering carrier.

Some notch filters can have effects similar to receivers which reduce threshold by narrowing bandwidth (see Figure 6-17). If the width of two filters located at 60 and 80 MHz is increased, the band between them becomes

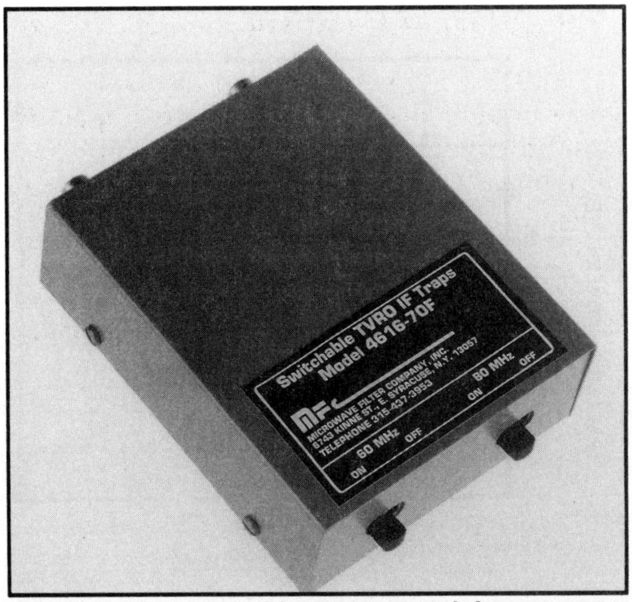

Figure 6-16. Switchable IF Trap Model 4616-70F. *This trap can be used with receivers having final IFs of 385, 400, 403, 510, 520, 560, 600, 610, 700 and some other special frequencies. The notch depth is 25 dB and the 3 dB bandwidth is 4 MHz. Input and output are via F-connectors. (Courtesy of Microwave Filter Company)*

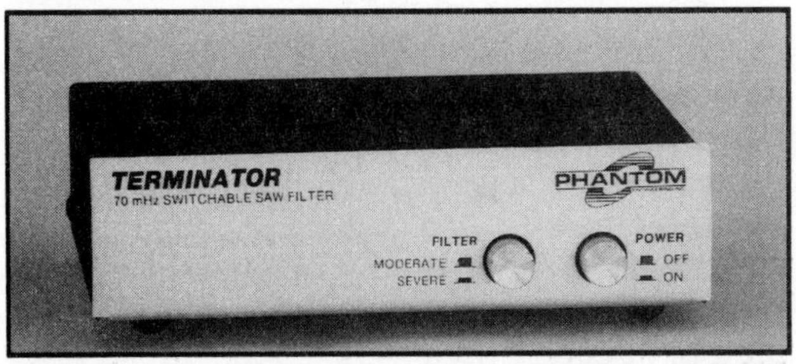

Figure 6-17. Adjustable Bandpass Filter. *This active filter features an adjustable 6 dB bandwidth that can be tuned from 32 to 12 MHz, as well as a variable gain for improving TI rejection. Maximum gain is 10 dB and maximum attenuation is 3 dB. (Courtesy of Phantom Engineering)*

squeezed down. For example, if the bandwidth of each trap were 4 MHz then the middle bandpass region would range from 64 to 76 MHz, a bandwidth of only 12 MHz.

Many modern analog receivers, especially higher end units, now incorporate built-in threshold extension as well as IF notch filters. These filters can be automatically switched in and out as channels are changed. In fact, only a minority of receivers that can be purchased today have IF loops for insertion of notch filters.

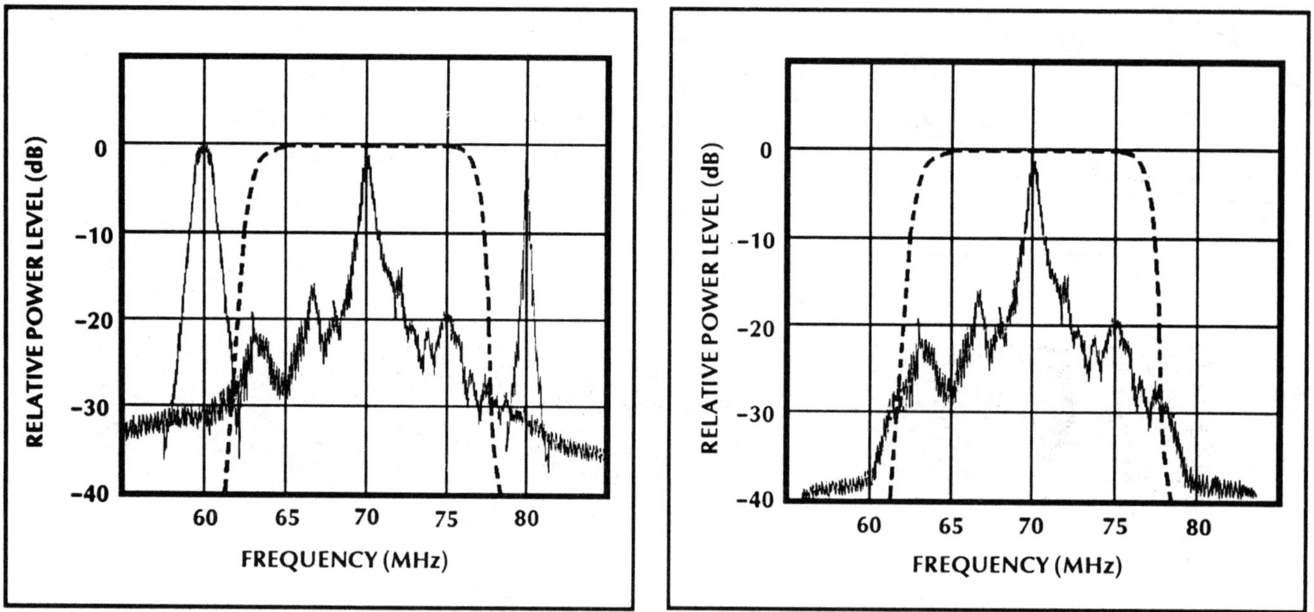

Figure 6-18. Eliminating TI with a Bandpass Filter. *The effects of both a wide band and a narrow band TI carrier may be eliminated by using a bandpass filter as shown here. In this case, the bandwidth is reduced to about 17 MHz. Picture fidelity is degraded by reducing signal bandwidth so severely.*

An adjustable bandpass filter can be used, if necessary, to combat standard TI carriers by narrowing the signal bandwidth to a range less than 20 MHz. Figure 6-18 illustrates a case whereby the effects of both a wideband and a narrow band TI carrier can be eliminated. This is certainly accompanied with a reduction in picture fidelity.

Other Filters

A variety of filters are available for use in combating TI at microwave frequencies (see Figures 6-19 to 6-22). For example, MFC's 4352 bandpass filter (see Figure 6-20) passes only 3.7 to 4.2 GHz. It directly connects into the line between an LNA and downconverter with male/female N-connectors so no jumpers are necessary. The 3966A can be adjusted to pass only one of any selected transponder frequency in the C-band. However, remember that most good quality LNBs built today feature built-in bandpass filters.

Figure 6-19. Tunable Microwave Bandpass Filter. *This bandpass filter can be tuned to any particular transponder frequency in the C-band. It has a 3 dB bandwidth of 40 MHz with under 1.5 dB insertion loss. If several are to be used in series, a 50 dB LNA should be installed. (Courtesy of Microwave Filter Company)*

Figure 6-20. Microwave Bandpass Filter. *The 4532 bandpass filter is designed to pass only those frequencies in the C-band range. It directly connects into the 4 GHz line between an LNA and downconverter so that no extra cable or connectors are necessary. This filter also passes DC power. (Courtesy of the Microwave Filter Company)*

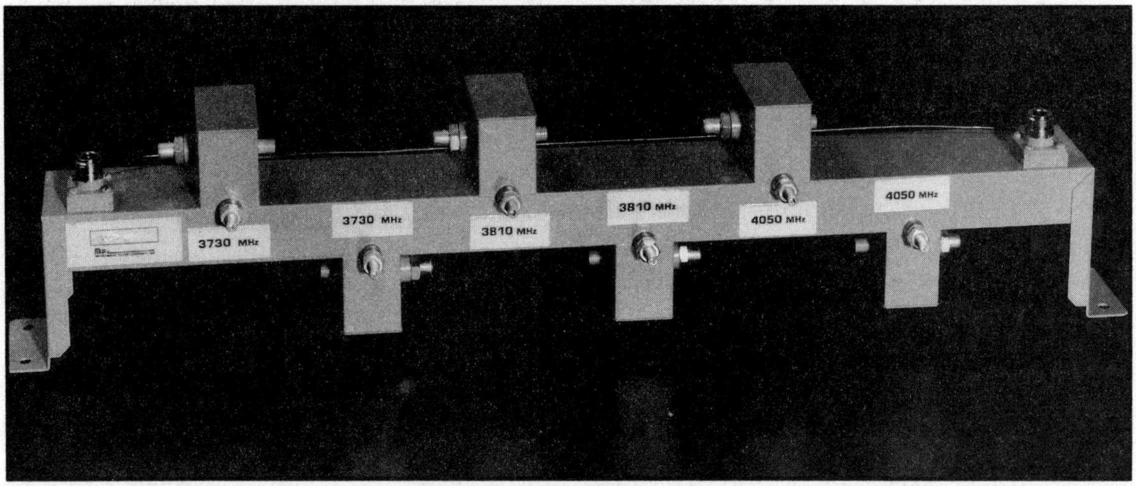

Figure 6-21. Microwave Notch Filter for 4 GHz Interface. *The model 5045-A 30 dB microwave notch filter suppresses 4 GHz interfering carriers. It is used to combat TI strong enough to completely wipe out a television picture. In these cases, the interference levels are usually too high for IF traps. (Courtesy of Microwave Filter Company)*

Figure 6-22. Block Downconversion Microwave Trap. *This filter notches out six frequencies in the 950 to 1450 MHz band. Each notch can be adjusted in depth to approximately 25 dB and the center frequency can be tuned over six 80 MHz adjacent bands to cover the entire 500 MHz bandwidth. (Courtesy Microwave Filter Company)*

Natural and Artificial Screening Methods

TI can often be overcome by a combination of natural and artificial screening methods and, if necessary, appropriate use of filters. Natural screens include trees, buildings, mounds of earth or any other structures already on the site. The best and generally least costly strategy is to make effective use of natural screens before resorting to filters or artificial screens. Natural screens would include using houses, garages or trees as protection (see Figures 6-23 and 6-24).

TABLE 6-4. MICROWAVE ABSORPTION ABILITIES of VARIOUS MATERIALS

Material	Absorption (dB)
Wooden Wall	5
Brick Wall	10
Concrete Wall (15")	30
Cinderblock Wall	15
4 foot Thick Earth	30
Evergreen Tree Cluster	25
Tree Cluster - no leaves	10

Materials have various abilities to absorb or reflect microwaves and to serve as either artificial or natural screens (see Table 6-4). Metal is by far the best reflector of microwaves. Most non-metallic structures are generally not very reflective but have varying degrees of absorbing abilities. Wood, for example, is a poor reflector with limited ability to absorb microwaves. However, an existing wooden structure such as a garage or barn can be made into a reflector of microwaves by lining it with a metal screen mesh. This can usually be accomplished for a relatively low cost. Bricks, cinder blocks, concrete and other masonry products are also poor reflectors but reasonably good absorbers of microwaves. The moisture content of green leaves causes them to be excellent attenuators of TI. But beware using deciduous trees as shields, because when they lose their foliage in winter so goes their value as protection against interfering signals.

In general, the higher a dish is installed above the ground the more susceptible it is to interference. In contrast, placing a dish in a depression or a valley is usually a safe bet in reducing or even eliminating interference. There is a case of an SMATV installation in a motel in California that was so plagued by TI that the least costly solution was to install the 15-foot dish at the bottom of an empty swimming pool. A new pool was constructed as a lower cost option than curing the interference by another method!

Figure 6-23. Using a Building to Screen Antennas from TI. *These dishes are shielded from interference by a brick building. (Courtesy of Brent Gale)*

Figure 6-24. Natural TI Screening. *Placing a dish in a protected area can shield it from wind as well as from interfering signals. Note that if deciduous trees are used, when leaves disappear in the winter, TI may return.*

Some materials are optimally designed for absorbing C-band radiation. An example is space cloth, a canvas-like fabric impregnated with carbon to make it resistive to current flow. A similar material coated with metal on one side could serve a dual purpose as an absorber in one direction and reflector in the other. Bulk absorbers can be attached to existing structures to prevent reflection of TI into a dish. Another example of such a material is a carbon-impregnated foam rubber having an array of small pointed pyramids on one side. (It is interesting to note that similar techniques and materials are used by acoustic engineers to absorb sound).

Figure 6-25. Using Wire Mesh Screens to Eliminate TI. *Appropriately placed mesh reflectors can protect a dish from very strong levels of TI. (Courtesy of Microwave Filter Company)*

Metal screens like builders' "hardware cloth" are adequate reflectors of microwaves. As long as the mesh openings are less than approximately one tenth of the wavelength of the radiation, the same amount of energy is reflected as it would be by a solid metal surface (see Figure 6-25). Since C-band signals have wavelengths of about 3 inches, openings less than 3/10ths of an inch are adequate. (This is why wire mesh dishes can reflect microwaves as well as solid dishes).

The first step in finding a screened location for a dish when TI is present is to draw a scale map of the site and note the composition of all natural and pre-existing structures. If the direction of the TI source is known, an ideal site may often be easily determined. Occasionally, extra man-made screens may be necessary. The least expensive solution is to attach wire mesh to an existing structure but be careful not to cause additional problems by bouncing more interference into the dish (see Figure 6-26). Alternatively, lightweight screens mounted on wooden frames can be constructed. The screen height should be at least three feet above the dish as well as wider than the dish. The ping-pong effect which causes interference to be reflected into the dish instead of away from it should be avoided by tilting these screens at a minimum of 30° or more away from the antenna. These man-made structures must be fastened down securely to prevent them from catching a strong wind and being blown into the antenna or causing other damages (see Figures 3-27a and 3-27b).

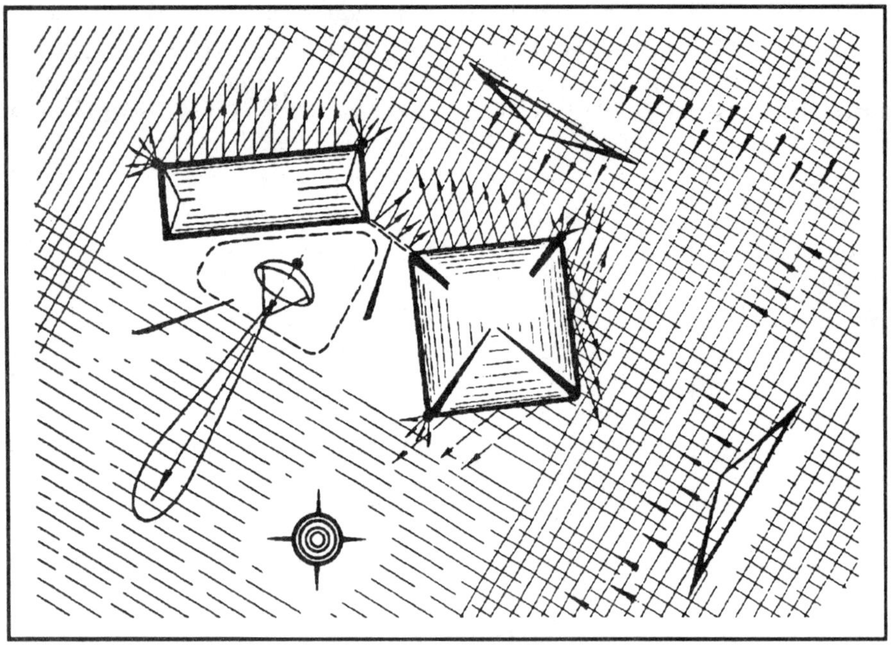

Figure 6-26. Scale Map of Installation Site. *This scale map shows interference originating from two directions. However, if the antenna would be located behind the two metal buildings and the fence a quiet zone would allow TI-free reception of satellite signals. (Courtesy of Microwave Filter Company)*

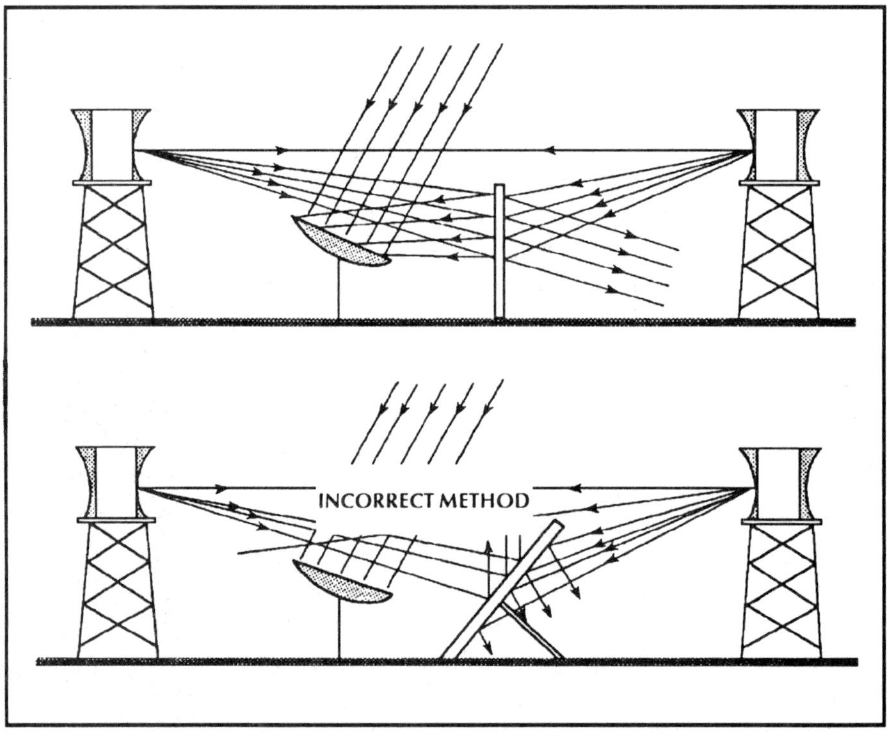

Figure 6-27a. Correct and Incorrect Use of Screens. *A screen should be tilted away from a dish by a least 30 to 45 degrees to avoid the "ping-pong" effect which can cause even more TI to be intercepted.*

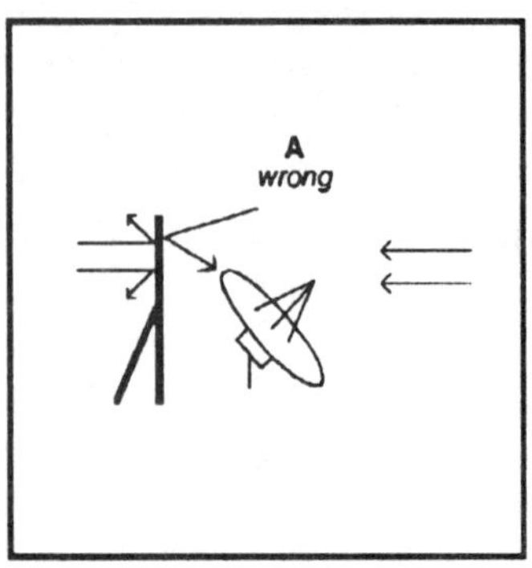

Figure 6-27b. Correct and Incorrect Use of Screens. *When correctly installed, metal screens are rolled at the top to avoid channeling TI into the dish via the edge effect.*

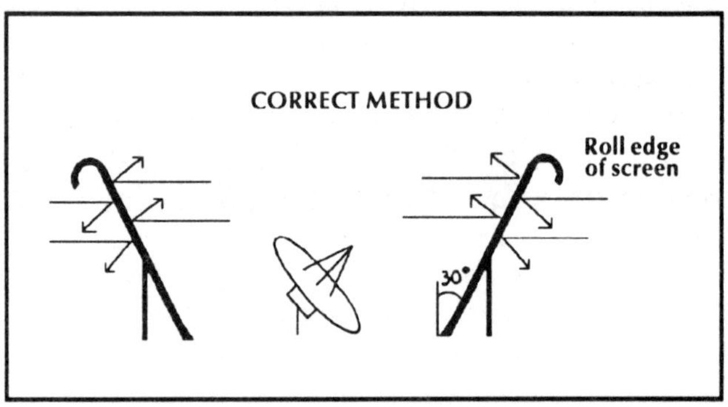

All reflecting barriers should be terminated at the edges with at least a six inch roll of the material used. This prevents creating a sharp edge from which microwaves could be diffracted and bent into the dish and feed.

Similar effects can be achieved with edge absorbers attached to the rim of a dish. These are made from absorbing materials and reduce antenna side lobes by intercepting some of the "spillover" energy from the surrounding terrain and interfering carriers (see Figure 6-28). Edge absorbers differ from antenna extenders or skirts which increase the reflective surface area of a dish. Since dish extenders reflect microwaves, they must carefully follow the parabolic contour to keep side lobes low while increasing gain. They are a "fix" that allows building a smaller into a larger dish.

Occasionally terrestrial common carriers relay full video signals occupying the entire 30 to 36 MHz band from sources such as a local TV station. These will completely wipe out satellite transmissions sharing a similar frequency band. The only completely effective method of shielding against such interference is to screen out these signals with artifical or man-made barriers before they enter the feed.

It should be noted that artificial shielding techniques, while generally more expensive than using filters, do not degrade the satellite signal. Whenever notch filters are inserted into the electronic system, some portion of the bandwidth is removed. This causes a loss of video information often resulting in noticeable picture degradation.

When traps in the microwave range are required, costs can escalate rapidly. If the interfering signal level exceeds 15 dB, a filter between the LNA and downconverter is required. At higher levels of TI, specially ordered, very expensive filters are required. One filter which may cost many hundred or thousands of dollars is then necessary for each affected transponder and must be inserted between the feed and LNA.

Charting Problem Sites

A simple method to combat TI even before visiting the site is by charting microwave routes on a local map. The required background information includes all present and planned locations of terrestrial 4 GHz microwave relay stations in the area as well as their transmission routes and frequencies. In many countries this information can be purchased from any one of the professional frequency coordination firms, usually for less than a few hundred dollars. It should be compiled at most once a year. Or it can be obtained with some more work from either the national telecommunications authority, such as the FCC in the United States, or local telephone companies.

A map of the region can then be marked with the location of each microwave repeater station. Straight lines are drawn between the sending and receiving sites. Any planned installation which falls too close to any one of these communication routes could be susceptible to TI. Since future sites of relay stations are also marked and since the chart will indicate even those potential sources of interference which might be intermittent, this method can be a very important complement to an on-the-site test.

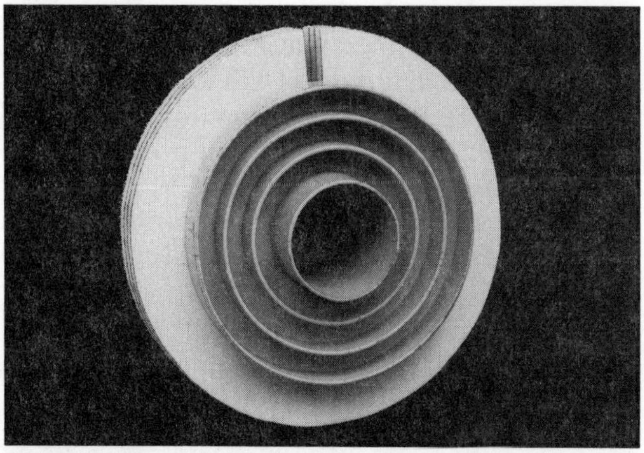

Figure 6-28. Microwave Absorbing Material. *A ring of microwave absorbing material shields the feed from interfering signals. (Courtesy of Microwave Filter Company)*

Similar results can be obtained at a reasonable cost from professional frequency coordination companies. For a fee, these firms will provide a computer printout showing the detailed TI environment of a satellite TV installation site as well as planned sources. This includes the expected worst-case levels of interference, their operating frequencies and polarities and their beam directions. Depending upon the services offered, prices range from $200 to $1000.

Other packages of information including a route map and an intensity map are also available for a nominal fee. The route map, a semi-transparent overlay used with a topographical map, identifies and links each relay tower by a line. The intensity map is similar but also includes the expected power of these common carrier relays.

Another option is to register a planned receive-only earth station site for a fee with the local telecommunications authority. This process may be necessary, for example, when an expensive SMATV (satellite master antenna TV) installation is planned for a condo, apartment complex or hotel. The registration protects this system from future terrestrial relays because it must be factored into their planning. Chances are low that the FCC would allow a common carrier to design a new relay which sends a beam of microwaves directly across an expensive, registered SMATV or cable TV installation. Frequency coordination firms also offer this licensing service and provide what they call "frequency protection" which warns both the licensed earth station and planned land-based communicator of a potential for interference.

It is important to realize that nothing can replace a site survey. An installation which may be far from the route of a land-based microwave communication path could easily be troubled by a signal reflected off a building or a metal billboard. Or a site directly between two relay towers may have just the necessary natural screening to avoid any difficulties. Any installation could also easily be susceptible to out-of-band or even ingress interference.

Frequency Coordination as Future Insurance

Until TVROs were deregulated in 1979, a license had to be obtained prior to installation. Today, uplinks are still regulated in many countries. These require careful planning and a process called frequency coordination.

The frequency coordination begins with a study of other nearby communicators sharing the same frequency band who might interfere with or be affected by the planned installation. A detailed map, usually created by computer from an extensive data base, shows where the potential sources of interference may lie. If all seem to be right, the license is granted. Any future sites which may affect this new one must likewise register. In this fashion, all are protected from each other.

Few receive-only stations bother to follow the frequency coordination or licensing route. Some of the larger SMATV (satellite master antenna TV) systems serving hotels or apartment complexes often take this safer route. Home satellite dealers should also be aware of these services.

CHAPTER 7

SELECTING SATELLITE TV EQUIPMENT

The home satellite TV business has evolved at an astonishing pace during its brief lifetime over just over fifteen years. Today, selecting equipment for personal or customer use can become a matter of choosing from among scores of dishes, receivers and other components. In 1980, only a few brands of equipment were available and actuators were virtually unknown.

In order to serve customers properly, a dealer must have a thorough understanding of the basics of satellite TV operation and should also keep track of new developments and equipment. He or she must know what systems are least susceptible to terrestrial interference, how to properly design a TVRo station having an acceptable performance margin and the effects reduced spacing between satellites and weather will have on picture quality, as well as a host of other issues. Learning all this information is simply a matter of consistent research. But by far the most important and enjoyable task is selecting and actually installing a system.

This manual is not written to advertise various brands of equipment or to list everything on the market. The trade journals do that perfectly well. Our intent is to provide the ammunition to allow a dealer or consumer to read and understand any trade journal or set of instructions in order to make well-informed judgments about the quality of equipment.

COMPONENT SELECTION

Among the criteria used in choosing components for satellite TV systems are performance, cost, features, durability and aesthetic appeal. Each of these must be weighed in selecting equipment. Some owners will be satisfied with an $800 system having acceptable quality pictures on most but not all transponders; others will pay $3000 for a TVRO having all the bells and whistles, will demand perfect video on all channels and then will purchase an elaborate home theatre system as a clincher.

Mounts

The mount, dish, feed and LNB are the most important parts of a satellite system and should be properly chosen. It all begins here. Spending extra money in the future on receivers or other electronics will usually not improve picture quality if the outdoors components have not been properly selected and installed.

A dish must be mounted on a stable foundation with a reliable mechanism for aiming the dish at the arc of satellites. A flimsy mount permits a dish to sway in the wind and may even collapse under the weight of a heavy snow. The reflector must have a secure attachment to the supporting pole and mount, and a method for setting and maintaining declination and polar axis adjustments. The bearings which allow tracking should be well designed to move freely and to resist excessive wearing. The points of attachment to the dish must be strong but should not cause the dish to warp or allow twisting or excessive movement under loads (see Figures 7-1 to 7-3).

By far the best method of evaluating a mount is to examine its structure in detail. Push the dish from its outer rim to see if there is any play or if the various adjustments may be upset too easily. A comparison with other mounts is usually instructive. There are differences between various brands.

Actuators

Actuator arms and horizon-to-horizon assemblies are subjected to mechanical stresses. Jack tubes and motors are especially prone to problems and failures. However, satellite TV technology has developed since its infancy to the point where reliable units are readily available (see Figure 7-3 and 7-4).

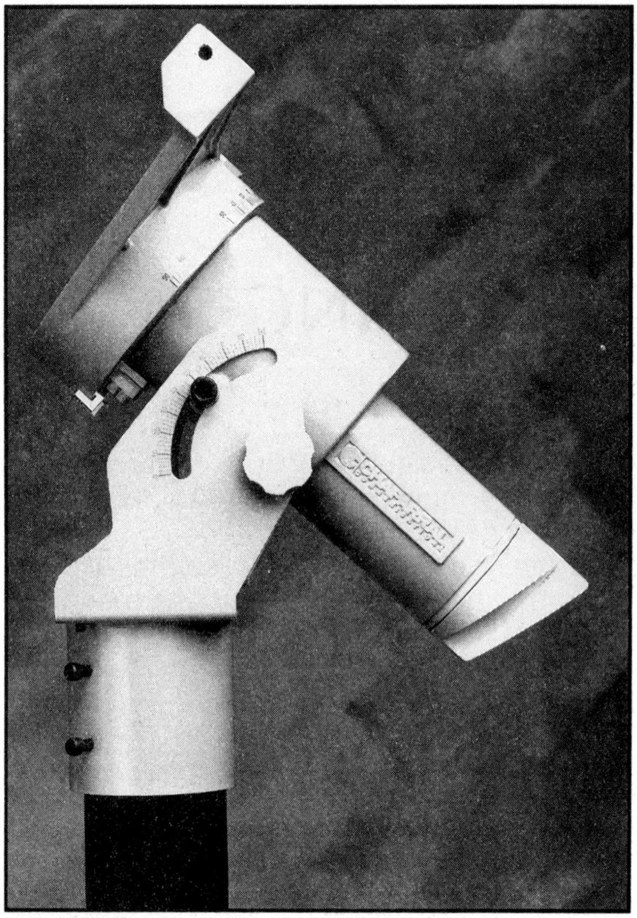

Figure 7-1. Polarmotor™ Horizon-to-Horizon Mount. *The Polarmotor™, designed for prime focus or offset fed parabolic dishes, tracks the full 180° arc in 50 seconds. It has a pointing accuracy of 0.2° based on 5 counts per degree of rotation and uses an optical encoder for counting that is compatible with either reed switch or Hall effect circuitry. (Courtesy of Chaparral Communications)*

Figure 7-2. A Gravity Mount. *This non-penetrating roof mount is designed to be an economical yet stable mount to support 6.5 and 7.5 foot dishes. It features adjustable brace legs to plum the mast pole when installing on uneven surfaces and allows a minimum of eight concrete blocks to into each of the three trays in order to provide stability in winds up to 100 miles/hour. Gravity mounts for 2.5 and 3.5 foot dishes are also available. (Courtesy of KTI)*

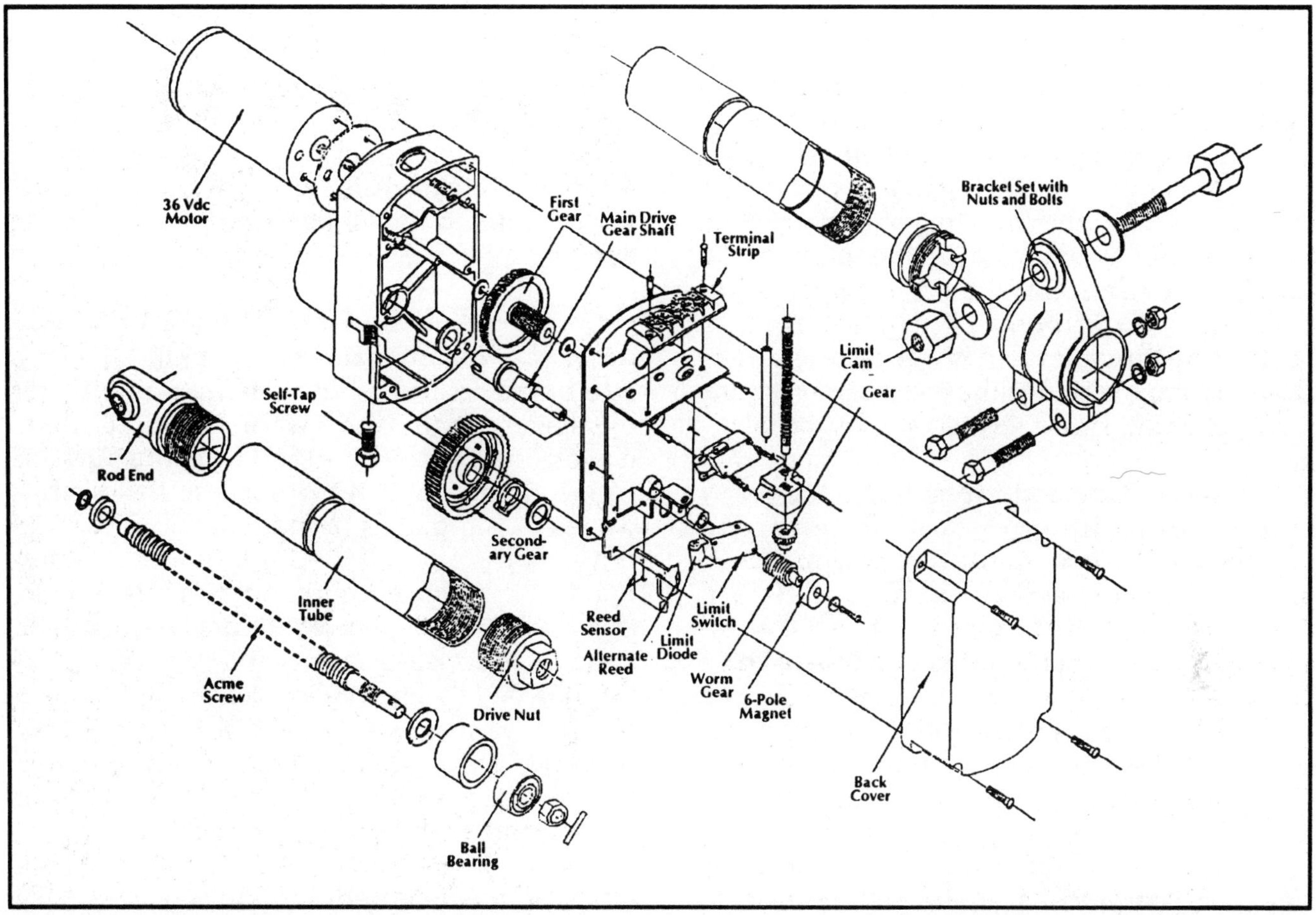

Figure 7-3. SuperJack II+ Construction. *This complete assembly of the SuperJack II+ is diagrammed here. (Courtesy of Pro Brand International, Inc.)*

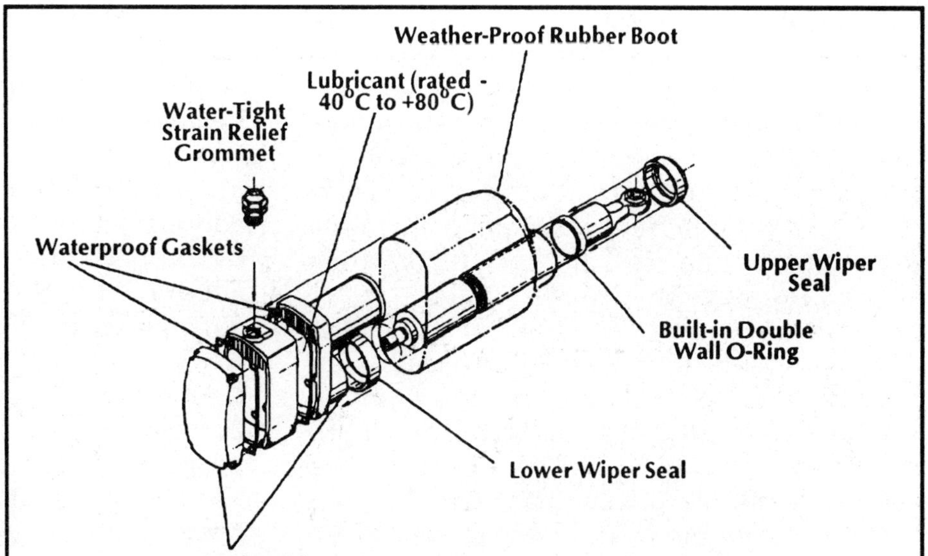

Figure 7-4. Waterproofing an Actuator Arm. *This drawing shows the points where an actuator can be protected from water entry. Drain holes must be installed and weatherproof covers as well as shaft wipes are necessary. (Courtesy Pro Brand International)*

Make certain that the actuator jack does not have excessive play. This could cause a dish to flop around in the wind. A longer arm is need for a larger reflector so that adequate leverage and force is maintained throughout its sweep across the entire arc. Seals against water entry should be thorough; drain holes should be drilled at the lowest point on the motor housing. Actuator arm units should be shipped with ball joints for attachment to the dish and mount. Also make sure that the hardware is properly protected against the weather; galvanized, stainless or zinc-plated steel is the best.

Actuators are controlled from indoors by signals from either potentiometers, reed switches, Hall effect transistors or optical sensors, each type having somewhat different characteristics. Most are built into satellite receivers or IRDs today. Some actuators take an undue amount of time to send a dish from one end of the arc to the other. Be aware of the programming requirements for the actuator controller: some brands are easy to install and program; others are more difficult and can be unreliable. Do not purchase an actuator which does not have mechanical limits or the ability to protect itself with devices such as slip clutches in the motor should some over-ambitious user or confused microprocessor try to drive it far beyond the end of the arc.

Dishes

The dish is the eyes of a satellite system. It must accurately concentrate and reflect the satellite signal as well as reject interference and noise. This is a tall order considering that it sits outdoors and is subjected to all the elements including corrosion, wind, ice and snow loading, hail, heat, cold and children. Of course, a reflector that may be perfectly adequate in Florida may not stand the test of time in Wyoming where snow and winds can be severe.

Judging which type of dish to use is based upon many factors including aesthetics, performance, weight, ease of shipping and assembly, wind loading, durability and cost. For example, spun aluminum dishes can have excellent surface accuracies but come in one piece and are not easily shipped. Wire mesh dishes which many find very attractive can be broken down and shipped in small boxes but may conform to a parabolic shape less accurately than other types if the panels are not held securely. Stamped aluminum dishes are very durable if their wall thickness is sufficient but are more easily buffeted by winds than a wire mesh reflector.

A simple physical inspection of a dish can be very revealing. First, when sighting along one rim, does the other side line up perfectly parallel? If not, the dish is warped. Second, does it feel smooth to the touch or is the surface rough or bumpy? If the surface has visible waves or has panels or mesh sections that do not line up accurately, beware. Third, how is the hub or central portion of the dish attached to the mount? Check that the mount is adequately strong and well attached to the dish to support its full weight without allowing the reflector to sag and warp in time. Some dishes are secured with only a few screws or bolts which may twist as well as rock when it is under stress. These may not withstand the test of time in very windy or snowy environments. Any rocking or twisting movement caused by wind will eventually produce metal fatigue and lead to failure and poor video quality.

The design and method of supporting the feed assembly are very important. Check how much weight is hanging out in front of the antenna. Some designs resemble a bowling ball on a bamboo pole. They certainly should be supported by guy wires attached to the outer antenna rim. The feed should block a minimum amount of the dish aperture or side lobes may be unnecessarily high.

The designs and materials of a dish must be adequate so expansion and contraction caused by temperature changes will not ruin surface accuracy or structural integrity. If a panel is damaged it should be inexpensive and simple to replace. If the dish is located near an ocean, the materials must be treated to resist the highly corrosive effects of salt spray. When a dish will be installed on a roof mount, it makes sense to use a lightweight mesh or spun aluminum reflector.

Dishes can be grouped into two major categories: mesh and solid. The solid ones are either fiberglass, or stamped or spun metal. Brands from each category can be evaluated by all the above criteria as well as by some which are specific to a given type of dish.

Mesh dishes can be made from expanded or stamped steel or aluminum. These materials are protected by painting, powder-coating or anodizing. Judging which methods are best for the given climate depends upon where the dish will be installed. A painted mesh dish may be perfectly durable in a desert environment but a disaster on the coast of Oregon because of salt corrosion. It is important that a mesh dish feel strong to the touch. A flimsy mesh will probably not withstand a hail storm without dimpling, deforming or having some panels badly damaged or destroyed.

How the shape of a mesh dish is maintained is crucial to its performance and is often related to the ease of assembly. As few as eight or as many as 24 supporting struts and panels of mesh may be used. Struts are usually manufactured from tubular steel, extruded or angular aluminum for structural strength in order to maintain their shape when under wind or snow loading. In some mesh dishes the actuator arm attaches to just one rib which certainly must be strong in order to maintain a true shape. If not, the reflector could be warped from one side. Mesh dishes often have one outer circular ring by which these supports are connected to the middle hub; others have one or more middle supporting circular rings as well. Some designs and materials yield an accurate and stable surface; others are poorly designed and manufactured. Common sense judgments of these designs are usually surprisingly accurate.

Solid dishes are usually stronger and often can be more durable than those having mesh reflective surfaces. They can be assembled more quickly. Snow and ice tend to slide off instead of accumulating as they do on some mesh antennas. But solid dishes usually weigh more and are more subject to wind loading and sagging under weight.

Spun aluminum or steel dishes do have the most accurate surface but are one piece and cumbersome to ship and to move to a site. This type of dish as well as the stamped or hydroformed variety must have their surfaces protected from rusting or oxidation. Of course, those that are made from stainless steel are free from this concern. But unpainted stainless steel dishes can have problems with heat build-up. In some brands, the mount is bolted to a back plate that is attached to the reflector surface. Others have mounts which are bolted directly to the back of the reflector and, if not carefully assembled, can distort the dish surface as the bolts are tightened. This is especially so if the mount does not properly match the reflector shape and if rubber washers, which can act as shock absorbers, are not provided for insertion between the mount and reflector. There can also be an electrolysis action between the different metals which can cause severe corrosion.

The quality of a fiberglass dish depends upon which manufacturing method was used (as outlined in Chapter 5). If they are cured and molded properly, fiberglass reflectors can have a very accurate, smooth surface. Improper techniques can result in warping, shrinking, surface deterioration and even separation of the protective gel coating from the underlying structure. In the worst cases, the embedded reflective material could delaminate. Fiberglass dishes can be very heavy and must have solid mounts attached securely at numerous points. Like other types of dishes, sections should have compound curves and rib structures to form a true parabola and these should fit snugly together. The surface should also feel smooth to the touch.

Fiberglass reflectors have the advantages of being corrosion resistant and durable. The best test of any brand of fiberglass dish is probably how it will maintain its accuracy and structural integrity over a period of time in the field.

Figure 7-5. The Effects of Solar Energy. *A highly reflective dish can literally melt a plastic feed cover, especially at solar outage times when the sun is aimed directly down the reflector boresight. (Courtesy WV Publications, London)*

Figure 7-6. Buttonhook Feed Support. *This 4-foot perforated aluminum dish has the advantages of lower wind resistance and see-through appearance while retaining an accurate reflector surface. (Courtesy of Winegard Satellite Systems)*

Remember that microwave dishes are also capable of performing as solar concentrators. So the choice of colors and surfaces can be critical. A brightly painted dish or a smooth metal surface can reflect enough solar energy to melt polarizer caps or fry cables (see Figure 7-5). Even rolled mesh and fiberglass surfaces are smooth enough to cause a problem unless they are properly painted. When painting a reflector, an optically rough paint must be used so that the sun's rays are scattered not focused to generate heat. It is also better to use paints which are non-metallic since rough spots or bumps in the coat could cause reflection errors.

Figure 7-7. Tripod Feed Support. *This dish supports the feed and LNB at its focus by a tripod. Both the feed weathercover and the feed bolt to all three support struts.*

Feed Supports

Prime focus parabolic dishes have three varieties of feed supports: buttonhook, tripod and quad (see Figures 7-6, 7-7, and 7-8). Whatever type is used, the feed must be stable under wind or snow loads and should not move when the dish tracks across the arc. Make sure that the attachments between the dish and support structure and the support and feed are strong. A feed and LNB mounted at the dish focus can easily stress their mounting hardware and cause the feed to move from the desired "sweet spot" at the exact focal point of the reflector.

Some dishes have feeds which simply pop into a preset position so no adjustment of focal length is possible. This is both an advantage and a limitation because it does not allow fine tuning of the feed position. The adjustable varieties are more difficult to work with but allow additional fine tuning which can optimize the picture quality received.

Figure 7-8. QuadPod Feed Supports. *This illustrates the additional feed/LNB support that can be achieved compared to a buttonhook or tripod design. However, the extra support structures block additional incoming signal and can result in an increase in side lobes.(Photo Courtesy of Brent Gale)*

Feeds

Feeds have the important job of properly illuminating a dish and of capturing as much of the incoming signal as possible. It is critical that the proper feed be chosen for each f/D ratio. Deep dishes (f/D from 0.32 to 0.25) will use different feed configurations than shallow ones.

Well designed and manufactured feeds have very low insertion losses, i.e. they transmit most of the signal into the LNB. The selection of a feed can be an important determinant of picture quality. The cost is a small enough portion of the entire system expense to warrant purchasing top of the line equipment.

Polarity selection is generally accomplished by servomotors, voltage switching or ferrite devices. Each type has its strengths and weaknesses. For example, a ferrite device will probably perform more reliably in very cold climates where motors may seize. While voltage switching and ferrite polarizers are the most common variety found in European installations, they are rarely used in North America where servomotors predominate.

Both Ku-band and dual-band feeds are available (see Figures 7-9, 7-10 and 7-11). C/Ku-band operation can be efficiently accomplished with some dual-band feeds that are installed and function as well or nearly as well as single band C or Ku feeds.

Figure 7-9. Masterfeed™ Orthomode Feed. *This component combines two LNBs and the feed into one compact housing. It allows reception of both horizontal and vertical channels for simultaneous viewing of any of 24 channels from two receivers. It therefore eliminates the need for an orthomode feedhorn and two separate LNBs. (Courtesy of California Amplifier)*

Figure 7-10. Corotor II™. *The Corotor II™ dual-band feed is designed to simultaneously receive both C- and Ku-band satellite broadcasts. Both the C- and Ku-band waveguides which are set at 180° from each other share the same probe. This feed has a maximum VSWR of 1.4:1, a cross-polarity isolation of greater than 30 dB and an f/D ratio adjustable from 0.28 to 0.42. (Courtesy of Chaparral Communications)*

LNBs

While an LNB is the most complex piece of a TVRO it is usually the most reliable, except when fried by lightning. Its noise temperature should be low enough to yield a good quality picture. C-band LNBs in the 15 to 25K range are available.

While a minimum LNB gain of 40 dB is sometimes acceptable, a 50 dB unit should be chosen to guarantee top performance.

Satellite Receivers

Sometimes it seems that there are as many receivers available as salesmen to sell them. Furthermore, each brand of receivers seems to have a different look and different features.

Video quality may vary substantially between receivers. Some units produce jumpy, flickering or grainy pictures, a reflection of receiver electronic design. If a low threshold has been obtained by the least desirable method, reducing the bandwidth, the result is often fewer sparklies but a picture with less resolution and smeared colors. Phase lock loop (PLL) receivers will normally have lower thresholds than coaxial delay line receivers. However, in more extreme cases when using PLL receivers, bright colors or high contrast scenes can tear or streak. But when using under-sized dishes, PLL receivers can have thresholds 1 to 2 dBs lower than coaxial delay line types and therefore better pictures.

Figure 7-11. Bullseye III™. *This orthomode feed is designed to simultaneously receive C-band polarities as well as higher frequency Ku-band broadcasts. Both feeds are located at the prime focal point of the reflector. (Courtesy Chaparral Communications)*

Audio quality also varies between satellite receivers. The audio should be easy to tune over the full 5 to 8.5 MHz range. There should be no buzzing or rasping on weaker transponders with bright colors on the screen. Check both the video and audio quality on transponders which are loaded with audio subcarriers. Each audio subcarrier should be heard clearly and crisply providing the signal strength is adequate.

Receivers also differ in the method used to tune between channels. Continuously variable tuning may be quick but it can be annoyingly inaccurate and determining what channel has been selected may be difficult.

When a remote control is desired, it should have the five main functions: channel selection, fine tuning, audio selection, volume control and polarization selection. If any of these is absent a user will have to walk over to the receiver to change the missing function thus eliminating the advantage of the remote control. Some of these functions could be automatically selected by the receiver.

Television Monitors

TV monitors accept raw audio and video signals from a satellite receiver and bypass the intermediate modulation step. As a result, picture quality can be better compared to conventional television sets. A satellite dealer may find a good quality monitor useful as a demonstration of how clearly satellite broadcasts can be received and as an installation tool.

Warranties

Even the best designed and manufactured equipment can fail. If the factory offers a warranty having too short a period, does not provide a fast turnaround on service or, even worse, is out of business when the equipment fails, a dealer can incur considerable costs.

The typical warranty for satellite TV components offers full coverage for at least 90 days and limited coverage for one year. But there are wide variations. A few manufacturers have extended warranties on electronic equipment lasting up to two years. Some dish manufacturers warranty their products for five years indicating either faith in their design or an unrealistic hope in the longevity of their product. In any case, responsible manufacturers should stand behind their products with prompt service. Some companies have gone out of business when high equipment failure rates overly taxed their financial resources. In these cases, warranties were only as good as the paper they were written on.

Therefore, a prudent course of action to follow should include careful selection of well-made equipment, a knowledge of the underlying warranties and an evaluation of the manufacturer's stability and reputation. Word-of-mouth information from knowledgeable sources can be invaluable in judging a manufacturer.

Selecting Receivers for International Broadcasts

C-band spacecraft such as Mexico's Solidaridad, Brazil's Brazilsat and the many North American satellites in orbit can be adequately detected by conventional reception systems. But Intelsat and other satellites which use circularly polarized signals, half transponder formats or unusual frequency allocations place additional requirements on the equipment. In addition, care must be taken to select an appropriate satellite receiver to match the broadcast format of a given country.

Circularly polarized broadcasts can be easily received by retrofitting a conventional feed with a dielectric insert as described in Chapter 8. This action causes a 3 dB or more decrease in signal power, an action which can be compensated for by increasing dish size.

In order to receive half transponder broadcasts a receiver must be tuned and centered onto the reduced bandwidth. If not, excess noise from those unused portions of the 28 to 32 MHz band may interfere with reception. If necessary, bandwidth reduction can be accomplished by using a 14 to 18 MHz bandpass filter. Remember that as receiver bandwidth is narrowed, C/N rises because the detected noise is reduced. So a smaller dish could be used to view these half transponder broadcasts. But as receiver bandwidth is reduced, picture fidelity worsens.

Receivers that do not have AFC defeat switches or that are permanently locked onto the standard C-band center frequencies cannot be off-tuned to broadcasts having different center frequencies. Therefore, synthesized tuned receivers which may perform excellently with conventional transmissions might be unusable in certain environments.

Most C-band broadcasts relay their audio information on subcarriers located on various frequencies in various parts of the world. Subcarriers of 6.2 or 6.8 MHz are common in North America. So it is advisable to have a receiver which is continuously tunable over the 5 to 9 MHz audio range for either stereo reception or for detection of audio carriers with uncommon center frequencies. Some international transmissions are relayed with their audio located on separate narrowly tuned carriers, known as single channel per carrier (SCPC) formats which must be received with a scanner. Some Intelsat spacecraft even relay this SCPC information on a transponder different from that which carries the associated video messages.

It is important to realize that both American and Canadian receivers have usually been constructed solely for processing the 525-line North American broadcast format. Numerous countries in Central and Latin America, Europe and other portions of the world use the higher resolution 625-line PAL or SECAM systems. The shaping and filtering of a broadcast signal by de-emphasis and low-pass networks can cause some noticeable picture streaking and color smearing when used with this higher-line format. Some manufacturers have designed specially altered receiver circuits to solve this compatibility problem.

A television or TV monitor specifically tailored to the particular NTSC, PAL or SECAM format is required for quality reception in any particular country. Excellent multi-standard components are now available if uncertainty exists or if the equipment will travel across international borders. Note that some multi-standard TVs or monitors are not yet capable of resolving problems with the particular variations of PAL used in Brazil and Argentina.

Decoders

In North America two analog scrambling systems have become standards, the VideoCipher II-Plus and the Oak Orion. The latter is used only in Canada and is not available to American consumers. The situation in other regions of the world varies. For example, in Europe a wider number of analog systems including Videocrypt, Filmnet and Eurocypher have been employed.

However, changes are in the wind with the introduction of compressed digital MPEG-2 broadcasts. In North America alone DirecTV uses Videocrypt, PrimeStar uses Digicipher and many business satellite television networks use the CLI or TV/COM systems. This wide variety of systems in use underscores why C-band analog TVROs have been termed "full view" systems. Most programming marketed to consumers is encrypted via the Videocipher II+ technology.

Most receivers available today in the United States are IRDs (integrated receivers/decoders). While stand-alone Videociphers are still available, IRDs have overcome any receiver/decoder compatibility problems that may have previously existed.

CHAPTER 8

INSTALLING SATELLITE TV SYSTEMS

One of the objectives of this manual is to instruct readers how to install home satellite systems that will stand the test of time. A theoretical knowledge, while academically interesting, is of little value unless the result is a system that works. Even the most sophisticated troubleshooting expertise may be overtaxed if the original installation had been seriously flawed and unacceptable shortcuts had been taken.

In this chapter, each installation procedure is examined in detail. For simplicity's sake, some steps are presented in a slightly different order than might occur during an actual installation. For example, the section on aligning the feed precedes the description of mounting actuator arms, although it is convenient to hold a dish in place by installing the arm before the feed assembly. In those situation where a fixed dish is being installed, the sequence would be different compared to a full-blown installation.

The techniques used to install smaller C-band dishes that have become possible as satellite EIRPs have increased differs slightly from those used for larger reflectors. C-band dishes as small as 5 to 6 feet in diameter can be more easily affixed to walls or roofs. Larger reflectors and their mounts most often have been supported on poles set in concrete on the ground. Smaller dishes are lighter, less susceptible to wind loading and are simply easier to install. The methods used to mount dishes that are fixed onto one target satellite are also explored. In such cases where tracking is not necessary, installation procedures are somewhat easier.

The Installation Process – Overview

A TVRO can be installed via a series of rather straightforward steps. Although the complexity of each step may vary between installations, most jobs are quite similar. This usual sequence of installing a tracking dish is listed below to allow the reader an overview of the process before diving into the details. Tables 8-1 and 8-2 also provide a point-by-point overview of the intallation process for both a fixed and a tracking dish TVRO. These illustrate that the general method is rather simple, although the details might require some time and experience to master. Tables 8-1and 8-2 apply to a ground-mounted, tracking and to a fixed dish system. In some cases installing some systems may involve steps from each sequence.

1. Site Survey

The goal of a site survey is to determine where the satellite dish will be located. Factors that must be considered include (1) finding a position with a clear view of the entire arc of satellites, (2) verifying that terrestrial interference, if present, will not be a serious problem and (3) planning cable runs from the dish to the indoors electronic equipment.

2. Installing the Dish Support

Satellite dishes are often mounted on poles that are set in concrete. However, other types of supporting structures can be successfully used. The critical element of this step is to be sure that the pole is mounted in a perfectly vertical orientation. Any tilting in the east/west direction can result in poor tracking. While setting a concrete support, install the section of conduit near the dish base.

TABLE 8-1. INSTALLATION PROCESS – GROUND-MOUNTED TRACKING SYSTEM

STEP	TASK
1. Conducting Site Survey	Locate a position with a clear view of the arc and plan the overall installation.
2. Installing Support Structure	Install the supporting pole in a perfectly vertical orientation.
3. Trenching and Cable Runs	Dig cable trenches, where necessary, and lay conduit. Avoid sharp bends in any cables.
4. Assembling Mount	Assemble mount and lift it onto the pole.
5. Installing Actuator	Install the actuator on the mount and set elevation so that it sits with the dish resting in either a vertical or horizontal position.
6. Lift Dish onto Mount	After assembling the dish, if necessary, either lift it onto the mount so it faces upward and can be bolted to the mount or roll it into position so it can be bolted in place.
7. Assembling LNB/Feed	Bolt the LNB onto the feed and attach both structures onto the dish.
8. Wiring	Attach the coaxial cables, polarizer wires and actuator wires.
9. Aligning Dish	Set the elevation and declination angles, turn on the power and align the dish onto the arc.
10. Fine Tuning	Complete the tracking and electrical fine tuning.
11. Waterproofing	Once the system is working faultlessly, waterproof all outdoors electrical connections.
12. Sealing Cable Runs	Bury the conduit.
13. Programming Receiver	Program the satellite receiver by following manufacturer's detailed instructions
14. Connecting Accessories	Hook up accessories such as VCRs and stereos.

TABLE 8-2. INSTALLATION PROCESS – WALL or ROOF FIXED SMALL DISH SYSTEM

STEP	TASK
1. Conducting Site Survey	Locate a position with a clear view of the chosen satellite and plan the overall installation.
2. Installing Support Structure	Install the supporting bracket.
3. Trenching and Cable Runs	Dig cable trenches, where necessary, and lay conduit.
4. Assembling Mount/Dish	Assemble the dish, mount and LNB/feed. Lift and attach this equipment onto the mount.
5. Setting the Angles	Set the elevation and azimuth angles.
6. Wiring	Attach the coaxial cables, polarizer wires and actuator wires.
7. Aligning Dish	Set the elevation and declination angles, turn the power on and align the dish onto the arc.
8. Fine Tuning	Complete the tracking and electrical fine tuning.
9. Waterproofing	Once the system is working faultlessly, waterproof all outdoor electrical connections.
10. Programming Receiver	Program the satellite receiver by following manufacturer's detailed instructions.

3. Trenching and Cable Runs

The route from the dish to the indoors equipment is then prepared. Conduit is recommended when cable is to be installed underground. Lay the conduit and cable in the trench but do not cover it up. This should be done only when the installation is definitely complete and successful. It is crucial that the conduit be leak proof and it is wise to use conduit of sufficient diameter so extra cable can be pulled in or the old cable can be replaced at some future date. Be sure to use 45° or 90° electrical sweeps on any sections where directions will change.

4. Assemble the Mount

Whenever possible, the mount should be assembled independently of the dish and then lifted onto the pole. Bolting the reflector and the heavy mount together and then lifting them onto the pole together might warp the reflective surface. This mistake should be avoided. It is also usually easier to set the declination angle on the mount at this stage.

5. Install the Linear Actuator

Installing the linear actuator onto the mount assembly prevents it from flopping about. At this stage, set the elevation angle and actuator position on the mount so it faces either directly upward or as closely to vertical as possible. The dish can then be easily lifted or rolled into place.

6. Assemble the Dish

After assembly, the dish can either be lifted onto the mount so it sits horizontally like a bird bath or it can be rolled into alignment with a vertically oriented mount. Secure the reflector to the mount. Then attach the feed support struts or buttonhook.

7. Install the LNB/Feed/Weathercover

After bolting the LNB, feed and weathercover together, lower the dish by adjusting the elevation angle and possibly by moving the actuator in its clamp so the LNB/feed assembly can be attached to its supports. Always install a weathercover to protect the LNB.

8. Electrical Connections

Complete the necessary connection to the LNB, polarizer and actuator at the dish and to the indoors receiver/polarizer/actuator controller. If the dish will be aligned with a test setup located outdoors at the dish, temporary connections should be made.

9. Power On and Align Antenna

Check that all wires are connected correctly, then turn on the power. Then carefully set the elevation and declination angles and follow the procedure that aligns the dish onto the arc of satellites.

10. Fine Tuning

Once a clear signal from each satellite has been detected, the alignment of the dish and feed can be fine tuned.

11. Waterproofing

All electrical joints should be waterproofed. It is also a wise practice to install an accordion sleeve on a linear actuator arm.

12. Seal all Cable Runs

Finally, bury the length of the conduit run. Be sure that weather covers are installed at the dish and at the entry to the building, if applicable.

13. Program Receiver

Program the receiver by following the steps in the manufacturer's instruction guide.

14. Connect Accessories

Hook accessories such as VCRs or stereos to the satellite receiver. The installation is now complete.

THE SITE SURVEY

Every job must begin with a thorough site survey, one of the most critical yet often most neglected steps in a satellite TV installation. All present and future conditions must be considered. The survey is the equivalent of a doctor's examination and diagnosis prior to surgery. No surgeon would ever open up a patient without a complete understanding of the task at hand. A competent installer would not cut corners in conducting a site survey, also a vital task.

Three important tasks are completed during a site survey.

- First, a location with a clear view of the arc of satellites is found
- Second, a check for terrestrial interference is performed
- Third, the entire installation is planned

Numerous facts must be established before beginning an installation. These include which satellites and transponders are to be received (see Table 8-3), what type of equipment should be used, where to place the dish, how much concrete or what type of wall fittings are required, whether or not to use a long pole, roof, tower or ground mount, what length, route and type of cable is planned, the distance from the dish to the satellite receiver as well as other similar questions.

TABLE 8-3. C-BAND SATELLITES in the NORTH AMERICAN ARC

Satellite Name	Orbital Location (°W)
Satcom SN2	69
Galaxy 6	74
Comstar 4	76
Telstar 302	85
Spacenet 3	87
Galaxy 7	91
Galaxy 3	93.5
Telstar 401	97
Galaxy 4	99
Spacenet 4	101
Anik E2	107.3
Solidaridad 1	109.2
Anik E1	111.1
Solidaridad 2	113
Morelos 2	116.8
Telstar 303	123
Galaxy 5	125
Satcom C3	131
Galaxy 1R	133
Satcom C4	135
Satcom C1	137
Satcom C5	139

The planning process, carried out with the cooperation of the customer, must deal with seemingly minor concerns as well as issues with potential legal ramifications. For example, cable paths must always be clearly marked because the installation crew might be unfamiliar with the site. Furthermore, permits for installing the system and receiving the programming may be required.

One central goal is to ensure that the strongest possible signal reaches the receive dish consistent with the customer's wishes about where to mount the dish and run the cables. It is wise to take the customer seriously so that all decisions are mutually acceptable. Tradeoffs may sometimes be made so a weaker but still acceptable signal is received. For example, if the customer does not want the dish next to her home but 150 feet away instead, a line amplifier and perhaps a slightly larger dish may be required. This may result in a substantial variance from the original bid. The customer's home and yard should be treated with respect. A wise practice is to respect the old adage that "the customer is always right."

Another objective is to ensure an environmentally sound installation. The dish should be well-protected from wind, hail and lightning as well as from other environmentally harsh conditions. It should blend as well as possible into the surrounding terrain and buildings. Bylaws from certain local authorities may actually make this mandatory, an advisable practice to follow in any case.

Fixed dish receive systems are generally easier to install that tracking dishes to receive more than one satellite. When just one spacecraft is targeted only a narrow portion of the arc must be seen. A small dish can also be more easily mounted in a wider variety of places than a larger reflector. Both of these considerations make the site survey easier.

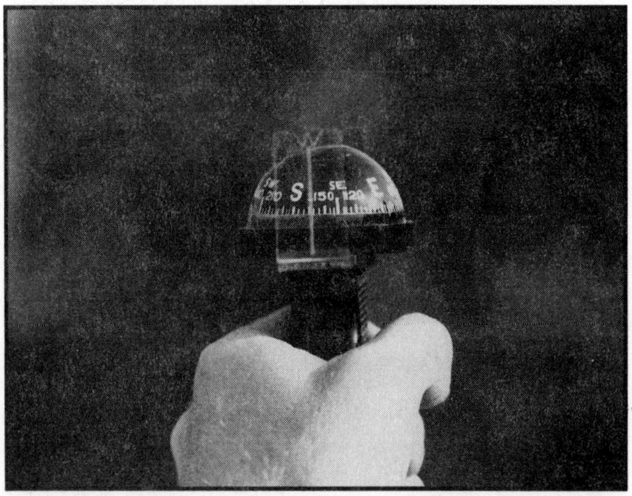

Figure 8-1. A Compass. *This hand-held compass allows the user to aim towards a particular heading while maintaining a view of the satellite dish. (Courtesy of Steve Berkoff)*

Ensuring a Clear View of the Arc

A dish must have a clear view of each satellite to be viewed because any obstruction can absorb or reflect microwaves and thus lower the signal power. Water is a particularly strong absorber of microwaves. Some reflectors installed in winter had perfect reception until spring when foliage returned to trees obstructing its view: then pictures either deteriorated in quality or disappear.

Two basic instruments are required to aim at the satellite arc: an inclinometer and a compass (see Figures 8-1 and 8-2). When a polar mount is installed, these instruments are used to aim the mount along a true north/south line and to set the polar axis and declination angles. When installing a fixed dish on an az-el mount, each satellite is targeted by adjusting the elevation and azimuth. Azimuth is measured with a compass in degrees of rotation from true north and the elevation with an inclinometer in degrees above the horizon.

There are many types of compasses on the market ranging from what are basically toys to very accurate and expensive prismatic types used by the army. Similarly, there are many types of inclinometers ranging from the simplest type that can be made with a protractor, a piece of string and a weight to instruments with digital read-outs.

When using a compass, it is important to realize that it can give inaccurate readings when used in the vicinity of iron-containing objects. Therefore, a bearing taken next to a large truck or any other large steel object could be many degrees off the local average magnetic north/south line.

Figure 8-2. Inclinometers. *An assortment of inclinometers is pictured here. The rounded instruments display the reading on an external scale. The smaller device is sighted through to the target. The instrument in the rear is a level/inclinometer with a digital read-out.*

Determining a True North/South Bearing

A compass points along the line between magnetic north and south. In most locations, there is a difference between the bearing to the north and south poles and the needle reading on a compass (see Figure 8-3). Both magnetic south and north are not located at the south and north poles but at the strongest magnetic center near these poles. Magnetic north is located near Bathurst Island off the northern coast of Canada. As a result, magnetic variation can exceed 30° in locations such as the American state of Alaska. However, magnetic variation is zero along the "agonic line" that runs just off the west coast of Florida, through Lake Michigan, over the Great Lakes, and to the magnetic north pole. East of this line, the north pole is east of the compass reading; west of this line it is west. Similar agonic lines are also found in other areas of the globe as the accompanying maps illustrate.

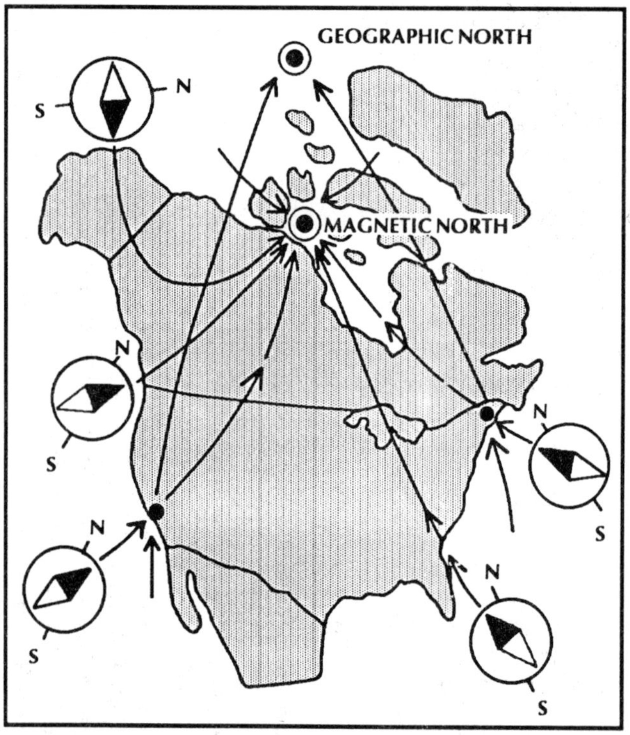

Figure 8-3. True North versus Magnetic North. *In both hemispheres magnetic north is quite far removed from the geographic north pole. In North America it is just off the coast of Hudson Bay in Canada. This illustration shows the difference between the bearing to true and magnetic north.*

A true north/south bearing can therefore be found by using a compass and correcting for magnetic variation. (see Figure 8-4). For example, in the American city of Denver, Colorado, magnetic variation is 12.5°E. So the arc of satellites centered on true south is found by rotating 12.5° east of the south reading on a compass. This is equivalent to a compass reading of a 167.5° (180° less 12.5°) as measured from true north. A similar procedure can be followed at any other point on the globe. A useful pilots adage to follow here is "west is best and east is least", that is add west magnetic variations to and subtract east magnetic variations from the compass heading.

Magnetic variation at any location in the world can be found from the charts shown here (see Figures 8-5 through 8-9) and, in many countries, by requesting the value from a local airport information service. Airline pilots, like satellite TV installers, must also take their bearings from true north and south. *Satellite Toolbox* software (see references section) can also be used to automatically determine magnetic variation from any site in the world.

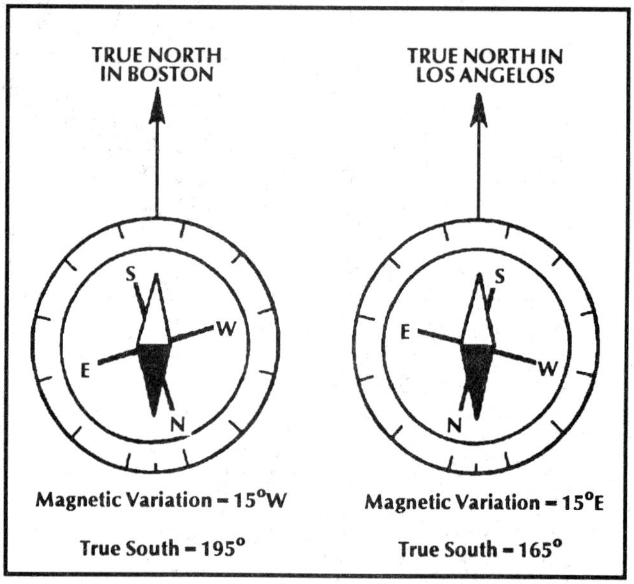

Figure 8-4. Adjusting for Magnetic Variation. *True south can be found by aiming a compass east of magnetic south when the variation is east (negative) and vice versa. Magnetic variation is positive when west of the zero-variation, agonic line north of the equator. South of the equator the settings are reversed.*

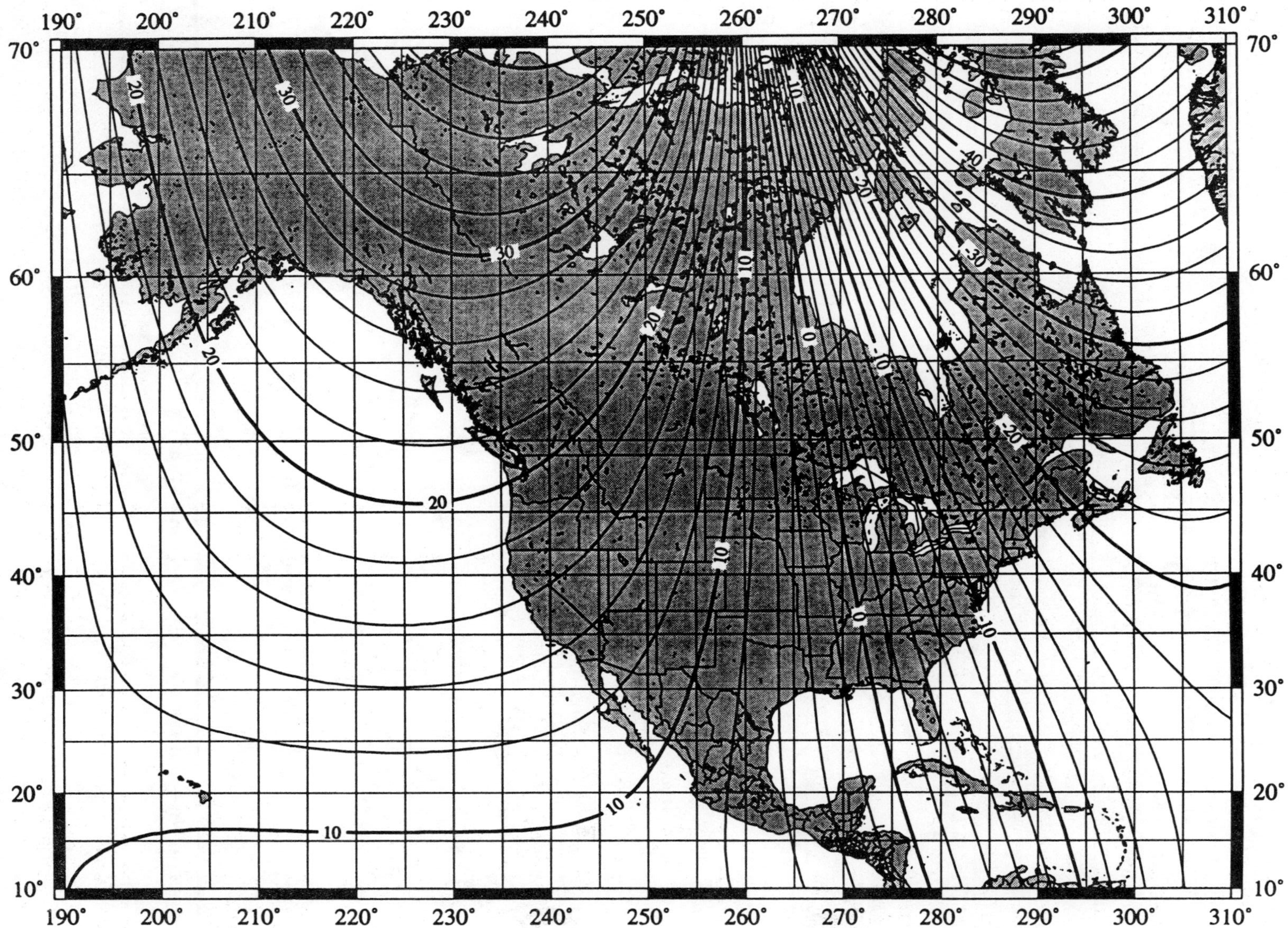

Figure 8-5. North American Magnetic Variation Map. *A compass points to magnetic north and not to true south except along the "agonic line" because of the difference in locations between the magnetic and geographic north poles. North of the equator, an east magnetic variation means that a reading taken with a compass must be corrected by rotating east to find true south. South of the equator the opposite is true. (Courtesy of National Geophysical Data Center – NOAA)*

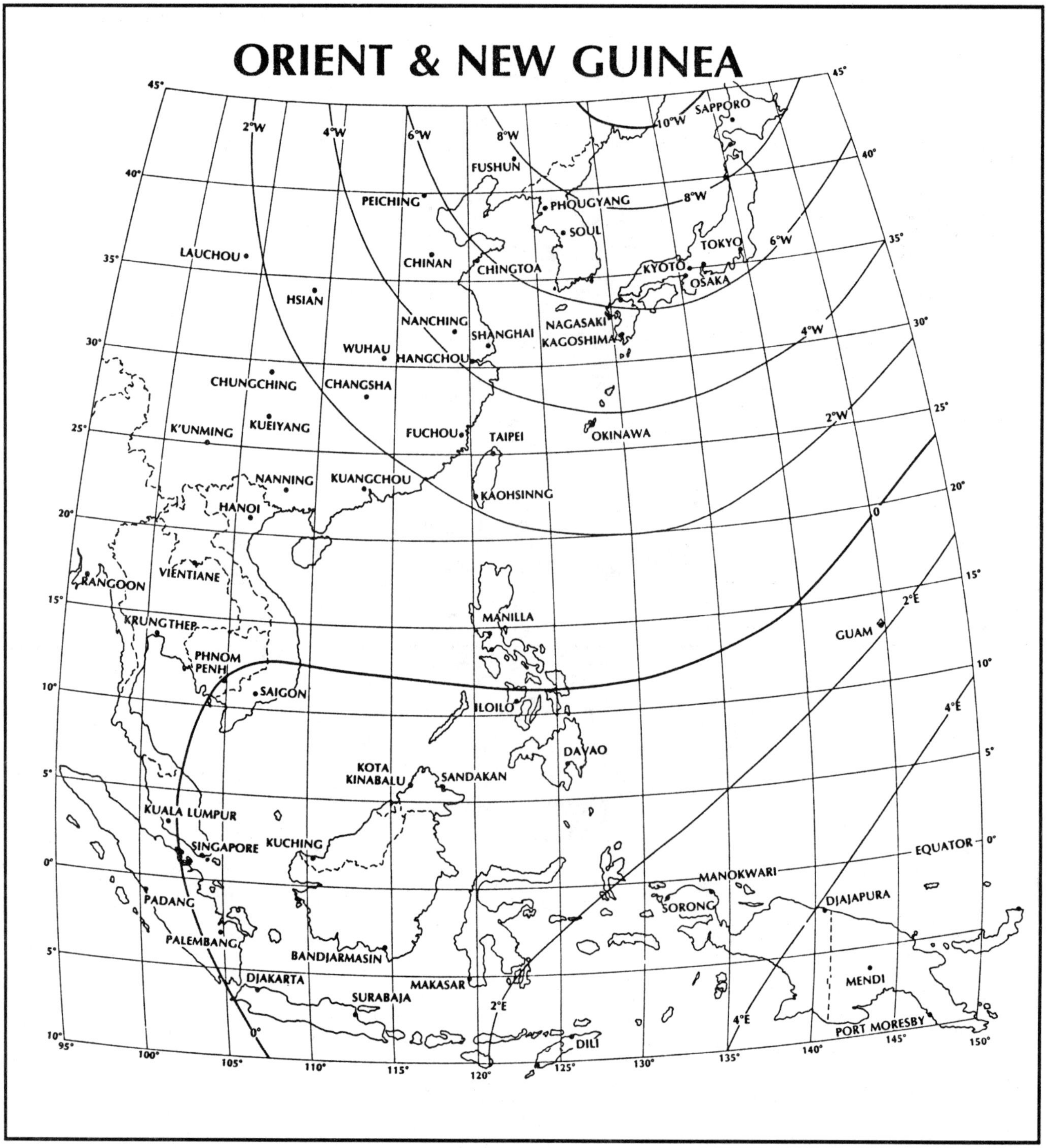

Figure 8-6. Orient and New Guinea Magnetic Variation Map.

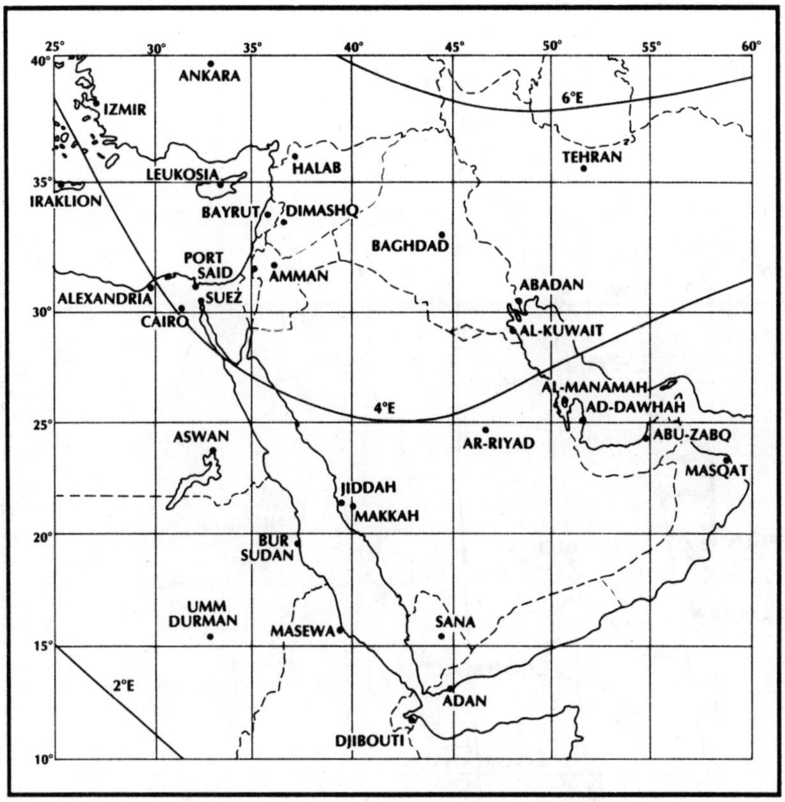

Figure 8-7. Magnetic Variation Map – Middle East.

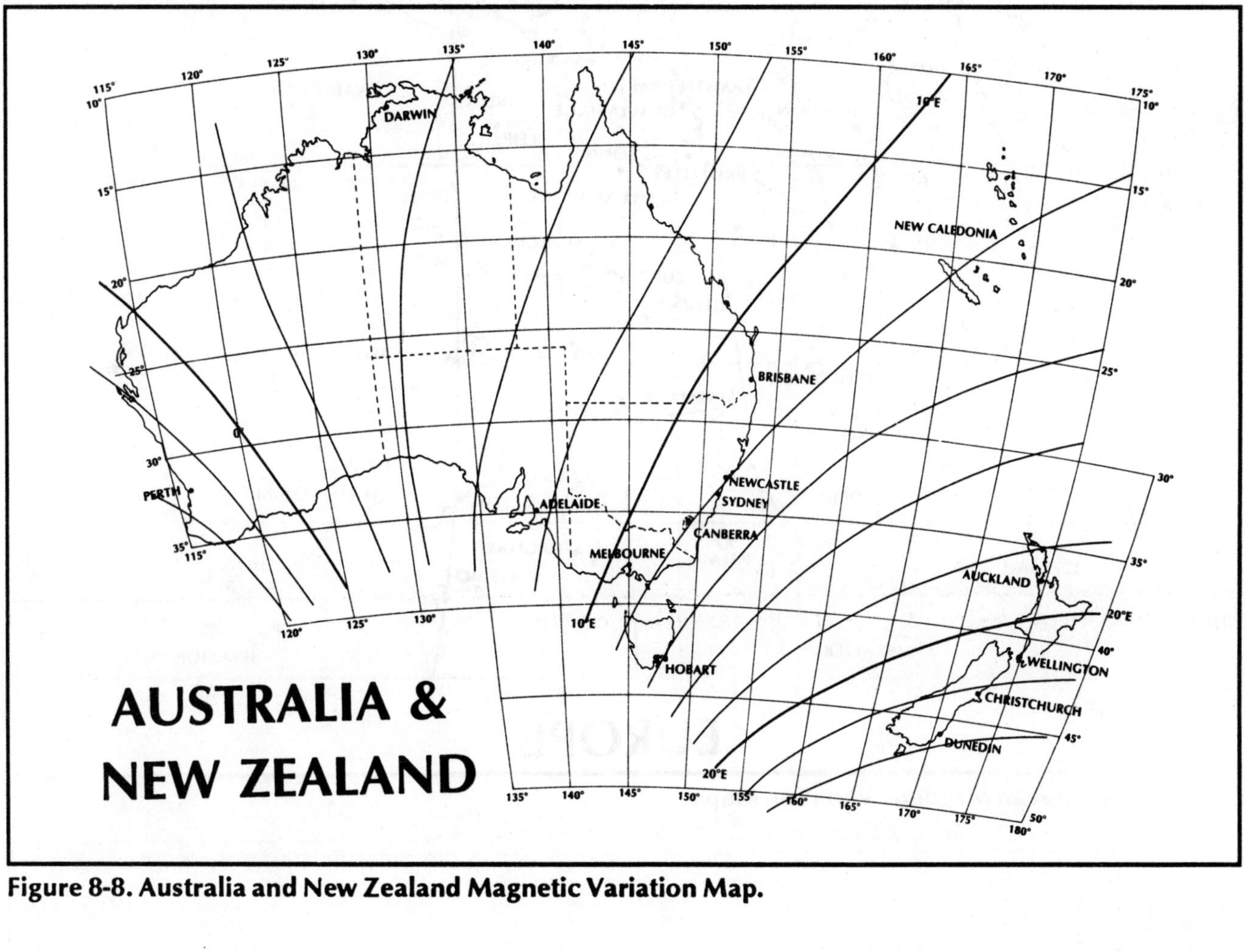

Figure 8-8. Australia and New Zealand Magnetic Variation Map.

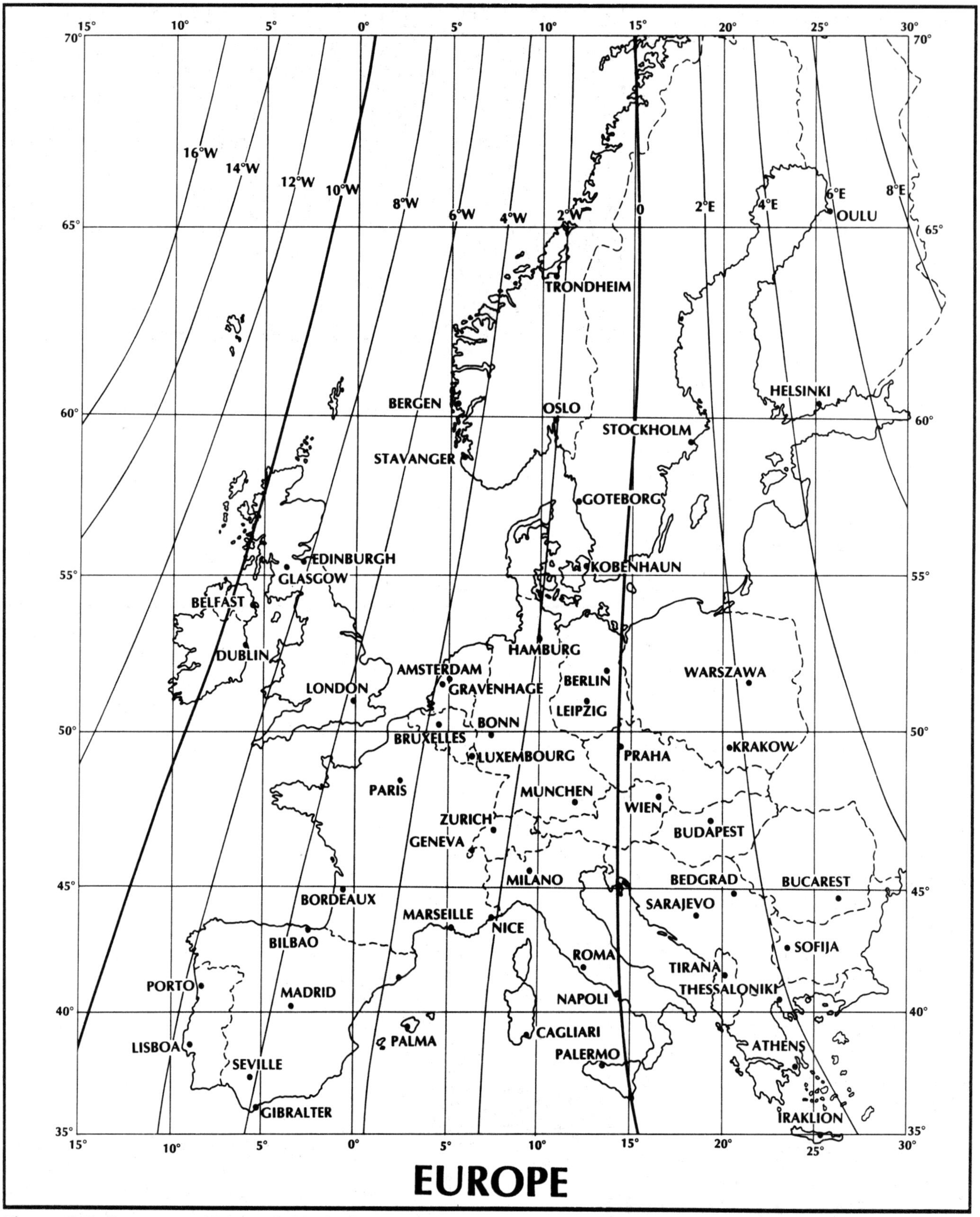

Figure 8-9. European Magnetic Variation Map.

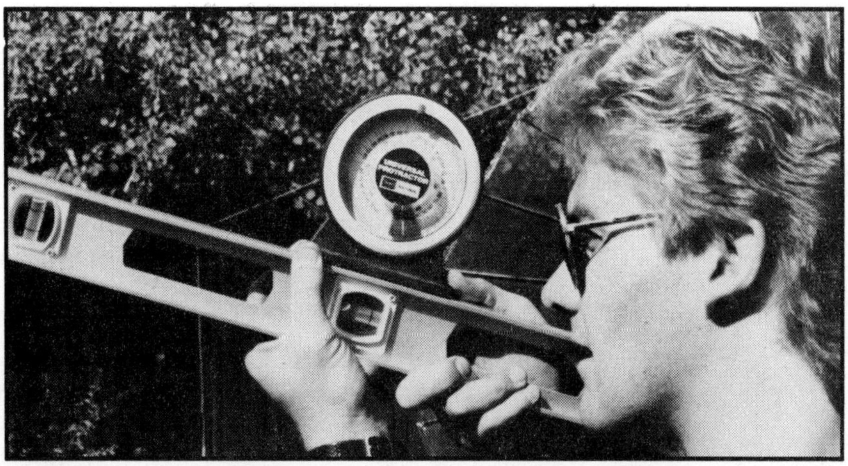

Figure 8-10. Measuring Elevation Angle. *Elevation angle can be measured by placing an inclinometer on a straight edge and sighting along it in the direction of the target satellite. (Courtesy of Brent Gale)*

Determining the Azimuth and Elevation Look Angles

The azimuth heading towards any chosen satellite can be found from a chart, from equations based on simple geometry as outlined in Appendix B, or from commercially available computer programs. Realize that a satellite whose angular location corresponds to the site longitude is found by aiming due south in the northern hemisphere and due north in the southern hemisphere. For example, Anik D1 at 104.5°W has a nearly due north azimuth from Boulder, Colorado at 105°W longitude.

Any other satellite can be targeted with a knowledge of its azimuth and elevation as well as the site magnetic variation. For example, the elevation and azimuth to Galaxy V as viewed from Dallas, Texas are 41.22 and 224.91 degrees. Thus this satellite would be targeted by rotating to a 218.61 degree compass heading (magnetic variation is 6.30°E so the compass heading is 224.91 less 6.3 degrees) and up to 41.22° elevation. If a tree or any other obstruction were blocking the view, this proposed installation site would have to be changed.

Elevation angles are easily measured with an inclinometer. This instrument can be placed on a long ruler or any other straight edge and used in a fashion similar to sighting a gun (se Figure 8-10). The ruler is raised until the desired elevation is reached. While sighting along its length, no object(s) should block a clear view to each satellite in the applicable portion of the belt. With some types of inclinometers, having a second person to help read the scale makes the process easier.

Various companies market instruments designed to make sighting the arc even simpler. Some of these products are see-through charts having satellite locations marked. These are mounted on a tripod and pointed due south for sighting the satellites. Others have telescopic view finders which are preset for any geographic location so a user can visually scan across the entire arc (see Figure 8-11).

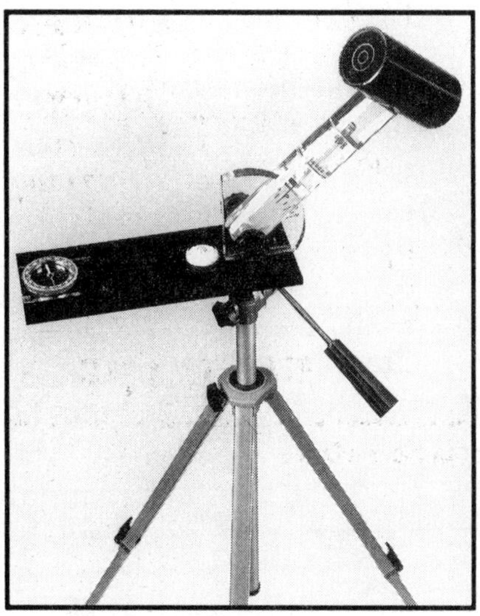

Figure 8-11. A Sighting Tool. *An installer can first set the azimuth and elevation angles to any satellite on the Viewfinder and then check for obstructions along the line to this satellite through the eyepiece. (Courtesy of Focii Manufacturing Company)*

Step-by-Step Procedure to Ensure a Clear View of the Satellite Arc

A rather simple procedure to ensure a clear view of the arc of satellites is outlined here and an example of this method is presented in Figure 8-12.

1. Use a map (Figures 8-5 to 8-9) to find the site longitude (its east/west location).

2. Use the map to determine the site latitude, its north/south location.

3. Find the azimuth/elevation chart that most closely matches the latitude and longitude of the site location (Table 8-4). This chart lists compass reading, the azimuth, and the elevation to the satellite closest to due south as well as a far easterly and westerly satellite. Note this table already includes the necessary correction for magnetic variation so the actual compass heading at each latitude/longitude grid is listed.

4. Correct the azimuth values in the azimuth/elevation chart for the magnetic variation at the site.

5. Use a compass and the magnetic deviation correction to find true south from the site. Aim up towards the most southerly satellite. Note its position.

6. In the same way, aim towards the most westerly and easterly satellites. In your mind's eye, connect an arc between the three satellites and note any obstructions to a clear view.

7. If obstructions to a view of the satellites exist, try the same procedure at another potential installation site.

Choosing the Best Dish Site

In cases where a dish that tracks the arc is being installed, at least three satellites should be sighted, the most easterly, the most westerly and the one highest in the belt. Then a line connecting these three points should be visualized in the sky. The entire arc should be free of obstructions to a clear view of the desired satellites. If there is a question about any other satellite in the arc being partially or fully blocked, it should also be targeted. If a clear view is not possible from one location, then another site should be tested. If no appropriate location which is acceptable to the customer can be found, raising the dish higher on a roof mount or long pole mount may be the solution. There is, however, a trade off to elevating a larger dish above the ground because the higher it is installed, the more susceptible it is to wind loading and interfering signals.

Power lines and other low frequency lines attached to telephone poles are not a concern unless they are blocking a large portion of the dish aperture. However, realize that such signals might be a source of low frequency ingress interference that leaks into poorly grounded or shielded distribution equipment (see Chapter 6 for more details).

Perfect pictures can be obtained when using correctly sized components only if the view to each satellite is unobstructed. If necessary, the dish may be installed in a location where the signal from one or more satellites is weakened or lost. Or, if there are minor obstructions to a clear view at a preferred installation site, a larger dish might be required to overcome this partial signal blockage.

The installation plan should include informing the customer of any necessary building permits and local ordinances. In some countries, distances from fences are usually set by zoning laws and easement restrictions. For example, in Canada the minimum distance to the dish edge is 5 meters although 4 or 6 meter clearances from the property line are not unusual. Changing an approved site may involve filing for a new permit. For example, in some European countries, local ordinances forbid installing dishes on roofs so eave mounts are suggested, since underground utilities such as water mains, telephone lines or electrical cables may be buried at these easement boundaries to conform to local building codes. Planting the pole a few more meters inside the property line than is mandated may therefore be an intelligent strategy.

The site should also be chosen with cable runs in mind. Unexpected difficulties may be encountered. For example, tunneling under a 10 foot wide concrete driveway may prove nearly impossible. In this instance, perhaps the dish might be located at a less desirable site where it is not hidden from view.

One additional important factor must not be ignored. A movable dish must have adequate freedom of motion to track the entire arc. If, for example, it is set too close to a fence, the reflector may either be bent or may demolish the fence before the end of arc limits are set. Nothing can replace careful planning.

SITE: Denver, Colorado
COORDINATES: Latitude - 40°North
Longitude - 104.5°West
OBJECTIVE: Determine look angles to middle, far easterly and far westerly satellites
PROCEDURE: Locate latitude/longitude grid from Table 8-4.

Latitude	**40.00°N**	**Magnetic Variation = 11.4°E**	
Longitude	**105.00°W**		
	Satcom C4 135°W	**Spacenet 4 101°W**	**Galaxy 6 74°W**
Azimuth	210.5	162.4	125.5
Elevation	34.4	43.5	33.8

Figure 8-12. Example of Step-by-Step Procedure to Locate Approximate Satellite Positions.

Table 8-4. Azimuth/Elevation Headings to Satellites in North American Arc

Galaxy 6 (74°W) — **Longitude (°W)**

	75	**80**	**85**	**90**	**95**	**100**	**105**
60	207.7/21.9	197.3/21.8	183.6/21.3	169.9/20.6	154.0/19.6	140.8/18.5	127.7/17.1
55	201.2/27.3	189.6/27.0	177.8/26.5	164.5/25.6	151.9/24.5	139.8/23.0	129.0/21.4
50	195.9/32.7	185.1/32.4	172.6/31.7	161.1/30.7	148.8/29.3	138.5/27.6	128.3/25.6
45	192.9/38.2	181.2/37.8	169.7/37.0	157.9/35.8	147.0/34.1	136.8/32.1	127.6/30.0
40	189.9/43.2	178.5/43.3	166.1/42.4	155.0/40.9	143.9/38.9	134.5/36.5	125.5/33.8
35	188.2/49.3	175.6/48.9	163.5/47.7	151.7/45.9	141.2/43.6	131.7/40.9	123.3/37.7

Spacenet 4 (101°W) — **Longitude (°W)**

	75	**80**	**85**	**90**	**95**	**100**	**105**
60	238.2/18.4	228.1/19.6	214.5/20.6	201.1/21.3	184.8/21.8	171.3/21.9	162.5/21.9
55	233.2/23.0	222.0/24.5	210.5/25.6	197.1/26.5	184.4/27.0	171.8/27.3	160.4/27.2
50	229.7/27.6	219.5/29.3	207.4/30.7	195.8/31.7	183.2/32.4	172.2/32.7	161.2/32.6
45	228.9/32.1	218.2/34.1	207.2/35.8	195.3/37.0	183.9/37.8	172.8/38.2	162.3/38.0
40	228.6/36.5	218.6/38.9	207.0/40.9	195.9/42.4	184.0/43.3	173.2/43.7	162.4/43.5
35	230.3/40.9	219.8/43.6	208.7/45.9	197.0/47.7	185.3/48.9	173.8/49.3	162.7/49.1

Satcom C4 (135°W) — **Longitude (°W)**

	75	**80**	**85**	**90**	**95**	**100**	**105**
60	272.3/5.9	262.0/8.1	250.2/10.2	237.3/12.2	222.0/14.1	209.1/15.8	162.5/21.9
55	267.1/8.1	257.0/10.7	246.7/13.2	234.4/15.6	222.7/17.8	211.1/19.9	200.5/21.7
50	263.3/10.2	254.7/13.2	244.1/16.1	234.1/18.8	223.0/21.4	213.4/23.8	203.4/26.0
45	262.1/12.2	253.3/15.6	244.4/18.8	234.7/21.9	225.3/24.9	216.1/27.7	207.2/30.3
40	261.1/14.1	253.5/17.8	244.6/21.4	236.3/24.9	227.3/28.3	219.1/31.5	210.5/34.4
35	261.7/15.8	254.1/19.9	246.5/23.8	238.4/27.7	230.6/31.5	222.7/35.0	214.7/38.4

Checking for Terrestrial Interference

Many dealers have learned about terrestrial interference (TI) the hard way, once the concrete had set and the installation was completed. This outcome can be avoided. There is no guarantee if an installation "down the street" had no TI. Probably the best advice about TI is "know your enemy."

The sensible approach is to check for TI before doing an installation, a protection from hidden costs. It is better to detect TI before than after an installation when filters may have to be purchased. In the worst cases, a system may need to be moved. It may be better to turn down a job than to have a dissatisfied customer. Bad news travels faster than good news. A few botched installations can certainly eliminate the word of mouth recommendations necessary for a healthy business.

By far the best strategy in dealing with terrestrial interference is to avoid the interference by correct dish siting if at all possible. Microwaves are very directional and can be reflected from metal objects such as a tin roof directly into a dish but also absorbed by many materials. So installing a dish behind natural shields such as trees or a building may be sufficient protection even in locations where TI levels are high.

If an interference-free site cannot be found, the next alternative is to use filters to "notch out" and suppress the interference. However, notching out the interference also eliminates part of the desired signal and can lower picture quality. A second alternative for suppressing TI is to construct the appropriate artificial shields, an expensive option required only in severe cases of TI (see Chapter 6 for more details).

Methods to Check for TI

By far the most effective method to check for TI is by testing a portable satellite system at the proposed site. A small-dish, low cost system can be easily transported to and quickly set up at the site (see Figure 8-13).

For a fixed-dish install, target the desired satellite. For a tracking-dish install, at least three or four spacecraft should be targeted since TI can be very directional. While interference may not show up on one satellite, it may on another. Sometimes moving a dish just a few degrees can cause a side lobe to point at a source of TI whereas before none was detected. The test dish should be placed near the same height the final one will be placed; if TI is found, moving it down only a few meters may suddenly recruit a fence as a good artificial screen.

All transponders on each satellite should be checked since some forms of interference often affect only a limited number of channels. If TI is present, it may be cured by simply relocating the dish. Placing an dish closer to the ground or, if possible, in a small depression in the ground, reduces its susceptibility to TI. Microwaves do not travel through the dense earth. If a location free of TI cannot be found, bandpass or notch filters can be inserted. Take care not to confuse a weak signal caused by the dish being blocked by an obstruction with interference. In this case, all channels would be equally affected. This could happen if a dish had been installed behind a deciduous tree in winter when no leaves were present to block the signal.

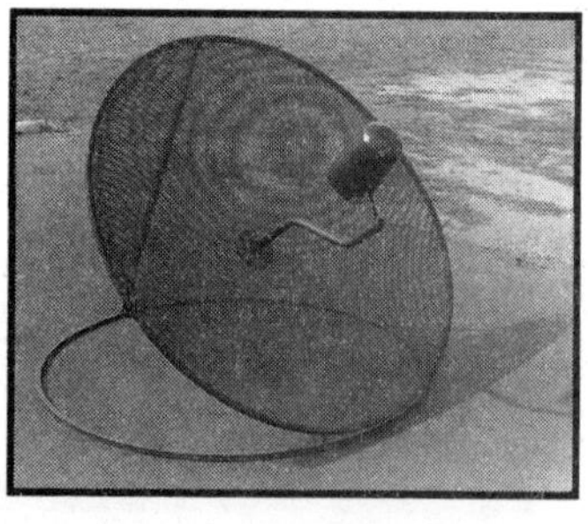

Figure 8-13. Portable Dish for Site Checks. *A small portable dish should be used if there is any uncertainly about obstructions blocking the field of vision of a dish or if TI has been detected by other methods. A 1.5 meter dish is pictured here. (Courtesy of DH Antenna)*

An alternative to testing a complete system is to use a hand-held LNB/feed connected to a satellite receiver, TV set or a monitor (see Figure 8-14). It is preferable to have a sensitive signal strength meter in addition to the one often built into the receiver. Turn on all electronics and then scan the test unit in all directions across the sky, not only at the satellites, since TI coming from off-axis directions can be detected by dish side lobes. The feed must be oriented in both horizontal and vertical planes so interfering carriers of both polarities can be found.

Watch both the signal strength meter and the TV screen. If the random white noise or snow on the TV remains unchanged and if there is little or no reading on the signal strength meter then there is no problem. If, however, the screen changes and becomes blank, black, or if white or black bars or lines appear and if the meter gives a positive indication then some form of TI is present. If the feed is covered up all these symptoms should disappear. If they do not, the test equipment is malfunctioning. When a TV set or monitor is not used, the signal strength meter can be used to detect moderate to heavy TI as indicated by a continuous fluctuation of its read-out. However, this method is not very sensitive to low or moderate levels of interference.

If interference is detected, the feed/LNB should be scanned back and forth to find the direction where the meter reading is highest. This process indicates where the interference is coming from. Orient the feed in all planes to check the polarity of this TI.

If TI is detected, move to another possible site, perhaps one better shielded by natural barriers, and do the same test over again. A logical first choice would be a location where there is an obstruction between the new site and the source of TI. If the interference is coming from behind the dish, it does not have as great an effect because the dish blocks out most of the microwave signal. If no location with a clear view of the satellite arc and free of interference can be found, it is time to bring in a portable dish to do a more sensitive test. Then microwave and IF filters can be inserted, if necessary, to find a solution to the problem.

Figure 8-14. Testing for TI. *TI can be diagnosed by using an alternative to either a portable dish or more expensive instrument such as a spectrum analyzer. An LNB and feed hooked to a satellite receiver and TV can be scanned across the sky to search for interfering signals as outlined in the text. (Courtesy of Steven Berkoff)*

The most accurate results can be obtained by using a spectrum analyzer. This device shows the frequency and power of microwaves coming from any direction at any chosen test site and is a very effective tool in checking for both in-band and out-of-band TI. A low cost analyzer rapidly pays for itself in saved time for an active installer.

Another option is to chart TI problem sites as described in Chapter 6. If a site happens to fall on the line between two microwave relay towers, be forewarned when conducting the site survey.

All these methods of testing for interference have their limitations. The information gathered applies only to TI sources that were operating during the site survey. The 5 to 7 P.M. time period is usually when microwave traffic is heaviest and is the best time for testing. However, none of these methods protects against common carrier relays that might be turned on at some point in the future.

Customers should be informed that even though TI may not be present today it is always a possibility in the future. Given this possibility, all contracts should have a clause that protects the dealer by either allowing cancellation of the sale if TI is discovered at the time of the site check or by passing along the cost of the necessary filters and installation time to the customer.

Planning the Installation

The installation should be planned as completely as possible during the site survey. The dish location must be chosen and coordinated with the cable runs to the receiver and all televisions. The cable entry into the building must be established. The type of cable and number of splitters, taps or connectors as well as the proposed placement of the satellite receiver must be determined.

Customers must be consulted during the planning process. Too many technicians have preconceived notions and may not pay close attention to the desires of their customers. Every satellite system should be installed as if it were for personal use.

When the installation has been completed have the customer sign a worksheet release form and a pre-installation worksheet. Examples of these forms are presented in Figures 8-15a, 8-15b and 8-16.

SITE SURVEY WORKSHEET

Job Number______________

Date____________________

Customer Name____

Address____

City_________________________________**State**______________**Zip**____________

Telephone (Home)____________________________**(Work)**___________________________

Lead Source___________________________**Salesman**________________________________

SITE MAP

Size and Make of Dish and Mount____

LNB Noise Temperature____

Style and Manufacturer of Feed____

Receiver Manufacturer and Model____

Length of Cable Installed____

Length of Conduit Installed____

Number of Wall Plates____

Figure 8-15a. Site Survey Worksheet. *By documenting everything about an installation and obtaining a customer acceptance signature, the job will be properly completed. In addition, information will be available for any future troubleshooting work that may become necessary.*

SITE SURVEY WORKSHEET (continued)

Make and Model of Actuator______________________________

Number of Extra TV Sets Installed______________________________

Description of Cable Routes______________________________

In Home______________________________

Outdoors______________________________

Obstructions to Cable Runs______________________________

VCR or Stereo Hook-Up______________________________

Type of Mount (pole, roof, etc...)______________________________

Types of City Ordinances in Effect and Permit Requirement______________________________

____

Type of Warranty______________________________

Status of Cable TV Disconnect______________________________

Type of Program Guide Requested______________________________

Type of Off-Air Antenna and Connection______________________________

Method of Waterproofing Used______________________________

Number of A/B Switches or Combiners______________________________

Date Installation Requested______________________________

TI Carriers Noted______________________________

Further Comments______________________________

____

____

____

____

____

____

Customer Signature______________________________

Figure 8-15b. Site Survey Worksheet. continued...

PRE-INSTALLATION WORKSHEET

Customer Name______________________________________**Customer Number**________________

Date____________________

Equipment List:

Number of Items	Item	Manufacturer	Model Number	Serial Number
	Dish			
	Mount			
	Feed			
	LNB/Feed Cover			
	LNB			
	Coax			
	Receiver			
	Actuator			
	TI Filter			
	Combiner			
	A/B Switch			
	Line Amplifier			
	Splitter			
	Wall Plate			

	Initials
All electronics have been checked before leaving office	____
Back-up electronics are on hand	____
Enough wire has been taken for installation of all TV sets	____
Proper tools are in installation vehicle	____
Customer contacted before leaving office	____
All lights and signals on vehicle and trailer have been checked	____
Correct customer invoice has been taken	____
Pole mount has set and has been checked	____
Connector kit has been checked	____

Time left office____________

Figure 8-16. Pre-Installation Worksheet. *This chart can be used as a guide to ensure that all necessary equipment has been taken to the work site.*

DISH SUPPORTS

A variety of methods can be used to support satellite receive dishes. The assembly and mounts can be supported on a pole, especially when more than one satellite is to be targeted. C-band systems to view exclusively one satellite, such as PAS-4 in Africa, feature fixed dishes usually on az-el mounts. These supports are often bolted onto an eave, a roof or secured to the ground by weights or concrete.

The methods for supporting a dish are broken down into two categories in this section: mounts at ground level and those above ground level. While dishes of any size can be mounted either at ground level or above, smaller fixed dishes are much more amenable to roof, eave or wall mounting techniques.

Pole and Pad Mounts at Ground Level

A variety of construction methods can be used to secure a pole to a ground-mounted site including pole supports, pads, pier foundations or any combination of these types. Most require the use of concrete or another strong binding material that will withstand the test of time.

When dishes are being installed on pole supports, the pole must be solid to secure the heavy weight that may be subjected to severe winds and other environmental stresses. It should be with a vertical orientation so all the remaining system adjustments can proceed smoothly. The three basic rules to follow when installing a TVRO, especially one which will also detect Ku-band broadcasts, are stability, stability and stability. Even minor instabilities in the support structure can degrade performance.

Preparing for the Job

Before work begins on the dish support, some preliminary steps should be taken. Gas, telephone, electric and cable television companies often have underground lines in unmarked areas. In many countries these utility companies gladly send personnel at no charge to locate and mark the route of their cables or pipes. If a utility line is severed during the digging and trenching portions of an installation, the costs in customer dissatisfaction, time lost and repairing the damage can be high.

Pre-Trenching for Conduit Placement

Once the cable runs have been mapped, a few feet of the trench should be excavated starting at the base of the mount. This allows the metal or PVC conduit leading in to a supporting pole to be set at the same time as other concrete work, such as a pad, is completed. If this step is overlooked, the cables would lie exposed near the base of the pole and could be susceptible to damage from gardening utensils such as lawn mowers, weed-eaters or shovels.

The Fundamentals of Concrete

Concrete is a building material that was invented centuries ago. Its properties are well known and new varieties are constantly being developed for a host of applications. When prepared and hardened properly, it has tremendous strength and is very durable. However, incorrect mixing or setting procedures can result in a weak base which can easily crack or disintegrate. If this were to happen, the dish base could move. Realigning this heavy block of concrete could be extremely difficult.

Concrete, a synthetic rock, derives its strength from its ingredients – gravel, sand and cement. Adding water to the cement mixture starts a chemical reaction. So setting of cement is not simply water drying out. The cement acts like a strong glue which powerfully binds the gravel and sand with other materials within the mold.

During the chemical reaction, heat is generated. If this heat is lost too quickly in cold weather, the concrete will not cure properly and perhaps lose strength and crack. If the surface of the concrete is covered with straw, blankets or other insulating materials during curing, heat will be retained and such problems can be avoided. Calcium chloride can also be added to speed up the setting process; generally 2% by weight of this salt is recommended. But too much calcium chloride can rapidly corrode metal structures such as the supporting pole.

Post type foundations can be made from ordinary Portland cement or standard premix. Slab foundations having larger areas of concrete should be made from air-entrained Portland cement and wire screen. This mixture contains additives that allow microscopic air bubbles to be entrained or trapped. These bubbles are like an internal lubricant which makes pouring and spreading easier. They also act like tiny shock absorbers that allow the larger piece of concrete to expand and contract without cracking. Both of these types of cement can be purchased in pre-mixed bags or they can be mixed on site. Proportionate ingredients for plain and air entrained concrete are listed in Table 8-5.

TABLE 8-5. INGREDIENTS for VARIOUS TYPES of CONCRETE

Type	Proportion	Ingredient
Plain Concrete	2.5	Portland cement
	3	Sand
	5	2 cm aggregate
	1.25	Water
Air Entrained Concrete	2	1A portland cement
	2.25	Dry sand
	5.66	Coarse aggregate
	1.25	Water

Freezing, Underground Water and Stability

Underground water is the chief enemy of any foundation. During winter, water at the base of sidewalks, patios or fence supports may freeze and generate tremendous forces which cause the surrounding rocks and soil to shift. When a thaw occurs, the water percolates away, soil and rocks collapse inwards and the concrete can twist or settle. This heaving effect is usually not too great a bother for fence posts or other such structures. However, if a pole supporting a dish is moved even slightly away from a vertical orientation, tracking of the geostationary arc of satellites can be ruined.

The frost line is the depth below which no freezing occurs. In North America, it varies from approximately one inch (2.5 cm) in southern Texas to over 10 feet (3 meters) in northern portions of the United States. In even more northerly locations such as Alaska or the Canadian Northwest Territories, special construction techniques are always required to manage the difficulties associated with the great depth of the frost line.

Pole or Post Supports

Pole or post type supports, commonly used for installing larger C-band dishes, provide strong, stable platforms. Such mounts cannot be used in locations where the ground is too rocky to dig the necessary hole or where the water table is too high. Because water is a good conductor of heat, a high water table may mean the frost line extends much deeper than normal in a given region. A pole support should also not be used for dishes larger than 4 meters (13 feet) in diameter since wind generated forces can be very high. A 3-legged tower which distributes the force over its whole base should be installed.

The hole for a pole support should be 2 to 4 times the diameter of the pipe. It should extend at least 6 inches (15 cm) below the frost line and 1 to 2 inches (2 to 3 cm) of gravel should fill the bottom to allow for proper drainage. Between 200 to 700 pounds (100 and 300 kg) of concrete are usually needed for an average pole support (see Figures 8-17 to 8-21).

Pole supports typically extend approximately 3 feet (1 meter) into the ground and 5 feet (1.5 meter) above. A general rule of thumb is to add 4 inches (10 cm) of extra length in the ground portion for every additional 20 inches (50 cm) of height. For example, a pole extending 16.5 feet (5 meters) above ground, 11.5 feet (3.5 meters) higher than normal, should be planted an additional 27.5 inches (70 cm) deep, for a total depth of 5.5 feet (1.7 meters). Poles set in very soft or sandy soil should be either planted deeper or have a wider concrete base.

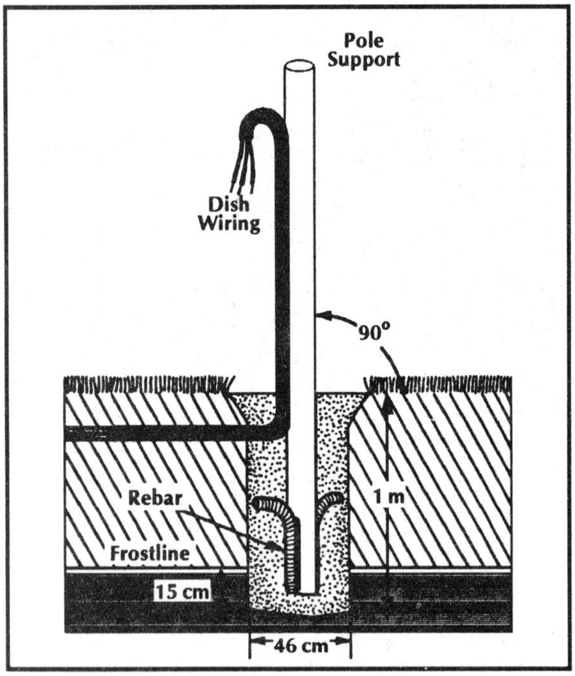

Figure 8-17. Ground Mounted Pole Support. *This illustrates all the necessary procedures required to install a pole mount for a receive dish. The rebar prevents the pole from twisting.*

Figure 8-18. Preparing Ground for Pole Mount. *After the site survey has been completed and all cable routes have been mapped, the hole for a pole mount can be dug. In general, a few simple tools including a post hole digger, a digging bar, a shovel, a garden hoe and a wheelbarrow are all that are necessary. The excavated soil should be loaded into the wheelbarrow and not thrown on a sod lawn. This simple step makes after installation clean-up much easier. (Courtesy of Brent Gale)*

Once the pole is placed into the ground, it can be temporarily supported by a few rocks and by wooden stakes or guy wires. It must be perfectly vertical so the dish can properly track the entire arc of satellites. A carpenter's level at least one meter in length or a plumb bob can be used to check this orientation from three or four vertical positions around the pole. These readings should be taken periodically as the concrete is setting. Note that after the concrete has been mixed it should have a thick consistency. Too much water that makes it "soupy" delays the curing process because all the excess water must either evaporate or drain off. The caustic ingredients in concrete kill grass so care should be taken to avoid spills. Some believe that concrete should also be added to partially fill the center of the pole for even further stability in high wind areas.

The pole must be strong enough to withstand both bending and twisting forces. At least schedule 40 steel poles should be used. Using schedule 80 or 120 steel is even better. (Schedule 120 is like pipe drillers' stem used in the oil industry). Most dishes typically require poles having a 3 to 4 inch (7 to 10 cm) outer diameter.

Figure 8-19. Using a Power Auger. *Digging holes with a power auger can save many hours of time and in the process save hands from uncomfortable blisters. Safety must be stressed when using this type of equipment in the vicinity of buried power and utility lines. (Courtesy of Brent Gale)*

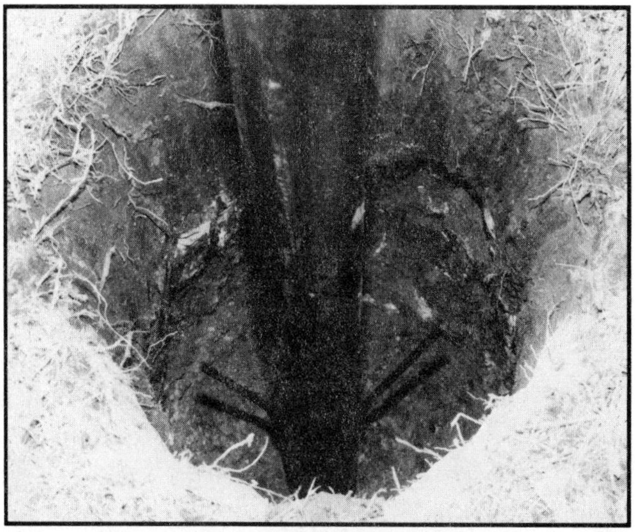

Figure 8–20. Pole-to-Ground Attachment. *It is recommended that construction rebar or a 6 to 8 inch (15 to 20 cm) bolt be welded to the bottom of the pole to keep it from twisting loose when high winds exert powerful forces on the dish above.*

Figure 8–21. Recommended Pole Depth. *The hole for a standard pole mount should extend at least one meter deep with a diameter of four times the pole diameter. In any case, the hole should be at least 6 inches below the local frost line. (Courtesy of Brent Gale)*

Figure 8-22. Pre-Trenching for Conduit Burial. *Before concrete is poured at least 2 meters of trench should be excavated and the conduit should be aligned within the trench and hole. When the concrete is poured the conduit will be completely buried and will extend parallel to the pole. If it were even partially exposed, cable near the base of the pole could be damaged by pets as well as by garden tools such as weed-eaters or lawn mowers. (Courtesy of Brent Gale)*

To prevent a dish from breaking free from the concrete form and rotating, a 1/2 inch (1 cm) metal rod should be inserted through the bottom part of the pole so that it extends out 3 to 6 inches (7 to 15 cm) on both sides. Or a piece of re-bar with similar dimensions should be welded onto it. An alternative to using one continuous pole is to implant a sleeve into which the long pole can be bolted. If designed properly, this type of assembly can allow a pole to be re-leveled in case the base ever shifts.

Most pole supports have no means to adjust the pole level once the concrete is set. This can be a concern because the effects of wind and water can, in time, cause the concrete base and thus the pole mount to shift. This effect is encouraged because a large dish angled towards the equatorial arc of satellites does not have its weight evenly distributed on the pole. This places quite a large bending force on the support. Without a leveling adjustment, the alternatives are to dig up the pole and replant it using additional concrete, install a new pole and cut out the original one, or attempt to level it with a piece of heavy machinery such as a truck or jack. None is an attractive option.

Figure 8–23. The Use of Leveling Stakes. *Leveling stakes serve to keep a pole in a vertical position as the concrete cures.. (Courtesy of Brent Gale)*

Figure 8-24. Stabilizing the Pole with Concrete. *Concrete can be poured down inside the pole to provide additional support. After mixing, concrete should have a thick consistency. Too much water makes it "soupy" and delays the curing process because all the excess water must either evaporate or drain off. Note that the caustic ingredients in concrete can kill grass so care should be taken to avoid spills. (Courtesy of Brent Gale)*

For the installer who regularly digs holes for planting poles, a gas-powered auger similar to those used to dig holes for fence posts is probably a good investment. This tool can save many hours of installation time.

A completed pole mount is shown in Figure 8-25. The section of conduit that is set in concrete with the pole must have a weatherhead to prevent moisture from entering the conduit.

Figure 8-25. Conduit and Pole in Place. *This photo shows the correct installation procedure with conduit extending into the concrete and secured in place with clamps. A weatherhead at the top of the conduit prevents water from entering the run. Conduit near the pole should be set in place when the concrete is poured. The concrete is poured to about 3.5 inches (9 cm) below the top of the lawn so that grass can be replanted adjacent to the pole. (Courtesy of Brent Gale)*

Figure 8-26. Pad Installed on Granite Rock. *This pad support was poured on a granite rock outcropping in the mountains of Colorado. A jack hammer was used to drill holes to insert rebar into the granite base. Over 4 cubic yards (3 cubic meters) of concrete were used to pour a flat pad and four 3/4 inch (19 mm) self-anchoring bolts secured the steel mount to the pad. (Courtesy of Brent Gale)*

Figure 8–27. Pad Support on Concrete Patio. *Approximately 3.3 cubic yards (2.3 cubic meters) of concrete were poured onto this concrete patio. The pad was secured by self-anchoring to prevent shifting. The 5 meter aluminum dish rests on an az-el mount. Cable runs are well protected by conduit. (Courtesy of Brent Gale)*

Concrete Pads

In those cases where the ground is too rocky or hard to allow digging of a narrow, deep hole, where the ground water table is too high or where an especially large reflector will be installed, a concrete pad can be poured on site (see Figures 8-26 and 8-27).

A pad is constructed by digging a shallow trench, building a wooden form to hold the concrete, adding gravel for drainage, embedding wire mesh for strength and pouring the concrete with the pole and the conduit in place (see Figure 8-28). It is best to recess the pad into the ground so that a lawn can be easily mowed and so the customer does not stub his or her toe on a concrete obstruction during the night. When installing a pad in a non-level location it is important to excavate the high ground and never to fill dirt into low terrain. Otherwise, the pad is likely to settle and cause the pole to tilt.

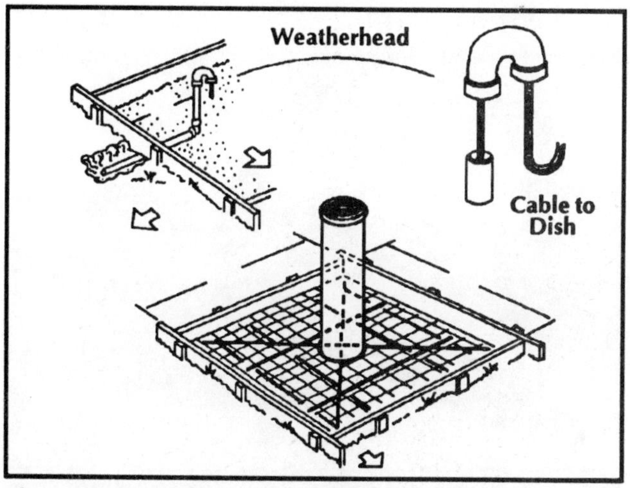

Figure 8-28. Schematic of a Pad Support. *A concrete pad support has steel rebar and wire mesh for strength and a gravel base for drainage. The pole has its weight distributed by extending rebar or channel iron from its base to near the edges of the pad.*

When concrete has set, the wooden form can be removed. These forms should be made from 2 by 6 inches (5 by 25 cm) green lumber because dry wood will absorb water from the concrete mix and hinder the setting process. Alternatively, the wooden form can be wetted or sealed with a water repellent material before pouring the concrete. Water should also be periodically sprayed onto the surface of newly poured concrete to improve hardening. This keeps the outside as wet as the inside and thus prevents cracking during setting.

A slab foundation supports a dish by virtue of its weight and size. Rules of thumb have been developed to size pads. A pad reinforced with mesh should have concrete at least as thick as the pole diameter. Without reinforcement, increase this width by 50%. The length and width of a pad should be at least half the reflector diameter or the measurement from the pad surface to the base of the reflector. For example, a 6 foot (2 meter) dish supported by a 3.5 inch (9 cm) pole should be supported by a pad measuring 3 by 3 feet (1 by 1 meters) that is from 3.5 to 6 inches (9 to 15 cm) thick.

Wire mesh used for reinforcement should be at least 8 gauge and have holes spaced so that openings are no more than 16 inches (40 cm) on a side. It should be located at a depth equal to about 1/3 the thickness of the concrete as measured from the pad surface.

Figure 8-29. J-Bolt Leveling Adjustment. *This 1.2 meter dish (left photo) is supported by J-bolts anchored in a pad. Additional thread extending above the surface of the pad permits re-leveling at any time in the future. (Photo courtesy of Northsat)*

A pad mount requires adequate drainage. Sandy soils require less drainage than rocky or clay soils which are both more impervious to water. The gravel bed under the concrete should range from a minimum of 50% to 150% of the concrete thickness. For example, a reinforced pad set on solid rock using a 3.5 inch (9 cm) pole should have about 5 inches (12 cm) of gravel and 6.5 inches (16 cm) of concrete.

As in a standard pole mount, the mounting pipe must have two pieces of rebar inserted through its base at right angles. When these are set with the pole in concrete, they spread the load over the whole base of the pad for stability and prevent rotation of the pole under high wind loads. This re-bar can be tied to the mesh reinforcement with baling wire for further strength.

Three Point Pads and Pier Foundations

Three point pads and pier foundations both bolt onto a tripod assembly that holds the dish support pole. The former is a concrete pad with three or more anchor bolts either preset into the concrete or drilled in after hardening (see Figure 8-29, 8-30, 8-31 and 8-32). Such bolts, also known as redheads, can most easily be installed with an electric hammer drill when the concrete is still soft. The latter uses three smaller pads called piers, each one having a small concrete foundation attached to deeply set reinforcement bars and anchor bolts (see Figures 8-33 and 8-34). While an az-el mounted dish is typically set on a three-point support, this differs somewhat from both of these mounting structures.

Anchor bolts should be made of galvanized steel, 3/4 inch (2 cm) thick by 25 to 35 inches (60 to 90 cm) long. After installation they should extend from 2.5 to 4 inches (7 to 10 cm) above the surface of the concrete. This design allows re-leveling of the pole when necessary. If the visible portion of the bolt is greased before pouring the cement, clogging of its threads can be avoided.

Figure 8-30. J-Bolt Leveling Adjustment. *This 1.2 meter dish (top photo) is supported by J-bolts anchored in a pad. Additional thread extending above the surface of the pad permits re-leveling at any time in the future. The bottom schematic illustrates the points of attachment for this mount. (Photo courtesy of Northsat)*

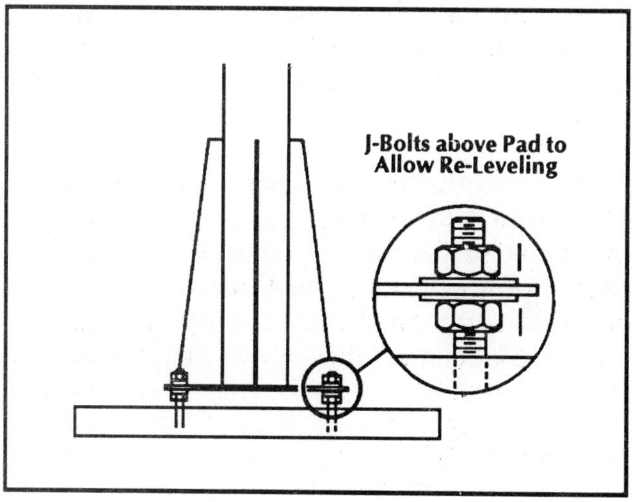

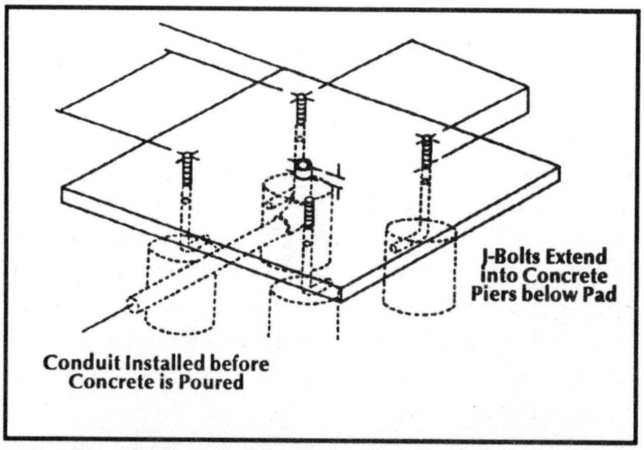

Figure 8-31. Pad Support Using J-Bolts. *This drawing shows the placement of J-bolts that extend into concrete piers below a pad mount. The conduit is also set in place to protect cable runs.*

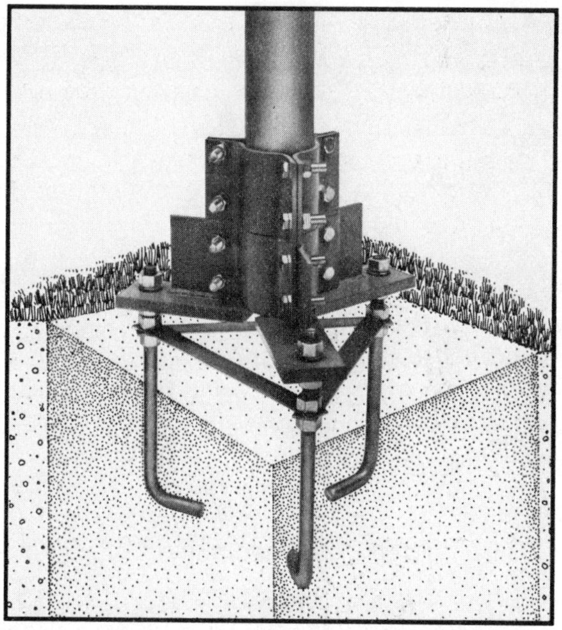

Figure 8-32. Level-Adjusting Pedestal Support. *This support is set onto a concrete base or even an existing solid concrete patio using either J-bolts or redheads. It can be quite useful in cases where the pole in a pad has become loosened. The pole would then be cut off at its base and this support would be installed in the pad near where the pole had previously been located.*

Figure 8-33. Three Pier Support. *This 20-foot dish with horizon-to-horizon mount is supported by a three pier pad foundation. Whenever such a tripod-supported polar mount is installed the back two legs should be set closely onto an east-west orientation. Fine adjustments to the mount are then made by sliding the telescoping legs to the final position (Courtesy of Antenna Development Corporation)*

Figure 8-34. Three Point Pad Mount. *A triangular az-el base can be installed with minimal concrete in view. Here, the excavated hole was 1.6 m deep and 1 m in diameter. (Courtesy of Brent Gale)*

A template of the base of the pole should be used to place the anchor bolts, generally J-bolts, before concrete is poured or while it is still "green" or soft. The bolts should be carefully kept away from the reinforcing mesh or rebar since any shifting of this supporting structure during pouring can cause them to move out of place. They should never be welded to the rebar since too much heat can cause them to become brittle. When the concrete is completely cured, 6 inch (15 cm) "redheads" having at least a 1/2 inch (16 mm) diameter can be inserted in a hole made by a hammer drill. Redheads, like some dry-wall screws or moly bolts, have a collar that folds up when the bolt is tightened, so they can support more than 40,000 pounds of force.

When the dish supporting tripod is mounted, two nuts with lock washers placed both above and below the attachment points allow a fine leveling adjustment during installation and at any time thereafter. When using either a pad or a pier to set the tripod, care should be taken to have the three supporting points as level as possible to facilitate making this vertical adjustment. It is also important to orient the legs of this tripod along the north/south axis during this procedure.

Some installers pour concrete at their shops and then truck the completed structure to a site. This technique has the advantage of lowering costs and making winter installations easier. However, slabs can settle in the spring, require special equipment to move them to a site and can be difficult to place once at the site. A 1500 kilo (over 3000 pound) trailer can crush sprinkler pipes in the ground and cannot often be moved into a customer's backyard. Nevertheless, such an approach can be very effective in maintaining business activity during spells of cold weather.

Figure 8–35. Setting of a Pad/Pole Combination Mount. *The 5/5" outer diameter pole could not be installed as deeply in the ground as necessary without conducting a major excavation job. Therefore, a pad mount measuring 10 feet in width was also poured around the top of the pole. Notice the conduit extending up against the pole to provide a protected route for the wiring. (Courtesy of Brent Gale)*

Figure 8-36. Combination of Pole/Pad Mount. *This illustrates a concrete pad poured with J-bolts inserted while the concrete was "green," before it had cured. A template of the dish base should be used to make certain that the bolts are in the correct pattern. If not, it may be necessary to remove them and re-insert redheads at the cost of additional labor and time.*

Combinations of Pole and Pad Supports

A creative installer can construct any variation of pole and pad supports that a given situation warrants (see Figures 8-35 and 8-36). For example, if an immovable rock is encountered half a meter below the surface when digging a hole for a pole mount, a combination pole/pad support can be created. The pole would still be set into the undersized hole. But a small pad would also be formed around the top of the pole to provide further support. Thus, less concrete could be used and the labor spent in digging the preliminary hole would not be wasted. In this case, the pole would extend further out of the ground than would normally occur and lifting the dish onto the pole might prove to be a little more difficult.

Ground Supports Not Requiring Concrete

Foundations that use a tripod support design but do not require concrete are also viable structures. For example, three or more rods can be driven deeply into the ground with a power auger. Then a tripod or even a four-legged assembly can be attached with weather resistant, adjustable bolts. It is interesting to note that this type of support is often not considered a permanent improvement by some building codes because no concrete is used. Thus, additional taxes can sometimes be avoided.

Above-Ground Mounts

Wall, roof or eave mounts are usually more feasible for mounting smaller dishes while above-ground towers or long pole supports may be required in certain difficult cases. Sometimes installing a dish at ground level is simply not possible because obstructions block a clear view to the satellite arc. In addition, in some countries and locales, real estate is a premium commodity and backyards are small or non-existent. This presents an installer with two options: choose a different site or mount the dish on a wall or above.

Roof, Wall and Eave Mounts

Although some of the same concerns including reflector stability apply to installing both small and larger dishes on roofs, walls or eaves, the supporting structures can be quite different.

Larger Dishes

Mounting large dishes on roofs should be regarded with caution, while mounting on waves or eaves is not feasible. Roofs are often not designed to withstand weights in excess of normal snow loading or the tremendous upward forces caused by winds acting on a large aperture dish. For example, an 50 miles/hour (80 k/h) wind blowing directly into the face of a 12 foot (3.7 meter) dish can generate a force of more than 1200 pounds (550 kg). This tremendous force can rip a roof off its rafters and cause even more damage when it finally lands!

Mounting a dish directly onto a roof has some potential disadvantages compared to ground mounting. Any play in the actuator or other components of a roof mount could create annoying sounds that are amplified by their direct connection to a roof, especially in high winds. Some dishes may even resonate loudly in some winds. A typical roof on a house that has some give built in can act much like the membrane of a loud speaker. This problem would probably not arise where a dish is solidly attached to a concrete or I-beam roof of a commercial building. An additional hazard is often not considered with dishes mounted far above ground: if the reflector is mounted on a long pole near an area where people walk, snow or ice resting on the dish may loosen, fall and cause injury to those below. This outcome is less likely when using offset fed dishes which are inclined more steeply towards the horizon than are the prime focus variety. In any case, if damage or injury does occur, both the owner and satellite dealer can be held responsible.

Anyone considering roof mounting a relatively large dish would be wise to hire a qualified structural engineer to analyze each particular situation. If a roof mount is judged to be feasible, the engineer takes responsibility for making this recommendation. This is mandated in those countries where a permit is required to change the structural integrity of a building.

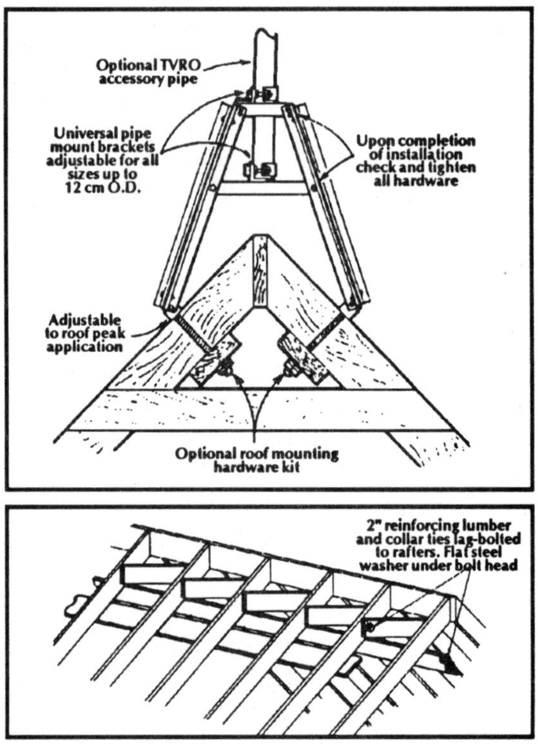

Figure 8-37. Peak Roof Mount. *An alternative to supporting dishes as large as 8 feet (2.4 meters) on a roof is a peak mount. Before installation a professional engineer and zoning authorities should be contacted and the appropriate license should be obtained. It is essential that as many as eight beams under the pitched roof be interconnected with additional steel or lumber members. Great care should also be taken to make certain that water cannot leak through the holes and cause damage to the inside structure and electrical wiring. (Courtesy of Rohn Tower)*

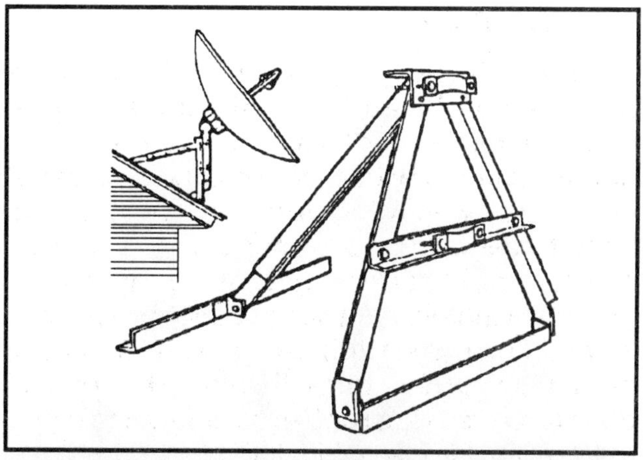

Figure 8-38. Pitched Roof Mount. *When using a side roof mount, the same care should be taken as with a peak roof mount. As many roof beams as possible should be used and additional support lumber should distribute the weight of the antenna/mount. Roof mounts, as in all satellite TV installations, should be well grounded. (Courtesy of Rohn Towers)*

The fundamental principle to be followed in designing and installing a roof mount is to tie the support structure to as many rafters as possible in order to spread the load. A number of companies now manufacture welded steel assemblies with built-in leveling adjustments designed to support poles on most types of roofs (see Figures 8-37 and 8-38). The support pads must be attached to a length of lumber or metal inside the attic which itself is secured to as many rafters as possible. If the roof has steel beams, it is highly recommended that the mount be tied in to these members. Two or three guy wires, preferably made from a material as strong as aircraft control cable, can be used to secure the dish for further stability. These also prevent the reflector or mount from landing hundreds of meters away if the worst should happen and it does break away from the roof.

Great care must be taken in breaking roof seals to prevent leaking (see Figures 8-39, 8-40, and 8-41). It is wise to plan this step with the help of a reputable roofing contractor. When seals are broken, the original roofers warranty is voided and a dealer can become liable for water damage to the structure or contents of a building. Even small unsealed holes can allow water to penetrate a roof and ruin a ceiling.

Figure 8-39. Flat Roof Pitch Pan. *Before beginning this installation, a roofing contractor was hired to lay down a 4 by 8 foot aluminum pitch pan which was filled with tar after the installation was completed. This sealed the roof and prevented leaks. (Courtesy of Brent Gale)*

Figure 8-40. Flat Roof Mounts. *Both these dishes were mounted on gravel-top roofs. After the gravel was cleared away a 1/4 inch (6 mm) steel diamond plate sheet was attached to the underlying steel roof members with four 3/4 inch (19 mm) double-nutted all-thread rods. Tar was poured onto the roof surface beneath the steel plate and on its top surface after all necessary tracking adjustments had been completed. This installation detail thoroughly sealed the roof and prevented water leaks. A professional structural engineer was consulted before the work was started. (Courtesy of Brent Gale)*

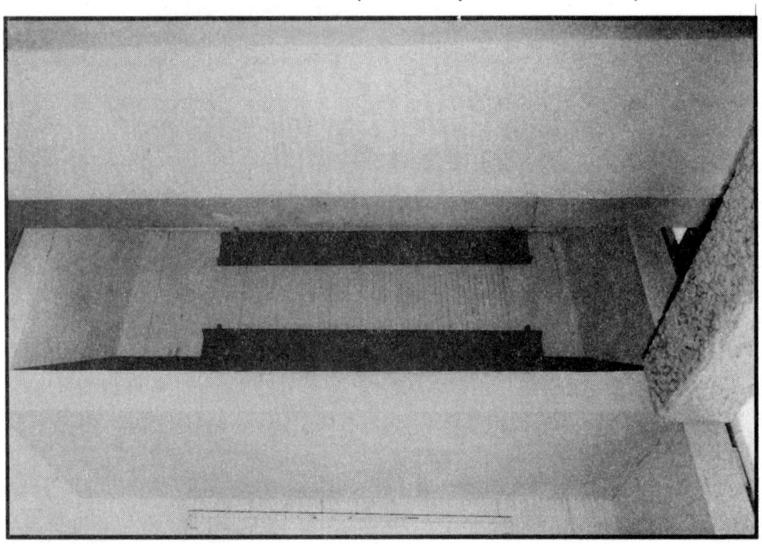

Figure 8-41. Under-side and Outside View of Dish on a Roof Mount. *This 8.5 foot (2.6 meter) dish was attached by spanning five rafters with two pieces of 8 inch (20 cm) channel iron both above and below the roof. The resultant "sandwich effect" helped in distributing the dish and mount weight over a larger surface area. Holes were caulked and then tar was poured on to the roof. If a pitch pan had been installed under the channel iron, water sealing would have been even more effective. (Courtesy of Brent Gale)*

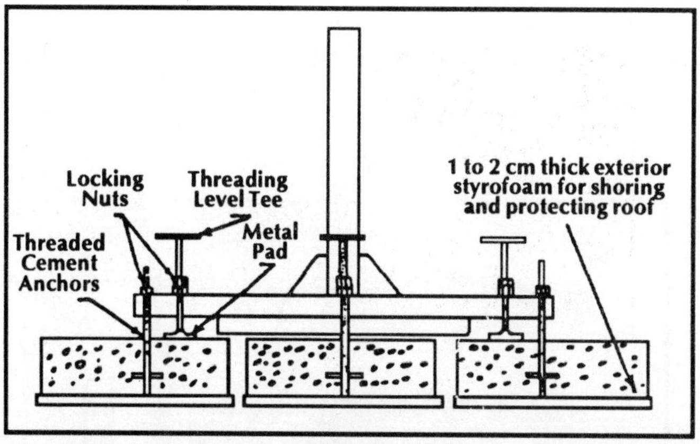

Figure 8-42. Concrete Flat Roof Mount. *This mount is bolted onto four concrete slabs which are spread over a 1 square meter area. "Tea handle" screws are used to allow leveling of a polar mount. (Courtesy of Upper Midwest Satellite Supply)*

Roof mounts do have the advantage of being easier and less expensive to assemble than most ground foundations. There are no problems associated with mixing and pouring heavy concrete and trenching through ground to lay cable runs. There is also a much lower chance of theft when the satellite equipment is mounted out of reach. However, winds are much stronger above a building than on the ground below and installing on a high or steep roof can be dangerous.

One type of roof mount, a gravity mount, does use concrete. Concrete pads or other supports secure the dish in place (see Figure 8-42). A similar mount can also be used at ground level.

Small Dishes

Smaller dishes, generally less than 6 feet in diameter, can be supported from walls, eaves or mounted on roofs (see Figures 8-43 and 8-44). However, even when installing reflectors as small as a few feet in diameter, wind loading and stability must still be considered. The weight of a dish as well as forces caused by winds and other loads can act to dislodge a bracket from a wall, an eave or a roof. An adequate safety margin must be factored into the choice of materials.

The mounting assembly should generally be secured to the wall or eave before the dish is attached. If this is not possible, first mark the position of the necessary holes using just the mount as a template. After drilling, the dish/mount assembly can be lifted into position and easily secured with the bolts. Although both steel or nylon wall plugs can be used on walls, nylon plugs are recommended when the dish is subject to vibrations caused by wind loading or when it is being mounted in solid masonry.

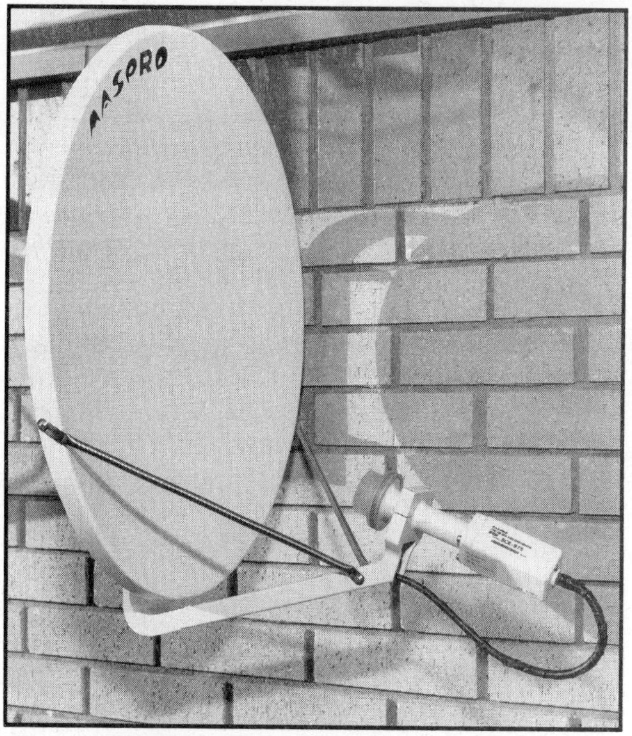

Figure 8-43. Wall Mounted Antenna. *This 3 foot (1 meter) stamped galvanized steel dish is attached to a wall for reception of European Ku-band broadcasts. (Courtesy of Maspro)*

Figure 8-44. Roof Mounted Dish. *This 6 foot dish is installed on the KTI universal mounting kit that can support reflectors up to 7.5 feet in diameter. (Courtesy of KTI)*

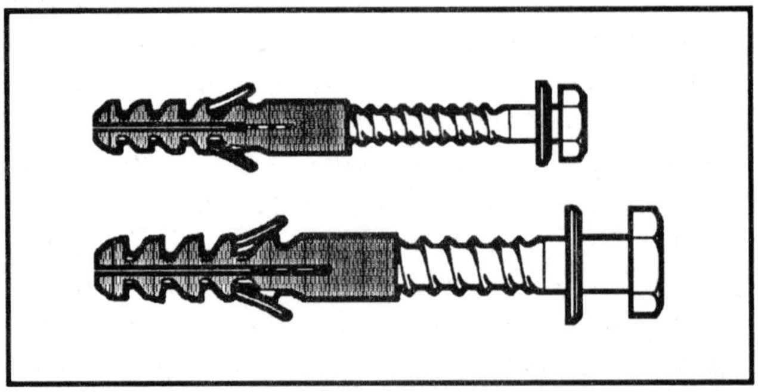

Figure 8-45. Wall Plugs and Bolts. *Wall plugs are used to attach a mount to a brick, concrete or concrete block wall. The plug inserts into a snug pre-drilled hole. When the bolt is inserted, the sides are forced to expand and make tight contact with the wall.*

Wall plugs can exert tremendous expansive forces that ultimately can crack concrete or brickwork. The bolts should be tightened to only the recommended torque. When wall plugs are being installed in bricks the bolts should be secured in the brick, not in the binding mortar between bricks which is much weaker. In situations where a weaker material such as aerated concrete serves as the foundation, expansion type fixings should be used. The sides of this type of device detach from its body and expand into a beveled hole when the bolt is tightened. When mounting to a wooden structure, a structural member such as a wall stud must be used. It is not sufficient to simply bolt the supporting bracket into a plywood wall.

Wall plugs and expansion bolts designed for masonry cannot be used with other building materials such as wood or shingles (see Figure 8-45). In these cases brackets have to be secured directly into the underlying wooden frame or onto a plate that rests inside the wall. Techniques similar to those described for mounting larger dishes can be used.

Figure 8-46. Free-Standing, Long-Pole Installation. *This dish is mounted on a free-standing pole. In such cases, the pole diameter should be roughly double the normal size and a pole reducer must be attached to the top of the pole where the mount attaches. (Courtesy of Brent Gale)*

Towers and Long Pole Supports

An alternative to a direct roof mount is to support a dish on a long pole or a tower, a method not generally used with smaller dishes (see Figure 8-46). This may sometimes be a somewhat easier install because the pole is secured in to the ground adjacent to a building and is more easily accessible. Furthermore, building permits are generally not required since no structural alterations other than attaching the pole to the adjacent wall are required. There should be at least one attachment point where the pole meets the top of a wall, the point of greatest torque (see Figures 8-47 and 8-48). It is better to weld tabs to the pole instead of using collars which may slip when high winds may twist the reflector. The structure of the wall must also be considered in attaching these tabs. Weaker parts of the wall could be damaged when the dish and pole twists or bends under forces of wind.

When installing large dishes on long poles, for every 4 inches (10 cm) addition to a typical 5 foot (1.5 meter) ground-mounted pole length, an extra 20 inches (50 cm) should be added to the normal 3 feet (1 meter) below ground segment. For example, if an 25 foot (8 meter) length were required above ground level, 51 inches (130 cm) 4 inches (10 cm) for each 20 inches (50 cm) of the extra 21 feet (6.5

Figure 8-47. Long Pole Supported by Wall. *This 10 foot dish is support by a long pole that is securely attached to the adjacent brick wall. The pole must either be welded or bolted to the points of attachment so that it will not twist in high winds. (Courtesy Brent Gale)*

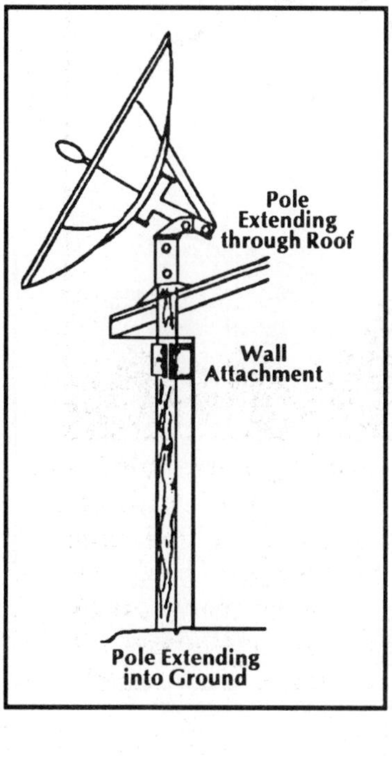

Figure 8-48. Exterior Long Pole Wall Mount. *Two methods to attach a long pole to an exterior wall, bolting the pole directly against the building and extending the pole through a hole in a pitched roof, are shown here. Both poles were secured at their bases in concrete filled holes. Care should be taken when attaching brackets because high wind forces can damage or detach some of the weaker types of wall materials. To determine how closely a support may approach a wall, a small hole should be dug at its base because some building footings may extend as far as 3 feet past the building edges. In these cases, a plumb bob can be extended from the roof to the center of the hole in order to determine the required length of brackets to secure the pipe. (Courtesy of Upper Midwest Satellite Supply)*

meters) should be added to the 3 feet (1 meter) normally below ground. Thus 7.5 feet (2.3 meters) would be underground and 17.5 feet (8 meters) would be above ground. Of course, this is a rule of thumb that does not have to be adhered to rigorously. Additional examples of long pole mounts are shown in Figures 8-49 and 8-50.

In addition to securely attaching the pole to one or two points on the adjacent building, it should be oversized in diameter by 2 to 6 inches (5 to 15 cm) for extra stability against twisting and bending. In general, this requires use of a reduction coupling at the top of the pole. Guy wires should be attached to the base of the mount and to nearby support points. A lesson can be learned from the designs of billboard installations which also use long poles. These supports begin with larger diameter poles at the bottom and step down to smaller diameter poles near the top.

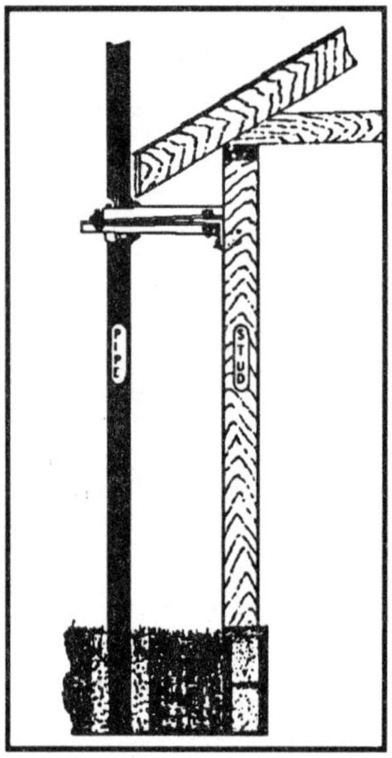

Figure 8-49. Adjustable Exterior Wall Mount. *This mount is designed so that its position can be adjusted beyond the eave of a building. It should be secured to at least two places as well as at its base in a hole filled with concrete. (Courtesy of Upper Midwest Satellite Supply)*

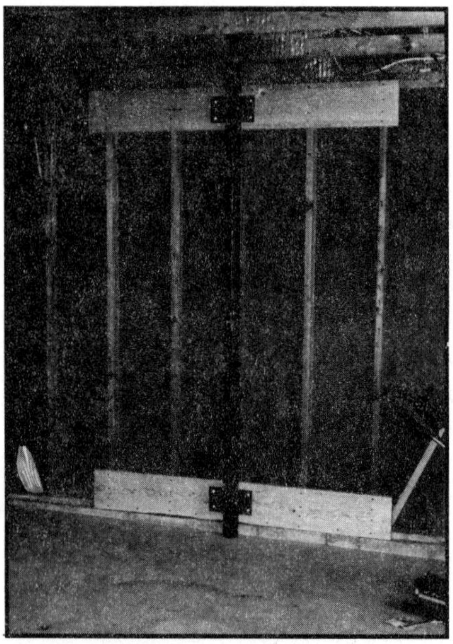

Figure 8-50. Interior Garage Support for Long Pole Mount. *This type of support is fastened to the inside wall studs with two headers located near the top and bottom of the pole. The headers connect to five 2 by 6 inch (5 by 15 cm) wall studs and pass the pole through a hole in the pitched roof. Holes must be drilled through the pipe and multiple bolts should be used to secure the pipe to the V-clamps to prevent it from twisting in winds. The manufacturer of this product recommends that dishes in excess of 6 feet in diameter should not be used in this type of installation. (Courtesy of Upper Midwest Satellite Supply)*

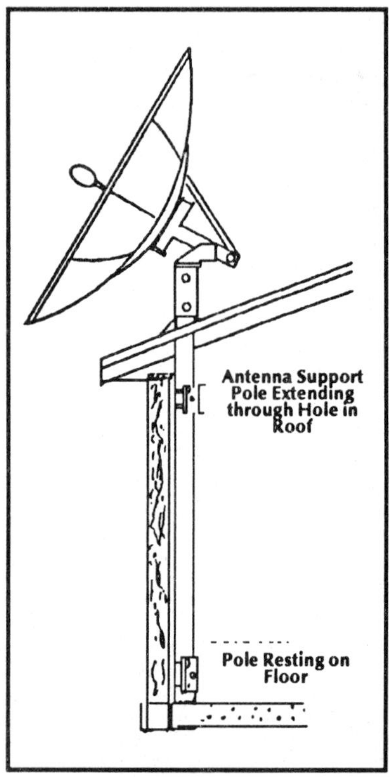

CABLE RUNS and TRENCHING

Cable runs between the dish and indoor equipment should be safely protected and neatly installed. Leaving unburied or unsecured cables lying around is an invitation to an animal or someone with a lawn mower or a destructive curiosity to cause damage.

A variety of techniques can be used depending upon the type of cable required, the length of the run, dish location, and, when applicable, ground conditions. For example, trenching is not necessary in conjunction with a roof or some types of long pole supports.

Digging the Trench for Ground Mounts

When a trench is necessary, before digging begins the first step is to determine where all water, electrical, gas and telephone lines are located. In many countries, utility companies gladly send a crew out to find these lines and even help dig to ensure that they are not severed. Customer owned underground systems such as lawn sprinklers are usually more difficult to pinpoint. It makes good sense to proceed with caution. In those regions where sprinklers are common, having parts on hand for repairing pipe can be a great time-saver.

The depth of the trench depends upon many factors. If rigid conduit is used there is less risk that an avid gardener will slice a shovel through the cable (see Figure 8-51). If the cable is running through an unused portion of the yard, it should be buried at least 6 to 9 inches (15 to 20 cm) and preferably 12 to 18 inches (30 to 45 cm) under the surface. Electrical codes in many countries specify required burial depths.

It makes good sense to keep a record of the length, route and gauge of buried cables. This information on a map of the site can be invaluable if an update or repair is required at some future date. This record is also useful if landscaping changes are made.

Whenever a trench is being cut across a lawn, care should be taken with the sod. It should be cut with a sharp knife or a flat shovel to a depth of at least 7 or 10 cm on both sides of the trench. The sod should then be carefully rolled in sections and laid next to the trench. All loose earth should be placed in a wheelbarrow or on plastic sheets since it is often difficult to clean loose fill from the surrounding sod with a rake after the trench is refilled. Any sod left above ground for periods longer than a day must be watered to avoid damage because it can easily dry out and wither.

Figure 8-51. Installation of Conduit. *This 2 inch conduit was placed in an 18 inch trench that had been excavated by a power trencher. Notice the weatherproof cap and the drip loop at the point of entry into the building. (Courtesy of Brent Gale)*

The trench should not be refilled until the installation is completed and it is certain that there are no defects in the cable. This precaution will prevent unnecessary re-excavation of a deep trench. Refilling the trench is one of the last steps in a successful installation.

The Use of Conduit

Many direct burial, multi-conductor cables do not necessarily require encasement in a rigid, protective conduit. However, conduit does provide extra insurance against damage. In certain situations using conduit is strongly recommended (see Figures 8-52 and 8-53). For example, it provides the necessary protection for runs under driveways or roads. If rodents are common, conduit can prevent them from chewing through cables. In areas with excessive ground water, conduits can protect against water damage. Be aware that there have been cases where improperly installed conduit has done just the opposite and has become a water trap. As a result, the only solution was to dig up the conduit to drain the excess water.

Conduit should be used to route cables up buildings or over roofs. While PVC conduit is fine for underground locations, it emits noxious fumes when burning. Metal conduit is therefore a better choice for locations on or within a building.

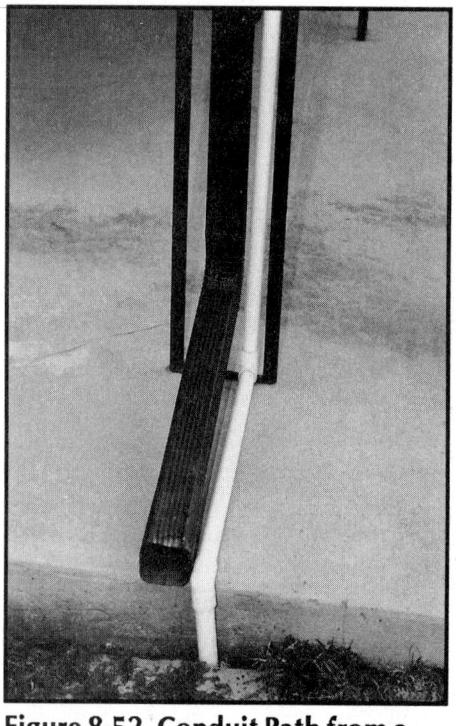

Figure 8-52. Conduit Path from a Long Pole Mount. *This conduit was routed from the roof along a gutter and then into a buried path. Its path along the drain was chosen so that passers-by would not trip over the conduit and perhaps damage the cable. (Courtesy of Brent Gale)*

Local building codes in some countries may require the use of conduit when cables carry power above a predetermined level. In these cases, white PVC water pipe or black rolled flexible tubing is not permitted. Only rigid varieties like gray PVC or aluminum are permissible. The reason for this regulation is that flexible conduit rusts and breaks through in a relatively short period of time. It should never be used in satellite installations.

Conduit should be a least twice the diameter of the enclosed cables. Installing sweeps instead of right angles corners makes pulling cables into and out from conduits easier. Since sweeps are not available in standard white PVC

Figure 8-53. The Use of Conduit. *Conduit protects cables from any damage that could be caused by rodents as well as from moisture ingress at points where the cables enter and exit the ground. A weatherhead must be installed to prevent moisture from entering the conduit at both the dish, as illustrated here, and at the entry to the building. (Courtesy of Brent Gale)*

used for plumbing lines, either gray PVC or standard metal electricians conduit should be used. Some types of conduit are designed to be bent into shape by applying heat when they are wrapped with a material similar to heat tape. It is important to realize that if more than four sweeps forming a full 360° turn are used throughout the cable path, it would be impossible to push or pull cables into or out of the conduit. If this happens, one major advantage of using conduit is eliminated. The proper installation of conduit allows removal and replacement of cables years after burial without the need for excavation.

Precautions should be taken to prevent water ingress, especially at the dish. Sections of conduit can be tightly glued together. Care must be taken when using PVC conduit, to avoid dripping solvent or cement onto cable insulation. Some types can be dissolved by these solvents possibly resulting in damaged coax. Weatherheads or 180° sweeps attached to vertical conduit should have removable screw fittings and should be used at all exposed ends. If the cable ever has to be removed from the conduit and replaced, these fittings can then be undone. The use of weatherheads as well as correct bonding at joints is very important since any water build-up can ruin cables and corrode connectors. In those cases where the conduit runs downhill from the dish to a building, improper installation could cause water to be channeled directly into its interior.

All cables should be attached to the supporting pole with wire ties or hose clamps. Drip loops can also be formed where the cable terminates to channel water away from the actuator motor. A drip loop is simply a loop in the cable which has its lowest point below connectors, a hole in the wall or an entrance in the bottom of a waterproof box. Any water from rain or melting snow and ice tends to collect at this low point and drip onto the ground (see Figure 8-54).

When pulling cable through conduit, do not lay the cable on the ground, as it can pick up debris and make the job more difficult. A "fish tape," which is a piece of rigid wire, can also be used to pull cables through conduits. It is inserted at one end and the cable is attached to it at the other end. Cables can also be lubricated with a liquid soap to allow them to slide through the conduit more easily. Of course, this soap must be cleaned off where the cable enters a building. If the cable is too short to span the entire length between the dish and receiver, never splice two pieces together. Start again. Run a complete and sufficiently long cable. Splicing would not only increase resistance to current flow, but would also present a perfect opportunity for cable breaks to occur in the future or for water to enter and short out the system. The splice is also likely to pull apart when cable is being fed through a conduit.

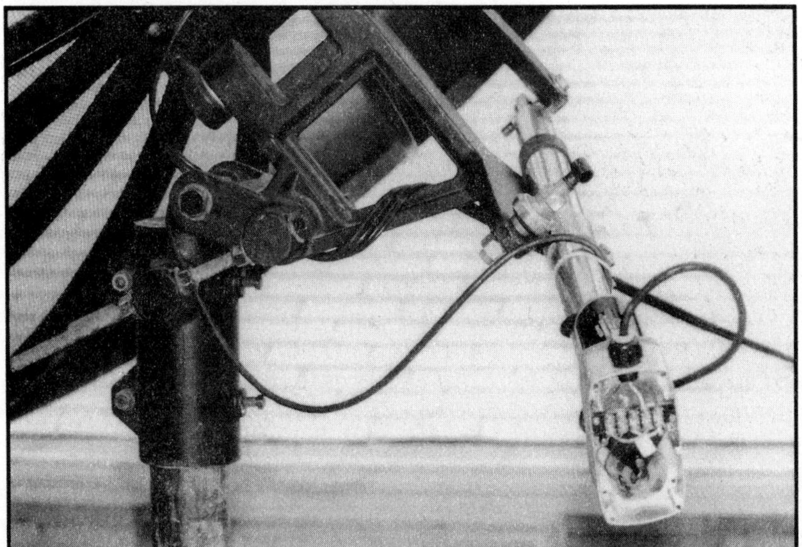

Figure 8-54. A Drip Loop. *A drip loop allows water build-up caused by melting snow, rain or condensation to drain away from components such as actuators. Notice how the cable here has its lowest point below the actuator motor.*

All of the cable used for both pole and roof mounts running from the dish to the building entry should be placed in electrical conduit. This prevents cable from being damaged by ultraviolet light and by other enemies such as birds and squirrels. Many building codes also mandate the use of conduit. Using well-installed conduit also results in a neater and more professional installation.

The Entry into the Home

When possible, cables should be hidden from sight. Holes drilled in both exterior or interior walls should be as small as possible and well caulked after the installation is completed. Multi-run cable can be balled up to minimize the size of the necessary hole. If the cable entry is through a crawl space, holes may be drilled below the ground level only if they are well sealed. In general, above-ground cable entries are recommended (see Figure 8-55). Always have sets of masonry and wood bits on hand. Many older homes have very thick walls so that extra long wood bits up to 18 inches (45 cm) should be available during an installation. Holes should be drilled at least 12 inches (30 cm) above or below and away from any electrical outlets to avoid hitting power lines. Exercise caution when drilling near bathrooms, kitchens or other places where plumbing may be close and where substantial damage can be done.

Drill from the inside to the outside of a wall, not the other way around, and only after all measurements have been taken and the path of the bit is known. When possible, use wall-plates similar to those on electrical outlets. A wide variety of brands is available at most electrical parts distributors. Cables fed either into an attic or crawl space can usually be fished up or down into a wall to make an unobtrusive entry. Difficulties can be encountered if the walls have cross-supports between studs.

When carpet must be disturbed, cut a small "X" with a razor blade or an "exacto" knife through both the carpet and padding so later if the system or cable is moved the rug can be easily repaired. A drill can often catch on the rug pile and tear a wide area open or ball up padding. Never run cables under a rug. The tacks used to attach the rug to the pad and floor can also easily short cable out. Cables when under rugs cause bumps and make tripping all too easy.

When cables are routed through either crawl spaces or attics, they should be attached to rafters or floor joists. Special staple guns are made to fit onto the cable surface to avoid damaging the conductor. Small plastic clips which are designed to be nailed onto a wall are another cosmetically attractive option.

One final note: never run cables parallel to electrical lines. Doing so can result in interference such as 60-cycle hum.

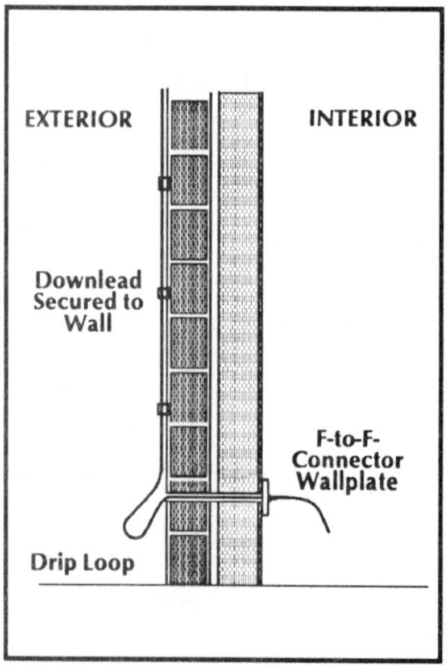

Figure 8-55. Wall Plate and Drip Loop. *Some fixed dish systems relay polarizer control pulses via the coax so just one run of coax is necessary. A drip loop should be installed on the exterior wall and a wallplate F-to F-connector socket on the inside wall. This is a good opportunity to make the transition from exterior block cable to another color that matches the interior decor.*

ASSEMBLING the DISH, FEED and LNB

The dish, feed and LNB are the most critical components of any satellite system. If they are not selected and installed properly, using the best and most expensive satellite receiver and modulator in the world cannot improve picture quality.

Assembling the Dish

Dishes generally are accompanied with a set of assembly instructions. These should be carefully read at least twice before beginning an installation to determine if any special tools or manpower are required or if there are any unanswered questions. Usually, most of these questions can be solved by simply beginning the assembly, but occasionally a call to the distributor or manufacturer is necessary. Be sure that all parts are accounted for before proceeding to the job site.

Smaller Dishes

Smaller dishes, and some larger brands, usually have a one-piece reflector and all that is required is to bolt the mount to its surface. Care should be taken to not attach these bolts so tightly that the dish could dimple. The mounts of off-set dishes usually have an integral support that holds the LNB and feed.

Large Dishes

Larger reflectors are generally not as simple to assemble as smaller off-set models. Each variety and type has some unique aspects to its assembly. Spun or stamped reflectors are most often delivered in one piece ready to be bolted to their mounts. In cases where holes have to be drilled to attach a mount, it is critical that they be centered. Do so by crossing two strings each tied to two opposite corners of the mount. Lining up their intersection with the dish center properly aligns the mount. Spun reflectors have been formed in a lathe and therefore always have a hole located precisely at their center. Stamped dishes often have a hole for the buttonhook or an identifiable center point remaining from the manufacturing process. It is good practice to sequentially tighten all bolts or screws on any dish only after all are in place to avoid warping the reflector. It is just as important not to over-tighten these bolts. This could cause a dimpling or indentation and a loss of gain. Rubber spacer washers or grommets are sometimes included to act as shock absorbers and should be used when provided. If the dish is aluminum and the mount steel, it is even more important to insert rubber grommets to prevent electrolytic action between these dissimilar metals. In time, this type of corrosion would weaken or even destroy the points of attachment.

Fiberglass dishes should be assembled on a flat surface so that all sections fit together with the rim lying perfectly flat. Again, do not tighten bolts until the assembly is completed. This procedure allows each section to be lined up to smoothly interface with the adjacent panels. Some fiberglass reflectors can be assembled on the mount. In this and all other cases, the bolts should be tightened progressively from the center to the outside to make the dish expand outwards from the center and maintain a true parabolic shape.

Mesh reflectors can be assembled on any surface but completing the work on grass or a soft material such as the cardboard boxes used for shipping avoids scratching the paint. It may also be convenient to rest the central hub on a stand such as a small table or a garbage can to make attaching all the ribs, panels and bolts or screws easier. If wind clips are provided they should be used, especially in areas having potentially strong and gusty winds. There have been cases of panels popping out partially or completely in a wind storm. Some installers run a small wire through each section near the rim to provide further protection against this type of damage.

There are two accurate and very simple techniques used to check that reflectors are not warped either during manufacturing, shipping or assembly. To "lip-sight" a dish, sight along one rim so that the other edge appears in the same plane. Both rims should line up as two straight lines one on top of the other. If not, the dish is warped (see Figure 8-56). The crossed string method involves using two or three strings taped or otherwise fastened over the outer edge of a dish (see Figure 8-57). If these strings either mesh together or are separated, the reflector surface is warped. They should just lightly touch one another. Using three crossing strings makes it easier to locate the center. If there is warpage, loosening and re-tightening the bolts which hold the dish structure can often be an effective method to realign the surface. A warped dish will not perform up to par.

Figure 5-56. Lip-Sighting a Dish. *Sighting along the edge of a dish to see if the line from the edge closest to the eye is parallel to the one farthest away can be an effective method to check that the dish is not warped. (Courtesy of Brent Gale)*

Mounting a Dish on the Pole

Except for bulky fiberglass dishes, most can be easily lifted by one or two people. Large fiberglass reflectors in excess of about 12 feet (4 meters) in diameter usually require four or more people to share the lifting. In some cases, a crane may even be required to complete the job. This extra weight has prompted some fiberglass manufacturers to design their products so they can be assembled directly on the mount after it has been set on a pole.

Other types of dishes can also be installed by first setting the mount on the pole (see Figure 8-58). Then the completely assembled dish can be lifted into the correct position where the attaching bolts can be inserted. Since the reflective surfaces of spun or stamped dishes are often easily deformed and can be dented or scratched these should be lifted into their final position after their mount has been attached to the pole. This avoids bending the rim at the points of lifting. Damage can also occur before the actuator arm is attached since the dish can often move freely and strike other objects.

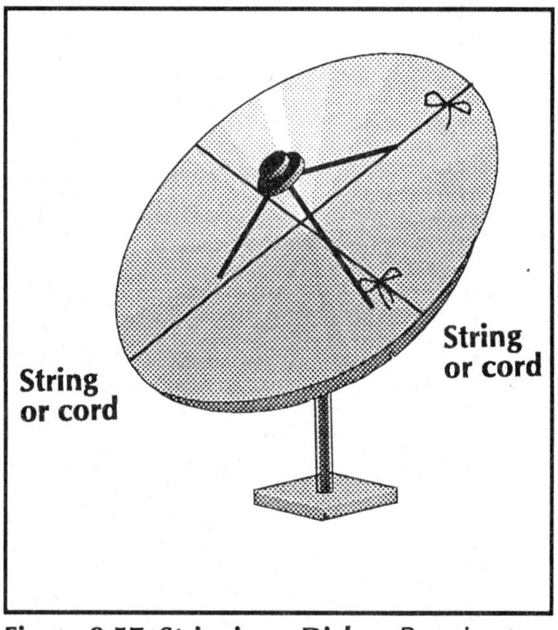

Figure 8-57. Stringing a Dish. *Running two or three strings attached to the edges of a dish is, like lip-sighting a dish, an effective method to determine if the reflector is warped. If these strings are either meshed tightly together or separated, the dish is warped.(Illustration Courtesy of MultiChoice)*

Assembling the Feed/LNB

The feed and LNB should be bolted together before being installed on their support. Two important rules should be observed. First, never touch or bend the probes on either component. They are finely tuned at just the correct location and any tampering will increase the VSWR and result in a lower C/N ratio. Second, always use the gaskets provided for insertion between the flanges. These protect against water ingress. But never use any sealant between these gaskets because there must be metal-to-metal contact in these joints. The distance between these joints should not be increased as this would upset the impedance match between the waveguides. The bolts around these flanges should also be evenly spaced. They should first be tightened by hand and then with a spanner.

Figure 8-58. Preparation for Lifting Dish onto Mount. *Here the mount is ready for the dish which can be rolled into place for attachment of the bolts. One technician can use this method to install even large reflectors. Alternatively, the mount could have been pointed directly upwards and the dish would be "bird-bathed" by lifting it onto the mount.*

Polarization Offset and Aligning Mechanical Polarizers

The probe in a waveguide must have the same orientation as that of the incoming signal polarization. Thus, for example, to receive vertically polarized signals, the probe must be in a vertical orientation. When a dish is aimed at a satellite on or near the site longitude, namely the highest in the sky or the one due north, the probe would be set in vertical position. (In this example we assume that the site is in the northern hemisphere. In the souther hemisphere the dish would point due south). However, when the dish is pointed toward either horizon the polarization is offset because the view towards that satellite is somewhat "twisted." This "polarization offset" increases the farther away from south that the dish is aimed.

When using a polar mounted dish, polarization offset is automatically set for all satellites if it had been properly set for the most southerly spacecraft. This should be done when aiming at the most southerly spacecraft. For voltage switched polarity feeds, first switch to either horizontal or vertical polarity. Then rotate the feed to a horizontal or vertical position and tighten it down.

For mechanically switched polarity feeds, set the probe halfway between the two mechanical limits. Then rotate the body of the feed until it is halfway between orientations of vertical and horizontal so the center of its range is also halfway. Although most mechanical feeds now have a full 270° range of motion, servo motors on older feeds rotate just 180° (see Figure 8-59). For many feeds that indicate the direction of vertical or horizontal, either on the feed itself or with a printed card that fits on the feed during installation, simply line up with vertical or horizontal when the dish is aimed due south (see Figure 8-60). For circularly polarized signals, the feed probe must be able to move at least 45° on either side of the dielectric slab as described below so this centering is even more important.

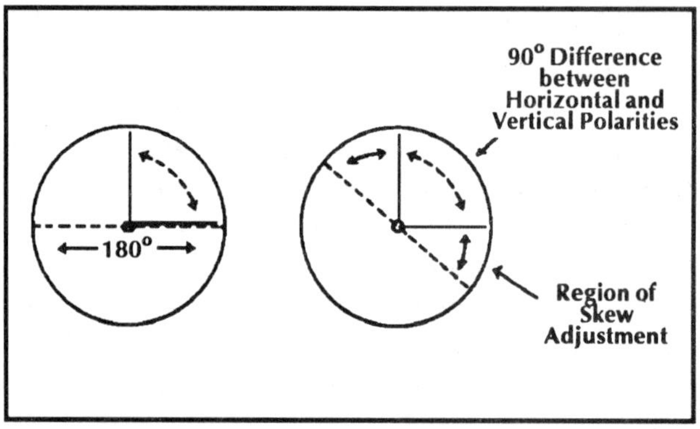

Figure 8-58. Servo Motor Mechanical Limits in Older Feeds. *Most older servo motor feeds have a 180° range of motion. If the mechanical limits do not lie past the orientation of both vertically and horizontally polarized signals, the probe will not be able to be properly aligned and the motor might be burned out in the attempt to move the probe to the correct position. Modern servo motors now have a 270° range of motion and the alignment problem is less critical.*

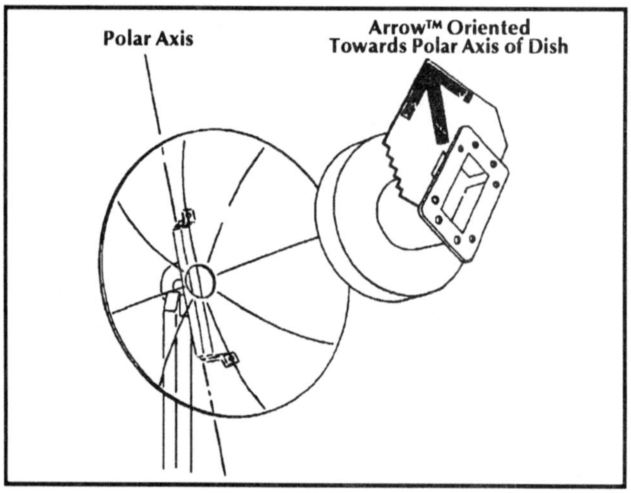

Figure 8-60. Orienting the Feed/Polarizer. *Chaparral Communications makes a useful device called the Arrow™ which aids in properly aligning a ferrite or servo motor feed. When the feed has been properly installed, the point of the arrow is lined up with the polar axis on the mount. This guarantees that the motor will not be driven up against a mechanical limit and be damaged in the attempt to line up the probe with the direction of polarization. The one shown here is used for orienting the Twister™ and Polarotors™*

The range of motion of a servo motor can be observed by hooking it to a test receiver when limits are set. When properly installed, the motor should have the freedom to rotate 90° between positions of horizontally and vertically polarized signals and still have extra room to move past either end of this range. This permits the necessary skew adjustment when tracking between satellites transmitting different orientations of linear polarization. Most important, it prevents the motor from running up against a limit and burning out either the motor or the timer in the receiver. This installation procedure also guarantees that the probe motor is functioning properly. Signs that a feed does not have an adequate range of motion are snowy pictures or floating lines (hum bars) that appear on the television screen.

Setting polarity offset is somewhat more involved for dishes on az-el mounts. Since a satellite is not targeted by rotating the reflector about the polar axis, polarization orientation must be set after a satellite has been targeted or "acquired." For voltage switched feeds this generally involves rotating the whole unit to maximize the signal level before securing it in position. For some offset fed dishes, this may be problematic (see Chapter 5 for more details). For mechanical feeds, care must be taken to ensure that the probe can rotate freely between vertical and horizontal positions once the dish is secured in place.

Retrofitting for Detection of Circularly Polarized Signals

Linear feeds can be retrofitted to discriminate between left and right-handed circularly polarized (LHCP and RHCP) signals by inserting a dielectric slab such as the teflon insert described in Chapter 5 (see Figure 8-61). The insert converts circular to linear polarized signals which are then detected as usual by the feed. If properly installed, losses incurred when detecting of the linearly polarized signals can be kept to an acceptable minimum, perhaps under 1 dB. Problems could arise with a reduced level of cross-polarization discrimination (also as described in Chapter 5).

In order to maximize reception of both linear and circular polarities by the same feed, it first must be set for optimal reception of either linear polarity. The slab is then inserted parallel with the direction of the probe to guarantee that linearly polarized signals are attenuated as little as possible. Picture impairments can be more easily seen if a weak transponder is being received. LHCP and RHCP satellite transmissions can then be received when the probe is rotated to a position 45° from either side of the dielectric slab. The feed must be oriented to allow the probe movement to 45° from both sides of the slab as well as to both horizontal and vertical polarity positions. This requires aligning the feed so that the dielectric slab is set within about 15° or 20° closer to what normally would be the center of the probe's range of motion.

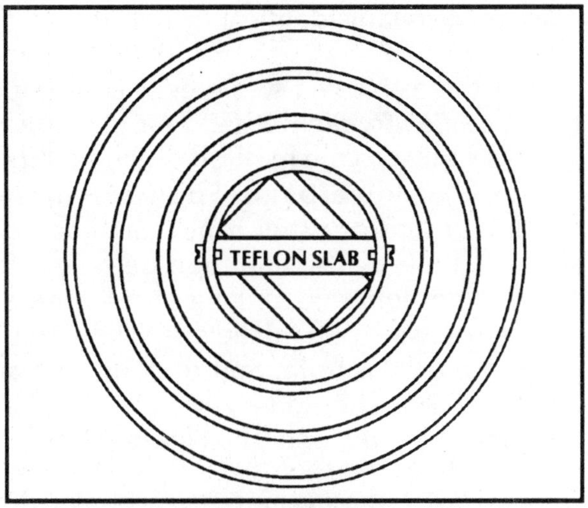

Figure 8-61. Feed Modification for Detection of Circularly Polarized Signals. *A linear feed can detect right and left hand circularly polarized signals if a slight modification is made in the circular waveguide in front of the transition to the rectangular waveguide. A teflon slab inserted at a 45° angle in the circular waveguide allows reception of circular polarity, albeit with losses on both circular and linearly polarized signals.*

Setting the Focal Distance and Centering the Feed

Installing the feed precisely at the phase center of either a prime focus or offset fed dish is critical to system performance. This involves correctly setting the focal length and, in the case of prime focus dishes, aligning the feed along the dish axis. In order to maximize the performance of the entire system, care should be taken to ensure that this is accomplished (see Figure 8-62). This is especially critical with Ku-band broadcasts where the wavelength is one third that of C-band microwaves. Small misalignments are three times as critical at Ku-band and cause three times the reduction in signal strength compared to that at C-band.

If the feed is not correctly located, gain will be lower and additional noise will potentially be detected because the feed would be illuminating more than its intended target, the dish surface.

Figure 8-62. Setting the Focal Length. *Setting the correct focal length is very important in optimizing reception of satellite broadcasts. This measurement should be taken from the dish surface to the front of the feed throat. The manufacturer of each type of feed generally specifies the precise point at the throat from which to measure the focal length. (Courtesy of Brent Gale)*

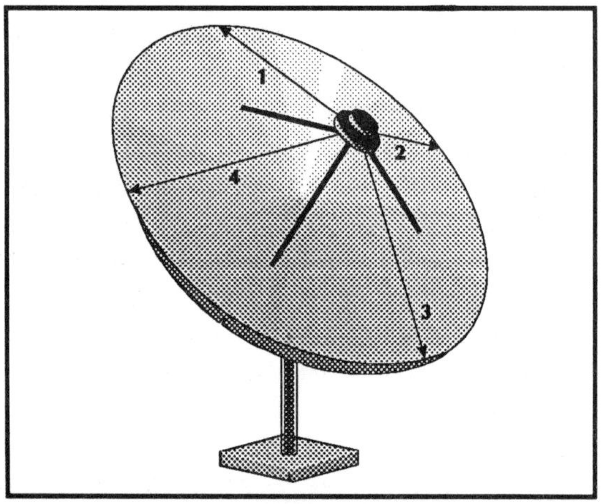

Figure 8-63. Centering the Feed. *A feed can be easily centered by making sure that three or four measurements from points around outer edge or rim of the dish to the feed are of equal distance.*

Always follow the manufacturer's installation instructions for measuring focal length. Each dish has a unique f/D ratio and focal length. Problems can arise with a poorly assembled dish. To illustrate, if a 1 meter reflector has a 12 inch (30 cm) focal length, the distance from the center surface of the dish to the front edge of the feed must be 30 cm. If the dish were warped this could change the position of the focal length and cause a marked impairment in reception.

Centering the feed is as critical as setting the focal distance. At least three methods can be used:

1. Measure from the outer lip of the reflector to the throat of the feed from four different points at 3,6,9 and 12 o'clock positions. The resulting measurements should be very similar in each case (see Figure 8-63). If not, the feed should be adjusted until the measurements are nearly identical.

2. Cross the dish using two pieces of string at right angles to each other over its surface. This is similar to the method outlined above to check for warpage. Spun aluminum, hydroformed or other reflectors having a hole in their center can then be tested for feed alignment by stringing two additional pieces of twine behind the dish. They should cross over this hole. Sighting along both cross- points should target the center of a properly aligned feed (see Figure 8-64).

3. Use an instrument called a focal finder. One end inserts into the mouth of a feed and the other end has a telescoping rod that should strike the dish at its center (see Figure 8-65). If the tip of the focal finder is not correctly aligned, it is a relatively simple matter to re-adjust the feed position until it does.

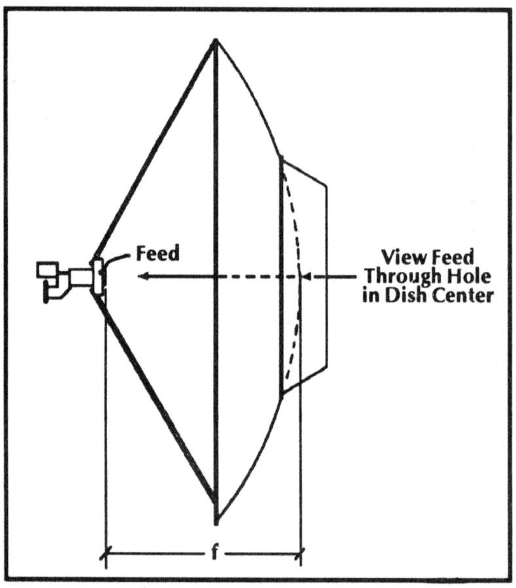

Figure 8-64. Sighting a Stringed Dish. *When looking through the hole in the center of many dishes to the point of crossed strings, the feed should be located directly on center. This is similar to sighting through cross hairs on a gun. It is crucial in this process that the strings span the full diameter of the dish so they will cross in its center.*

Centering a feed can be simplified when using a buttonhook feed support. Three small cables with S-hooks can be attached to the feed at the 12, 4 and 8 o'clock positions for stability. The other ends of these cables are then attached with metal clips that slide onto the outer lip of the reflector. This "wind kit" makes centering the feed by tightening or loosening the different threaded eye hooks easier.

All these efforts could be in vain if the correct feed had not been chosen. Each feed is designed to function optimally at a rated f/D ratio. Tests have shown that a mismatch between feed and dish could cause over a 1 dB drop in performance. That extra gain can be the difference between having a picture or no picture in a digital system or a good picture or a sparklie picture in a weak footprint area or when using a small aperture dish.

When all these adjustments are completed the plastic cover that fits over the throats of the feed must be installed. Such covers that have small holes that allow any condensed water to escape and protect the throat of a feed, should always be used. If a feed is not capped, wasps or other creatures may build nests inside and ruin reception.

It is also wise to use an LNB/feed cover when the installation and all waterproofing has been completed. This protection against the elements provides extra insurance against water intrusion.

Figure 8-65. Using a Focal Finder. *This photo shows a focal finder inserted into the throat of a feed. It points directly at the center of the dish as it should when the feed is correctly aligned. If a tripod instead of a button hook feed support had been used, the focal finder telescoping arm would have extended all the way to the center point of the reflector. (Courtesy of Brent Gale)*

INSTALLING ACTUATORS

Step number one in installing actuators is to read instructions. No one knows better than the manufacturer how many problems a poorly installed actuator can have. The instructions clearly spell out what mistakes to avoid as well as correct installation methods.

Mechanical Assembly

Once the dish is mounted on its pole, the actuator arm can be attached. Both ends of the arm must be bolted to the reflector and mount pivot points with ball joints in order to allow for lateral movement and to avoid binding. This is crucial because lateral forces on the inner tube can make it bend and seize up. These forces can also cause deterioration and potential failure of the internal O-ring seal that protects against water and dirt entry between the inner and outer tubes.

The arm should also be attached so it forms a minimum of at least a 30° angle with the rear surface of the dish at all points in its sweep. Having a smaller angle causes the motor to work harder because the arm has less leverage. Actuator arms vary in length from 12 inches (30 cm) and less to over 48 inches (over 120 cm). The larger and heavier the reflector, the longer the arm required to provide adequate leverage to easily move it. This arm should be installed as parallel to the plane of rotation as possible to minimize forces which could cause it to bend (see Figure 8-66).

Be aware that the installation of actuators on the west coast of North America differs from systems on the east coast. There are few C-band satellites west of Satcom C5 at 139°W while many satellites that relay video signals are clustered in the east. For example, the longitude of Los Angeles is 120°W, a difference of only 19° from Satcom C5. New York on the opposite side of the continent is located at 74°W, a full 65° from the same satellite. By comparison, a site in New York is only 5° removed from Spacenet II (Satcom SNII) at 69°W.

An actuator installed on the west coast should be attached to the left side of a dish from the perspective of an observer at the back. During installation, the clamp on a fully retracted actuator should be tightened only when the dish is swiveled as far west as possible. The reflector should then track from the far eastern satellite to a westerly position about 25° past true south. When installing an actuator on the East Coast, the actuator should bolt to the right rear side of the dish and the end of the fully retracted arm should face the reflector as far east as possible. If the actuator is installed on the wrong side in either the far east or west, the end result could be a bent tube or a system that cannot receive all satellite transmissions.

Install the actuator with the above details in mind. First, make sure that the actuator tube is fully retracted. Next the actuator clamp should be positioned so that the antenna can be aimed just past its most westerly or easterly target, depending upon which side the actuator is mounted. This is accomplished by allowing the actuator tube to move in its clamp while the antenna is rotated into position. Then tighten the collar bolts after the antenna is properly positioned.

Figure 8-66. Attachment of an Actuator Arm. *An actuator arm attaches to the mount at one end and to the dish frame at the other. As shown here, both joints can freely rotate and the upper joint has lateral movement so binding during motion is prevented.*

There is an exception to the above procedure. A movement called rock-over can occur if an oversized actuator has been installed or if the actuator tube is accidentally fully extended as a result of action by a user or the wind. Rock-over prevents the actuator from retracting and will usually bend the tube as well as a portion of the dish or its mount. During installation, align the dish beyond the lowest satellite and make sure that the actuator has mechanical limits.

Most ball jack actuators can exert 1500 pounds (700 kg) or more of force. Acme arms can generate about 750 pounds (350 kg) force. This force applied via the reflector outer rim could easily cause damage to surrounding objects. Be sure that the dish has a full range of movement without coming into contact with any obstructions such as trees or fences. In addition, the actuator must not bind on any part of the mount or reflector over its full range of movement, even past the preset electrically programmed east and west limits. Enough slack cable should be left on the actuator so that the antenna can move through the entire arc without pulling this cable and so that a drip loop can be created.

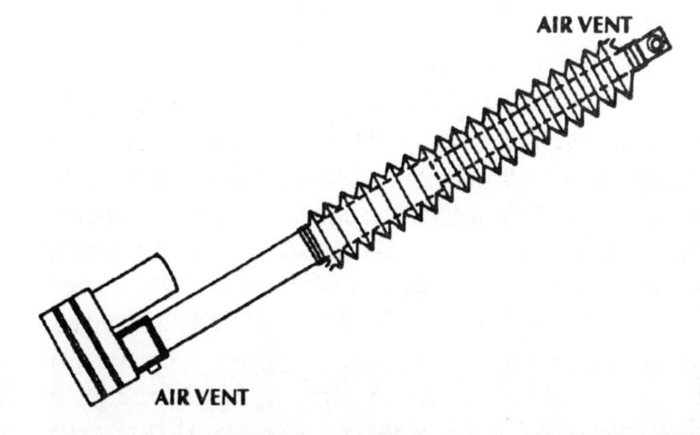

Figure 8-67. An Actuator Boot. *A neoprene actuator boot should have air vents at top and bottom and should attach securely to both the inner and outer actuator arms.*

Preventing Water Damage

Water is the chief enemy of actuators. If it manages to accumulate in the motor housing or in the gear box it can cause corrosion. More immediate damage could result if water freezes so that motors and gears seize.

Actuators should be mounted according to the manufacturer's instructions so that the motor is facing upwards and the drain holes under the housing are facing down. If there are no drain holes, drill one or two 5 mm openings into the housing at the lowest point. This will allow water which enters by condensation to escape before it accumulates.

Another place where water can enter is the opening between the outer and inner sleeves. Most manufacturers provide a rubber fitting, called a shaft wipe, to protect this opening. Further protection can be realized with a neoprene accordion sleeve. Because neoprene is transparent to sunlight, it permits any moisture to evaporate. There should be a vent hole at both ends of this sleeve to allow water and air passage so that before any water accumulates it can evaporate and escape (see Figure 8-67). If a rubber boot is used to cover the motor housing, make sure that it also has large drain holes. If not, it can do more damage than good.

Actuators are responsible for a great deal of mechanical work. If not properly installed, failure is likely. An estimated 75% of all service calls on larger dish systems have been for actuator related problems.

Horizon-to-Horizon Mount/Actuator

A horizon-to-horizon mount has a built-in actuator. This combination of mount and actuator is much simpler to install than the linear type and is also much less susceptible to water damage. The reflector can first be affixed to the mount which is then lifted onto the pole. In this case the weight of the mount must be supported so the dish will not be distorted. Alternatively, the mount can be directly lifted onto the pole and then the dish can be attached to the mount.

Wiring the Actuator

Most direct burial, multi-run cables provide the necessary five-wire conductor needed to connect actuators. This includes two #14 gauge wires carrying typically 36 Vdc to power the motor. If the east button is pressed on the internal control box and the dish moves west, or vice versa, reversing these wires will correct the situation. Three shielded #20 gauge wires are used to send the counting pulses between the actuator sensor and indoor control box (see Figure 8-68). These wires are shielded and grounded in order to prevent spurious voltage spikes from being mistaken for counting pulses when using controllers that expect similar pulses from Hall effect transistors or reed switches.

It is important to use the seals provided to protect the points of cable entry against water intrusion. These should be waterproofed with a material such as coax seal.

Safety Considerations

Installing and repairing actuators is a potentially dangerous activity. Actuators operate with the most current and voltage of any TVRO component, from 1.5 to 6 amps of current and typically either 24 or 36 Vdc. If a live circuit is touched by an installer who, for example, has wet feet and happens to be at ground potential, the shock will certainly be uncomfortable and possibly deadly. It takes only one milliamp of current to override the heart's electrical pulses.

The key to avoiding electrical shocks is to take steps to not be at ground potential. Always carrying a rubber mat to work on is a good habit to develop. If the old rule "always keep one hand in your back pocket when working on live circuits" is followed, the chances of receiving a fatal shock will be reduced to near zero. Never work on the electrical components of a system if an electrical storm happens to be nearby.

Using an ac power strip with a built-in circuit breaker known as a ground fault interrupter is an additional safety precaution. The satellite receiver, actuator and TV should all be connected via this outlet. If a ground problem such as a short circuit through a person's body develops, the breaker will open before any harm is done to either the person or the electronic equipment.

Actuators also have the potential of being a fire hazard because they draw a relatively large amount of current. It is generally a wise decision to install actuators as well as satellite receivers in a non-flammable environment having adequate ventilation. Actuators have been known to overheat and melt carpet fibers.

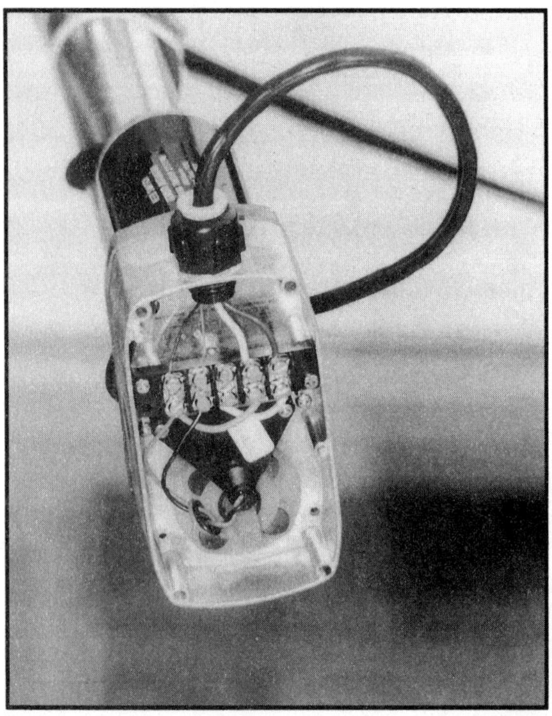

Figure 8-68. Wiring an Actuator. *Five wires connect this actuator motor/sensor to the indoors controller. Two of the heavier gauge wires on the right are for motor power. If they are interchanged, the west/east movement will be reversed. The ground and two lighter gauge wires on the three left terminal strips are used for control voltages or pulses. Notice how the incoming cable is sealed against water entry by a tightly fitting plastic cap. (Courtesy of Brent Gale)*

Retrofitting for Inclined Orbit Tracking

Both polar and az-el mounts can be retrofitted with actuators to track inclined orbit spacecraft. The simplest and most economical method is to motorize the declination of a polar mount. A single-axis system in which the elevation axis on either a polar or az-el mount is varied is also relatively simple but subject to certain geographic restrictions. Two-axis motor systems function well and accurately in any situation. These could take the form of two actuators on an az-el mount, one on each of the azimuth and elevation axes, or a second actuator that varies the dish's declination on a polar mount.

The tracking movement required depends upon the diameter of the receive dish as well as the difference between the satellite and receive site's longitudes. Smaller diameter dishes have wider beamwidths and so are somewhat less sensitive to satellite motion. When targeting an inclined orbit satellite, a dish with as small as possible a diameter will require less movement but at the expense of gain. The objective in making these choices is, of course, to receive a signal of adequate strength.

ELECTRICAL CONNECTIONS

Poor wiring techniques or improper connections can ruin an installation. A poorly grounded cable or a leaky connector may be all that is necessary sufficient to ruin reception of a satellite broadcast. Installers should be familiar with the correct techniques for installing all connectors that may be encountered. Fortunately, only a few types of connectors and cables have become standard in the satellite TV industry, so acquiring the necessary tools and knowledge can be relatively simple.

Connector Types and Techniques

The most familiar types of connectors encountered include F-connectors, RCA or phono jacks, moly plugs, the solderless lug terminal, Skotch Locks and SCART connectors. Other varieties often encountered in video accessory systems include K10, K14, BNC and DIN connectors. Each of these has certain uses and may require special tools to be properly installed.

F-Connectors

F-connectors are the industry standard used for attaching coaxial leads to the LNB, satellite receiver and television sets (see Figure 8-69). There is a variety of types used for both RG-6 and RG-59 coax. Some come with the crimp ring as a separate part. These connectors must be correctly installed, particularly when working with higher frequency signals.

F-connectors are attached by first stripping off about 3/4 inch (2 cm) of the outer insulating layer. The ground braid is next trimmed to a length of about 1/8 inch (3 mm) and folded back over the outside insulation past the 3/4 inch point. Then about 3/8 inch (10 mm) of the white center insulating jacket is removed from the inner wire. Use a knife and scrape the inner conductor wire clean to guarantee that a good electrical contact will be made. The crimp ring, if a separate part, is twisted on before the body of the connector. The connector should be twisted onto the coax so that the inner white dielectric jacket is flush with the inner shoulder of the F-connector. Finally, a crimp tool is used to bind the crimp ring onto the ground wire and outer sheath. This process is outlined in Figure 8-70.

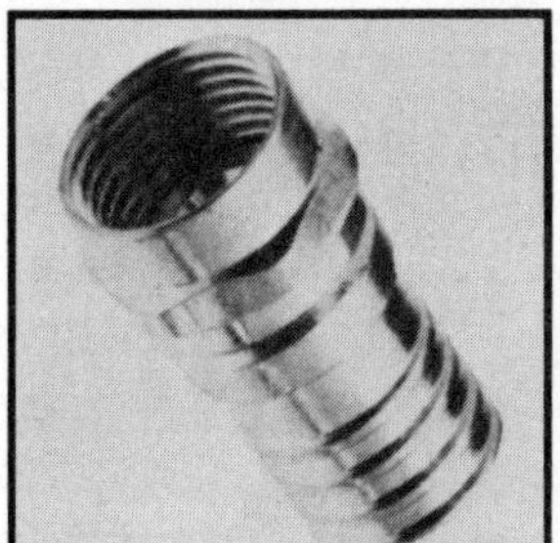

Figure 8-69. F-Connector. *F-connectors are used with coaxial cables such as RG-59, RG-6 or RG-11 which are rated for operation at frequencies below 2,000 MHz.*

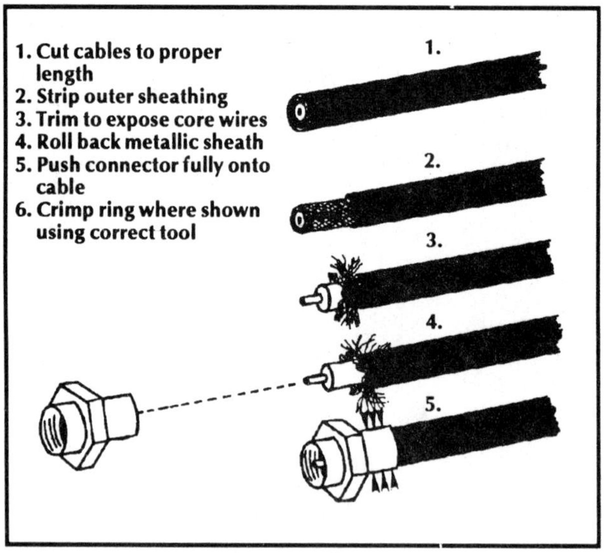

Figure 8-70. Installing F-Connectors. *This illustrates a step-by-step method to correctly install F-connectors. During this process, it is important not to score the center conductor wire because this would affect the current flow at the high frequencies used in satellite TV.*

When installing an F-connector, it is crucial that a crimp tool and not simply a pliers be used for the final step. This special tool compresses the connector onto the jacket equally on all sides. A pliers would distort the shape of the inner insulating jacket, alter the electrical characteristics of the cable and result in signal losses due to impedance mismatches.

The use of push-on F-fittings should be avoided. They do not provide as tight an RF connection as threaded connectors and can potentially let radiation leak into or out from the attachment point. They are also easier to pull off accidentally than the threaded type, for example, when furniture is rearranged during house cleaning.

F-connectors can be joined together with couplers called F-81 barrels that have a female lead at either end. This is not a recommended practice except when absolutely necessary as it causes some loss of signal at the junction and creates another possible route for moisture entry.

Tools Required: Scalpel, Wire Strippers, Cutters, Crimping Tool

Phono Connectors or RCA Jacks

RCA jacks are often used for relaying audio and video baseband outputs from receivers to VCRs, TV monitors, or external modulators. They should never be used for higher frequency signals because they are not shielded as well as F-connectors. Most phono connectors must be soldered onto a dual run wire or coax. A simple way to avoid the difficulty associated with soldering is to use phono-to-F-connector adapters. Varying lengths of cables having pre-attached phono connectors can also be purchased at audio or electrical supply stores.

Tools Required: Scalpel, Wire Strippers, Cutters, Soldering Iron, Solder.

Scotch Locks

Scotch locks are another similar type of connector for joining two wires together. These crimp-on connectors are filled with a flooding compound which is squeezed out during the crimping procedure. A good mechanical and well-protected electrical connection results. These connectors are ideal for use in attaching polarizer wires.

Tools Required: Stripper and Crimper

The Solderless Lug

Solderless lugs are designed to be used with screw-on terminal strips that are often used as rear panel outputs on receivers. The wire fits in one end and is either crimped down with a standard tool or soldered in place (see Figure 8-71). Such lugs are recommended for attaching wires to screw-on terminals. Bare wires can fray and, in time, even short out to adjacent terminals. As a minimum, if lugs are not used, wires should be twisted and tinned before being screwed onto a terminal.

Tools Required: Wire stripper/cutters

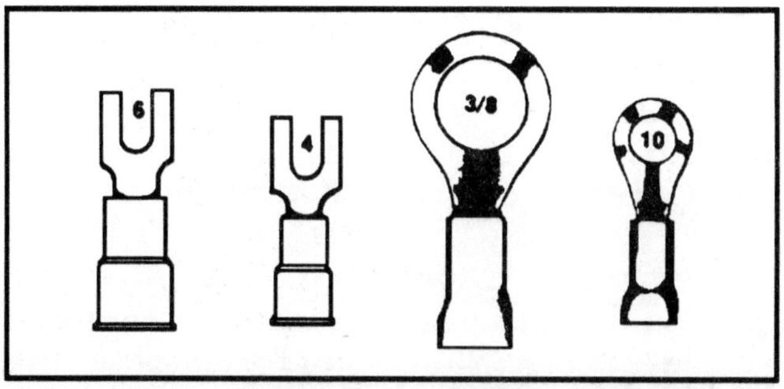

Figure 8-71. Solderless Lugs. *This type of connector can be used on any equipment that has a screw-type barrier strip. Attaching conductors to this connector will prevent stray wires from touching an adjacent terminal and shorting out.*

SCART Connectors

The SCART or Euroconnector ("Peritel" in France) is used on virtually all new European television receivers, other video equipment and some satellite receivers, especially those manufactured in Europe. The pin assignment is uniform among manufacturers, although in some cases not all pins are used.

The SCART fitting can be used to interface between decoders and televisions. The unclamped baseband video from the TV set is routed through the decoder and back into the television.

Tools Required: Stripper, Soldering Iron, Solder

Other Connectors - DIN, BNC, PL259

DIN connectors, soldered to very thin wires on a multi-wire cable, are standard audio/video fittings. BNC connectors are commonly used with video recorders, generally as video in and video out. PL259 connectors have been used on older brands of video recorders.

Tools Required: Scalpel, Wire Strippers, Cutters, Soldering Iron, Solder.

Additional Pointers on Installing Connectors

When stripping insulation or ground braid from cables and wires, be careful not to score the internal conductor. This would weaken it and could cause a break at some later time. Also, be careful, especially when peeling the outer insulation off coaxial cables, that the tiny pieces of wire do not fall into a vent in the actuator controller or receiver. Such wires could easily short an internal component and burn out the unit.

As a general rule, never splice cables. However, some cables can be extended with the appropriate adapters joined to fittings on both pieces of cable. If the point of union is exposed to weather, use both coaxial sealant and a rubber boot or shrink fit tubing.

Adapters are also available to join different types of connectors. For example, an RCA to F-connector adapter has a male F-fitting at one end and a standard phono jack at the other. This particular adapter is often needed when hooking up external modulators. It is much easier to make a cable with F-fittings at either end and use an adapter than to install a phono jack at one end and a F-connector at the other.

Whenever cables are terminated in connectors or adapters care should be taken to eliminate any forces between the cables. This can be accomplished by either leaving slack cables on both sides of the unions or by using devices such as wire ties to hold the cables and counteract these forces.

The Final Wiring

The wiring sequence to be followed depends upon how the final adjustments of dish position will be made as well as the type of system being installed. If a test setup is used at the dish for tracking adjustments, temporary wiring is required. If the dish is to be aligned using walkie-talkies to communicate from the indoors receiver to a person outdoors, the final wiring can be completed without an intermediate step.

An alternative to using walkie-talkies while still being able to avoid wiring a temporary set-up at the dish is to use an remote control extender when a two-run coaxial ribbon is installed. Two-run ribbons have an extra coaxial cable that may later be used to hook up a Ku-band LNB for dual-band reception. This extra run can be attached to the TV output on the satellite receiver and to the remote extender and a small TV at the other end.

In those cases where the dish will be adjusted using only the indoor satellite receiver, the LNB and polarizer should be connected before it is raised into position to track the arc. The electronics at the focus are more easily reached with the dish angled down to either horizon.

All receivers have an F-connector input for the IF signal from the LNB which falls in the range of 950 to 1450 MHz (950 to 1700 MHz in European systems). Every coaxial cable connector should be carefully examined prior to mating to ensure that the center pin has not been broken off and that it is centered and properly extended. There would be no electrical contact if the pin does not extend far enough, and damage may occur if the pin is not centered or is extended too far. Usually LNBs damaged by off-center pins are not covered under warranties. It can be very difficult to locate a poor connector once an installation is completed. One final point to remember. The F- connector output from the LNB should be waterproofed with a material similar to coax seal only when the installation is completed.

The rear panel usually has a terminal strip with three connections, +5 Vdc, pulse control and ground, for mechanical polarity power and control (see Figure 8-72). The three polarizer leads at the dish should be twisted tightly together and joined with Scotch locks. If not, they should be soldered and then covered with wire nuts or other types of crimp-on connectors. These are then covered with "shrink spaghetti" which, like shrink tubing, contracts when heated by a small flame or hair dryer.

Actuators are attached to either a separate controller or directly to receivers having built-in actuators. The five-wire actuator lead should be hooked up and double checked to ensure that a proper connection has been made.

Experience suggests that any spare or extra wires at the feed assembly, actuator motor or satellite receiver should not be cut off. These can be taped to the multi-run jacket and kept as spares for possible use at a later date.

Safety Considerations

Satellite receivers typically draw up to a maximum of one ampere of current to provide a range of regulated power outputs. A mechanical polarizer usually requires +5 Vdc while internal circuitry most generally requires a regulated +12 Vdc, +18 Vdc and often +23 Vdc. Other voltage outputs are not uncommon.

While it is rare that these voltages are sufficient to cause serious harm, the same precautions taken when installing actuators should be followed. Better be safe than sorry. Many receivers often have built-in actuator controllers and, therefore are potentially dangerous.

Protection Against Power Surges and Voltage Fluctuations

A direct lightning hit is a rare occurrence that may not only destroy the electronics in a satellite system but can also easily set a house on fire. Luckily, most hits are indirect ones that only cause power surges. Standard line power provided to many rural regions is often not very "clean" and produces similar but much smaller voltage transients that can cause a receiver or actuator microprocessor to lose its memory.

Power or voltage fluctuations are common in rural areas where the majority of receivers are used. These can range from typically 10 to 20 Vac but can be as high as 30 to 40 Vac during storms or times of peak summer heat. These variations in mains voltage are more prevalent if a system is near the end of a power line or if there are high-current loads between the installation site and the power source.

Whatever the source of voltage spikes or fluctuations, it is better to take precautions during installation than to be forced to repair components at some later date. While most receivers and actuators have minimal built-in protection in the form of voltage regulators and filters. it is best to take some preventive measures during installation.

Grounding Methods

A satellite receive system must be grounded not only to protect it from power surges but also to avoid problems in video signals that can crop up with poor grounds. When a dish is mounted on a pole and at a distance from the home, it is best to connect to a ground rod near the pole. When a dish is mounted on a wall, roof or eave or next to the home, the optimal strategy is to connect to the building ground, assuming it is well made.

Ground Rods

Ground rods are constructed from long narrow poles of copper-clad steel that are driven into the ground to provide an escape route for lightning. The effectiveness of "lightning rods" depends upon the ability of the soil to pass current, the soil resistance, and details of their connection to the dish and mount. A 6 foot (2 meter) ground rod is usually sufficient to handle most lightning strikes (see Figure 8-73).

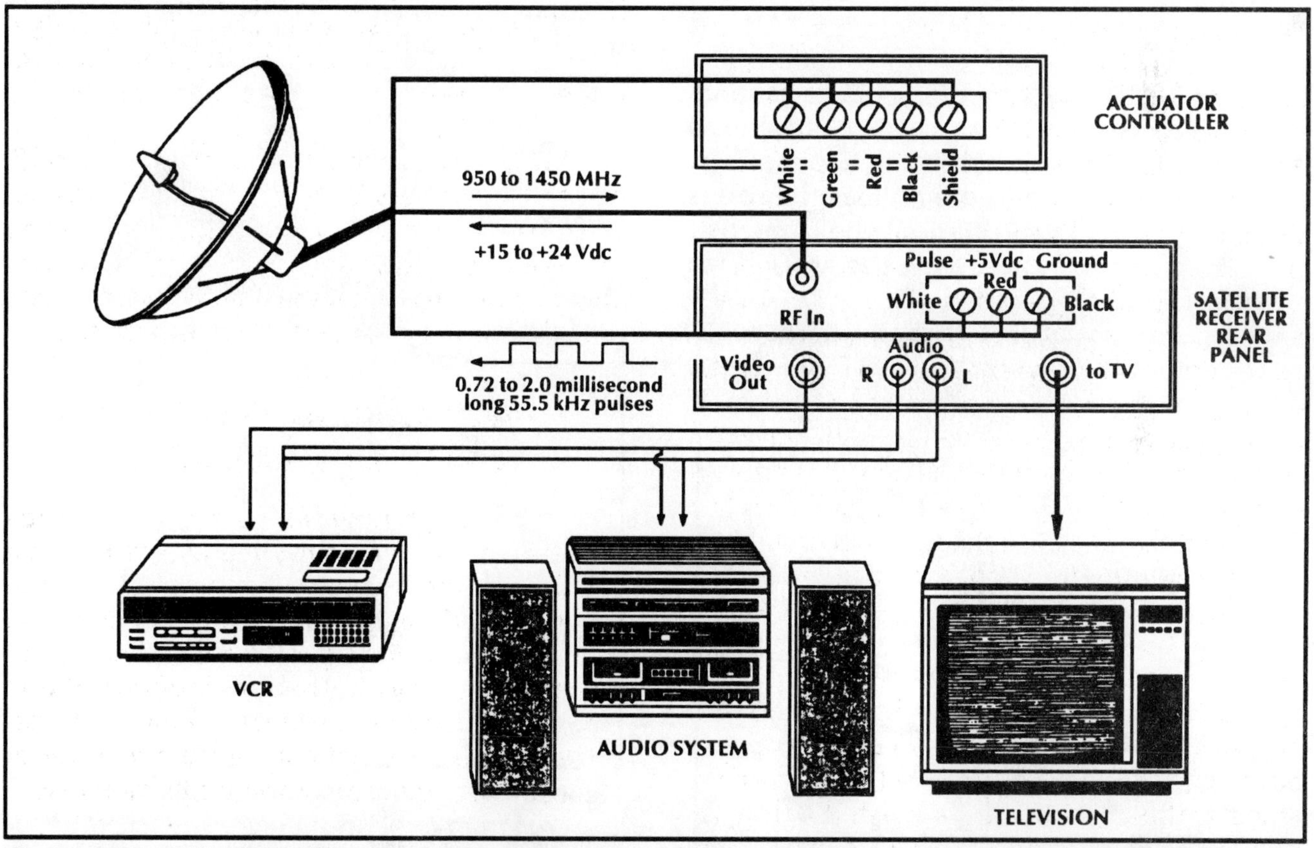

Figure 8-72. Typical Satellite TV System Wiring. *A satellite receiver can be connected to a complete audio/video home entertainment center. This illustrates how a system is wired. The receiver relays voltage to the LNB and receives RF signals on a single coaxial cable. Three wires send 5 Vdc pulses every 18 milliseconds (which equals 55.5 kHz hence the name 555 timer chip) to the polarizer. The position of the servo motor which moves the probe is controlled by the width of these pulses which vary from 0.72 to 2 milliseconds. The actuator controller transmits 6 Vdc to and receives pulses on the three sensor wires. It also sends 36 Vdc via two motor wires to control movement of the actuator. The receiver here has a built-in stereo processor which hooks directly into the home audio receiver via its output audio terminals. Alternatively, the baseband output could be used to drive a stereo processor. The VCR can be connected to either the TV set or video output ports.*

Most soils have sufficient conductivity so that a well-made ground connection will effectively shunt voltage transients caused by nearby lightening strikes. Soils which have little clay or loam but are mainly rocky may need to be treated with materials like rock salt, copper sulfate or magnesium sulfate to increase their ability to pass current.

A 6 to 8 foot (2 to 2.5 meter) lightning rod should be driven into the ground about 30 cm away from the supporting pole foundation. A 1/4 to 3/8 inch (6 or 9 mm) minimum diameter copper, brass or aluminum wire is then bolted onto the ground pole. A tight clamp, often provided with the rod, is used for making contact at the top end. This wire should follow a smooth curve; any sharp bends will impede current flow.

An effective method of planting a grounding rod is to use a pipe with a cap on one end as a driving tool. If the soil is difficult to penetrate, water can be added to the hole as the rod is driven. The top of the rod should be planted at least 6 inches (15 cm) below the ground for best results. This also protects the customer's lawnmower as well as the customer's feet from damage. A driver sleeve is also available for use with a standard jackhammer to plant the rod. This is a required tool in many cases.

Probably the best alternative to using a ground rod is to run a heavy wire, #8 gauge or larger, from the pole or dish into the building's ground point (see Figure 8-74). This should effectively shunt any excess currents to ground. Paint must be removed from these points of contact prior to attaching the ground strap so that a good electrical contact is made.

Grounding Roof or Wall Mounted Dishes

A ground block is used to connect to the building ground (see Figure 8-75). One side of the block is connected to the LNB coax, the other side to the dish. In addition, a ground wire should run from the block to the dish mount structure and the building ground point. All connections of this wire should be free of paint or any other materials that would impede current flow.

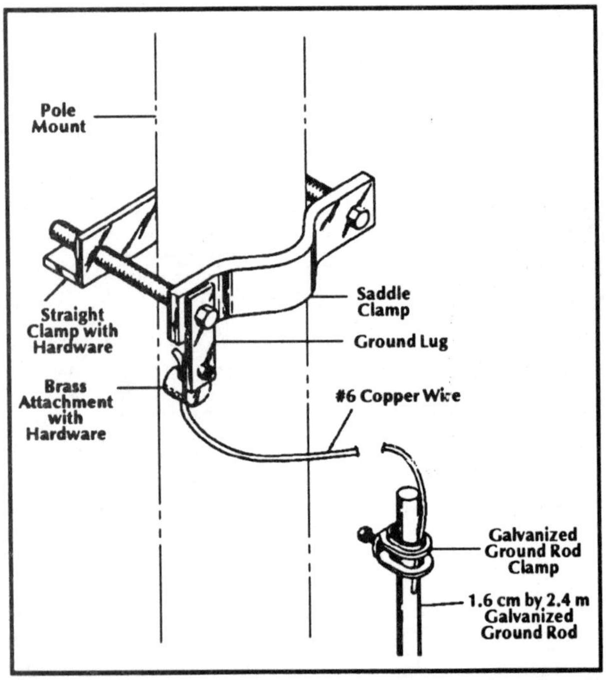

Figure 8-73. Grounding Kit. *Grounding hardware is available at many electrical supply houses. Number 4 or 6-gauge solid copper conductor is recommended. It is attached with brass hardware to the ground rod.*

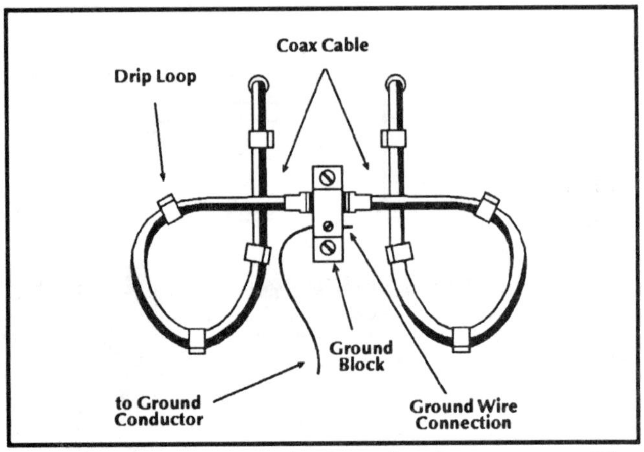

Figure 8-74. Ground Block and Drip Loop. *The coax run from the LNB should be attached to one side of a ground block. The other side connects to the satellite receiver. Each side has a ground loop that allows any moisture to drip away from, not towards, the ground block and F-connectors.*

Figure 8-75. A Surge Protector. *This type of surge protector has an internal power break that can be reset in case a strong voltage transient appears on the power line and trips the unit.*

Surge Protectors

Surge protectors can also be used to shield electrical components from voltage spikes (see Figure 8-75). There are many varieties available ranging from simple units which plug into a power line to more expensive ones requiring the installation talent of an experienced electrician. Any microprocessor-controlled receiver or actuator is similar to a home computer since its memory can be erased, interrupted or altered by power surges. A surge might cause channels to change, the dish to move unpredictably or a carefully programmed memory to fail.

Some surge protectors can handle up to 50,000 volts and are able to respond in microseconds. However, the least expensive brands may lose their ability to operate after only a few large spikes.

All satellite reception components should be joined together to one common ground as further protection against voltage spikes. Unless a separate ground line connects all components to a common point at the dish, the ground on a three-prong power cord must always be used. In those cases where only two-prong outlets are available, a ground relief adapter can be used. Paying good attention to ground procedures will also minimize the possibility of an annoying type of interference known as hum bars (see Chapter 14 for more details).

Constant Voltage Transformers

When a receiver often overheats and shuts down, a constantly high or low line voltage may be the cause. It may be necessary to install a constant voltage transformer to counteract this problem. This component can also be useful in improving the performance of TVs or monitors.

Protection Against Overheating

Actuators and satellite receivers generate enough heat to cause problems when not properly vented. At the very least, the electronics may not function at peak performance or could shut down or blow a fuse. In the worst situation, a fire may result. These components should always be placed on their feet above a flat surface, and a 3 inch (7.5 cm) air space should be provided above and to their sides. Care should be taken never to cover vent openings or install the receiver or actuator in a non-ventilated wall location or directly on a thick pile carpet.

DECODER INTERFACING

While hooking up a descrambler or a decoder to a receiver is not a difficult process, the interface between decoders and receivers can sometimes cause problems. Some details must be understood.

Clamped Or Unclamped Input?

The majority of descramblers need an unclamped composite baseband input signal to operate properly because receiver clamp circuitry can distort a scrambled signal. In most receivers, the clamp/unclamp switch is located on the back panel. In others, the clamp can be selected from a front panel switch or via the remote control when programming the channel parameters.

Baseband Or Composite Video?

In most cases, descramblers can decode either the unclamped baseband or composite video. The composite video is generally a better choice. The baseband signal is unfiltered and has not been de-emphasized. If a descrambler does not have any filtering and deemphasis circuitry, the colors and definition of the descrambled picture signal would be nonlinear and thus the picture would be somewhat too "sharp" in appearance.

ALIGNING a DISH onto the SATELLITE ARC

A dish can be aligned to receive signals from one of two locations: indoors or outdoors. When the final wiring is finished, a set of walkie-talkies can be used to communicate between one person at the receiver/decoder and another at the dish. Alternatively, a test setup can be wired at the dish, a recommended method if weather permits since there is less potential for mis-communication and just one person can complete the job (see Figure 8-76).

In either case the initial test setup should include a satellite receiver and a small TV set or monitor. While the dish and feed can be aligned using the receiver that is being installed, disconnecting a customer's TV and lugging it out to the site is not recommended. It is generally best to use an "old reliable" receiver that is known to be working to eliminate the possibility of having to troubleshoot new components. A more costly alternative is a test set, a combination TV set and receiver in one case.

A portable spectrum analyzer or lower cost signal detection instrument can be used to align the dish. Either of these can also be invaluable when fine tuning the system. A supply of 75 to 300 Ω transformers may be necessary if a customer's TV set is an older model that does not have F-connector attachments.

Aligning Fixed Dishes

Fixed dishes can be aimed at just one satellite knowing its elevation and azimuth bearing. The elevation and azimuth angles to any satellite from any location in can be determined from aiming software or calculated from the equations presented in Appendix B. Azimuth which is measured on a compass in degrees east of a north heading, must then be corrected for magnetic variation.

The process of aiming a dish at a satellite that broadcasts digital signals is different from targeting one that relays analog signals. An analog signal first appears as a slight alteration in the uniform pattern of white noise that appears on a TV screen. As aiming is improved, the signal becomes stronger and the picture more clearly defined. A digital signal is either strong enough to be "acquired" by a receiver or simply does not appear. It is an all or nothing situation.

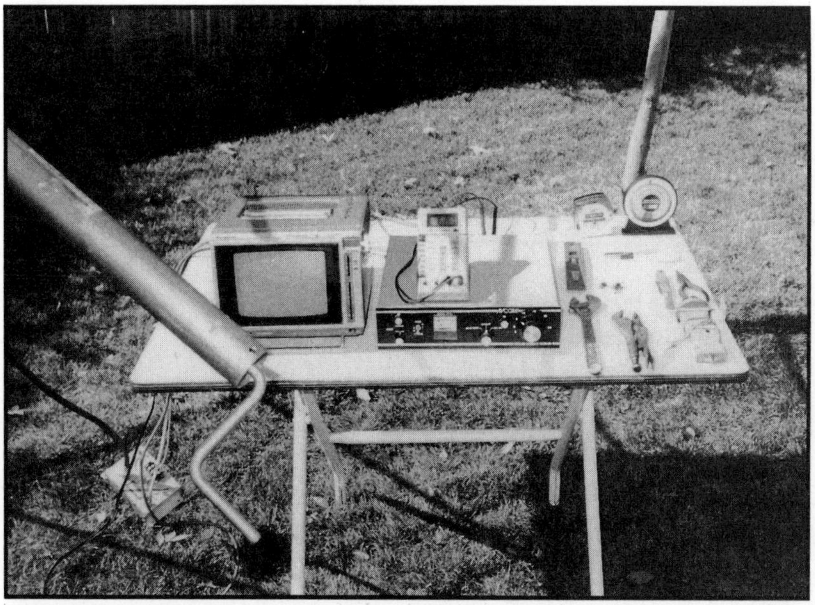

Figure 8-76. An Outdoors Installation Set-Up. *All the equipment required to align a dish onto the arc of satellites is set up on this table next to the dish. It includes a portable TV set, a satellite receiver (preferably one with a built-in signal strength meter), a volt/ohm meter, an inclinometer, a compass, a crescent wrench and assorted tools. During alignment the TV screen can be viewed for direct feedback on dish movements. (Courtesy of Brent Gale)*

The best strategy to follow when detecting either an analog or digital broadcast is to set the azimuth and elevation angles as well as the skew angle as accurately as possible before powering on. Skew angles for any site can be read from the graph in Figure 8-77. All that is required to read this graph is the site latitude and the difference between the satellite and site longitudes. (Remember that, as explained in Chapter 5, if the feed has been properly aligned onto the most northerly or southerly satellite in the arc when installing a polar mounted dish, the skew is automatically adjusted as the dish is tracked across the arc).

Then move the dish slightly back and forward and up and down until even the faintest picture is seen. For an analog transmission, adjust the aiming and then skew as well as possible to optimize the picture. A signal strength meter is useful in this process for pulling in that last decibel. When using a digital IRD an on-screen or stand-alone signal strength meter is a more crucial part of this process. If a meter is not used, it is virtually impossible to ascertain whether or not the dish has been "peaked" to a maximum C/N ratio because once the signal has been acquired the picture is nearly as perfect as it can become for the transmission at hand. This might be acceptable until a small pointing error creeps in or until rain fade knocks the C/N ratio below threshold.

When installing a fixed dish, it is still important to be familiar with methods of tracking the arc. In some situations where locating true south may prove to be difficult, the pointers below may be useful. Furthermore, many of the techniques and suggestions apply to both fixed and tracking systems.

Testing Components Before Setting Angles

By far the easiest method to test electronic components is to check each on a working system back at the shop. Taking the necessary time to do so in a familiar environment can eliminate many installation headaches. It is also wise to conduct a few pre-alignment tests to determine if the equipment being installed is working as expected. The most obvious is to ensure that the television set must be tuned to the same channel as the modulator output. Its quality can be checked by viewing a local over-the-air channel.

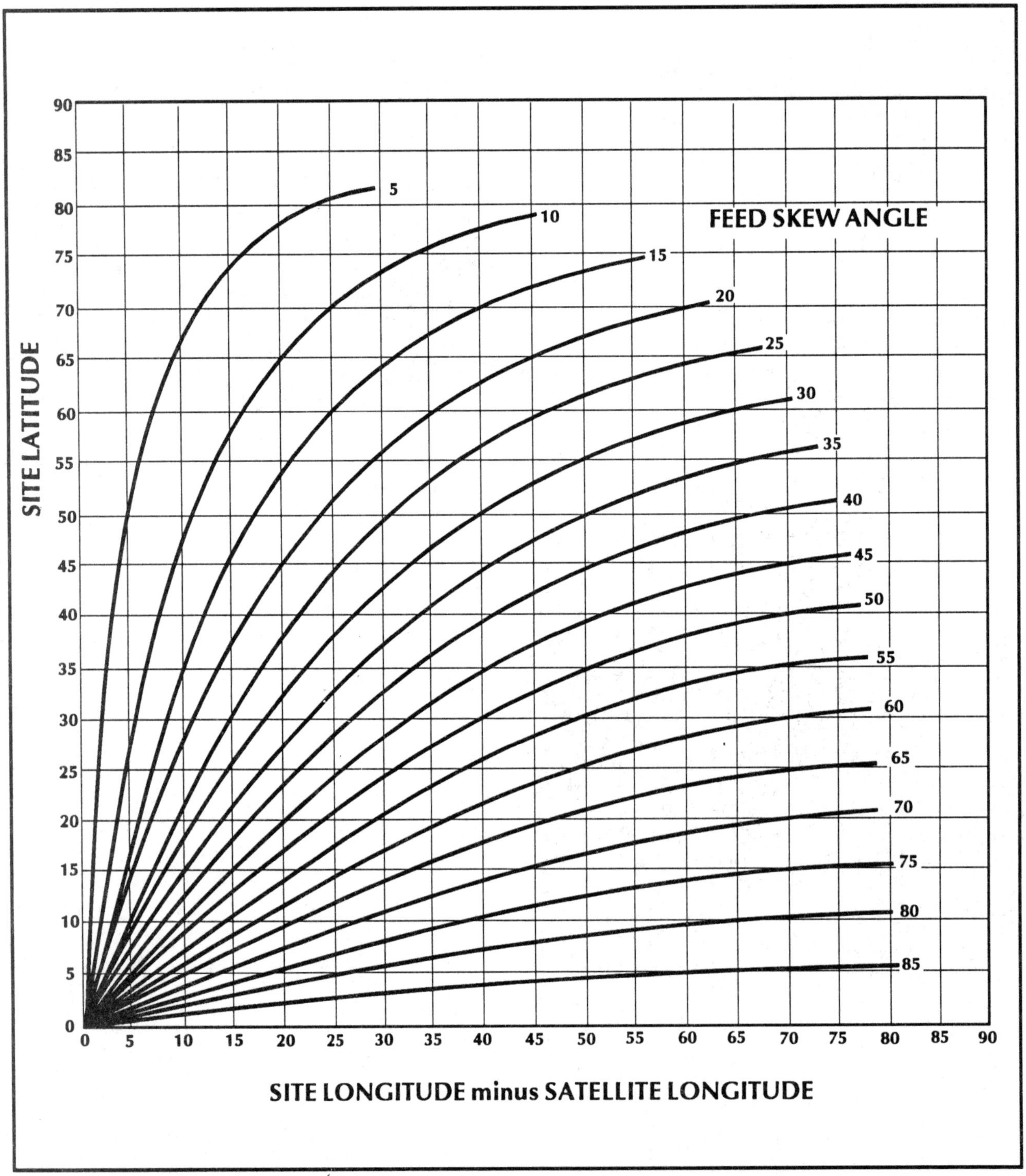

Figure 8-77. Feed Skew Angle versus Site Latitude and Difference between Site and Satellite Longitudes. *Skew angles are measured from the local horizontal plane. Sites in the southern hemisphere: feed rotation is counter-clockwise looking over the feed into the dish for satellites in the western sky and clockwise for satellites in the eastern sky. In the northern hemisphere reverse the sense of rotation. (Courtesy of Jim Roberts, Gourmet Entertaining)*

Next, attach a coaxial cable between the TV set input and the RF output on the satellite receiver. Set the audio mode selection to the mono position if there is a choice between mono and stereo. If the cable from the LNB to the receiver input is disconnected and if the receiver and TV are on, the screen should be black or solid white. If not, the receiver has a short or the cable between it and the TV is open. It is important to always turn the power off and unplug a receiver before disconnecting or hooking up any connectors since damage can be caused by shorting a center lead to ground.

Then hook up the LNB. White noise should be seen and a loud hiss heard. If not, the LNB is either defective or is not receiving power from the receiver. Alternatively, use a voltmeter to check for the necessary 15 to 24 Vdc at the receiver output and LNB input. Next check that the voltage at the LNB input is adequate. A 2-way, dc-passive power splitter that allows current to pass in both directions must be installed. Or disconnect the cable and measure the power between its ground and center conductor. If power is not reaching the LNB, the cable or one or both connectors are faulty.

Aiming a Tracking Dish

An actuator is required to align a tracking dish onto the arc of satellites. However, a simple alternative during this process is to remove the motor and its housing from the actuator arm and then to use a small hand crank to scan between satellites.

Three adjustments are required on polar mounted dishes to ensure proper tracking of the arc: north/south heading, declination offset angle and polar axis angle. These parameters should be set when the dish is at its highest position, namely aimed towards the most southerly or northerly satellite at the center of the arc.

North/South Orientation

A polar mount must have its axis aligned with the north/south axis of the earth in order to be able to detect all satellites in the viewable arc. This is easily understood by visualizing a dish at the equator. If it rotates in an axis aligned with the center of the earth, it will correctly scan the circle of satellites in the sky directly above from horizon to horizon.

Most dishes have a flat plane on the mount which can be used as a sighting reference. A hand-held compass with cross-hairs is the most effective type for lining up the mount with the north/south plane. Remember that a correction for magnetic variation is necessary. Add magnetic variation to the compass heading for west magnetic variations; subtract for east.

The Polar Axis Angle

The polar axis angle is set according to values outlined in Table 8-6. These values are within fractions of degree of the site latitude. Setting the polar axis angle equal to the site latitude would target a dish directly out into space along a plane parallel to one passing through the equator (see Chapter 5 for details). Using the "modified" polar mount angles yields better tracking of the arc.

Most mounts have one or two long threaded rods that are used to adjust polar axis angle. An inclinometer resting on the axis bar or back part of the mount is used to set this angle (see Figure 8-78).

Declination Offset Angle

The declination offset adjustment lowers the "sight" of a dish to the arc of satellites. Declination is greater in locations farther away from the equator. Table 8-6 shows how declination varies with latitude. These values can be used to set declination in either the southern or northern hemisphere.

TABLE 8-6. POLAR MOUNT AIMING ANGLES

Latitude	Polar Axis	Declination Angle	Apex Elevation
0.00	0.00	0.00	90.00
0.50	0.51	0.08	89.41
1.00	1.02	0.16	88.82
1.50	1.54	0.22	88.24
2.00	2.06	0.30	87.64
2.50	2.57	0.38	87.05
3.00	3.08	0.46	86.47
3.50	3.60	0.52	85.88
4.00	4.11	0.60	85.29
4.50	4.62	0.68	84.70
5.00	5.13	0.76	84.11
5.50	5.65	0.83	83.52
6.00	6.16	0.91	82.93
6.50	6.67	0.99	82.35
7.00	7.18	1.06	81.76
7.50	7.70	1.13	81.17
8.00	8.21	1.21	80.58
8.50	8.72	1.29	79.99
9.00	9.23	1.36	79.41
9.50	9.74	1.44	78.82
10.00	10.26	1.51	78.23
10.50	10.77	1.59	77.65
11.00	11.29	1.65	77.06
11.50	11.80	1.73	76.47
12.00	12.31	1.80	75.89
12.50	12.82	1.88	75.30
13.00	13.32	1.96	74.72
13.50	13.84	2.03	74.13
14.00	14.35	2.10	73.55
14.50	14.86	2.18	72.96
15.00	15.37	2.25	72.38
15.50	15.89	2.32	71.79
16.00	16.40	2.39	71.21
16.60	16.91	2.47	70.62
17.00	17.42	2.54	70.04
17.50	17.93	2.61	69.46
18.00	18.44	2.68	68.88
18.50	18.95	2.76	68.29
19.00	19.46	2.82	67.72
19.50	19.97	2.90	67.13
20.00	20.47	2.97	66.56
20.50	20.98	3.05	65.97
21.00	21.50	3.11	65.39
21.50	22.01	3.18	64.81
22.00	22.52	3.25	64.23
22.50	23.03	3.32	63.65
23.00	23.54	3.38	63.08
23.50	24.04	3.46	62.50
24.00	24.55	3.53	61.92
24.50	25.06	3.60	61.34
25.00	25.57	3.66	60.77
25.50	26.07	3.74	60.19
26.00	26.58	3.80	59.62
26.50	27.09	3.87	59.04
27.00	27.59	3.94	58.47
27.50	28.10	4.01	57.89
28.00	28.61	4.07	57.32
28.50	29.11	4.14	56.74
29.00	29.62	4.20	56.18
29.50	30.13	4.27	55.60
30.00	30.63	4.33	55.03
30.50	31.14	4.41	54.46
31.00	31.64	4.47	53.89
31.50	32.16	4.52	53.32
32.00	32.66	4.59	52.75
32.50	33.17	4.65	52.18
33.00	33.67	4.71	51.62
33.50	34.17	4.78	51.05
34.00	34.68	4.85	50.47
34.50	35.18	4.91	49.91
35.00	35.68	4.97	49.35
35.50	36.18	5.03	48.78
36.00	36.69	5.09	48.22
36.50	37.19	5.15	47.66
37.00	37.69	5.21	47.10
37.50	38.20	5.27	46.53
38.00	38.70	5.33	45.97
38.50	39.20	5.39	45.41
39.00	39.70	5.45	44.85
39.50	40.21	5.51	44.29
40.00	40.71	5.56	43.73

Latitude	Polar Axis	Declination Angle	Apex Elevation
40.50	41.21	5.62	43.17
41.00	41.71	5.67	42.61
41.50	42.22	5.73	42.05
42.00	42.72	5.78	41.50
42.50	43.22	5.84	40.94
43.00	43.72	5.89	40.39
43.50	44.22	6.00	39.83
44.00	44.72	6.00	39.28
44.50	45.21	6.05	38.73
45.00	45.71	6.11	38.17
45.50	46.21	6.17	37.62
46.00	46.71	6.21	37.08
46.50	47.21	6.27	36.52
47.00	47.70	6.32	35.98
47.50	48.20	6.38	35.42
48.00	48.70	6.42	34.87
48.50	49.20	6.47	34.33
49.00	49.70	6.51	33.79
49.50	50.19	6.57	33.24
50.00	50.69	6.62	32.69
50.50	51.19	6.67	32.14
51.00	51.69	6.70	31.61
51.50	52.18	6.76	31.06
52.00	52.68	6.80	30.52
52.50	53.18	6.85	29.97
53.00	53.67	6.89	29.43
53.50	54.17	6.94	28.89
54.00	54.67	6.97	28.36
54.50	55.16	7.02	27.82
55.00	55.66	7.06	27.28
55.50	56.16	7.10	26.74
56.00	56.65	7.14	26.21
56.50	57.14	7.19	25.67
57.00	57.64	7.22	25.14
57.50	58.14	7.26	24.60
58.00	58.63	7.30	24.06
58.50	59.12	7.35	23.53
59.00	59.61	7.39	23.00
59.50	60.11	7.42	22.47
60.00	60.59	7.47	21.94
60.50	61.08	7.51	21.41
61.00	61.58	7.54	20.88
61.50	62.07	7.58	20.35
62.00	62.57	7.61	19.82
62.50	63.06	7.64	19.30
63.00	63.55	7.68	18.77
63.50	64.05	7.71	18.24
64.00	64.53	7.75	17.72
64.50	65.02	7.78	17.19
65.00	65.52	7.80	16.68
65.50	66.01	7.84	16.15
66.00	66.50	7.87	15.63
66.50	66.99	7.90	15.11
67.00	67.48	7.93	14.59
67.50	67.97	7.96	14.07
68.00	68.47	7.98	13.55
68.50	68.96	8.01	13.03
69.00	69.45	8.04	12.51
69.50	69.95	8.06	11.99
70.00	70.43	8.09	11.48
70.50	70.92	8.12	10.96
71.00	71.41	8.15	10.45
71.50	71.90	8.17	9.93
72.00	72.39	8.19	9.42
72.50	72.88	8.21	8.91
73.00	73.37	8.23	8.40
73.50	73.86	8.25	7.89
74.00	74.35	8.27	7.38
74.50	74.84	8.29	6.87
75.00	75.33	8.31	6.36
75.50	75.82	8.33	5.85
76.00	76.31	8.35	5.34
76.50	76.80	8.37	4.83
77.00	77.29	8.38	4.33
77.50	77.78	8.40	3.82
78.00	78.27	8.41	3.32
78.50	78.75	8.43	2.82
79.00	79.24	8.45	2.31
79.50	79.73	8.46	1.81
80.00	80.22	8.47	1.31
80.50	80.71	8.48	0.80
81.00	81.19	8.50	0.30

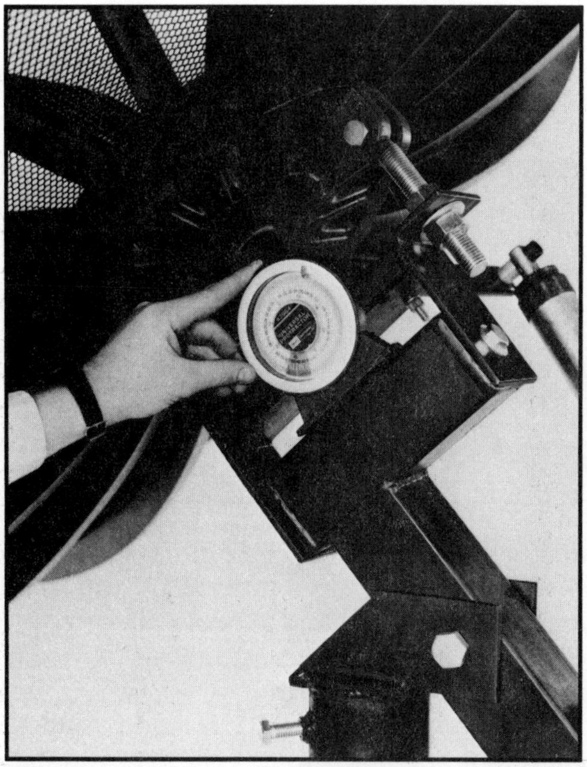

Figure 8-78. Setting the Polar Axis Angle. *The polar axis angle is set by placing an inclinometer on the polar axis bar and then raising or lowering the dish so the angle equals the site latitude. (Courtesy of Brent Gale)*

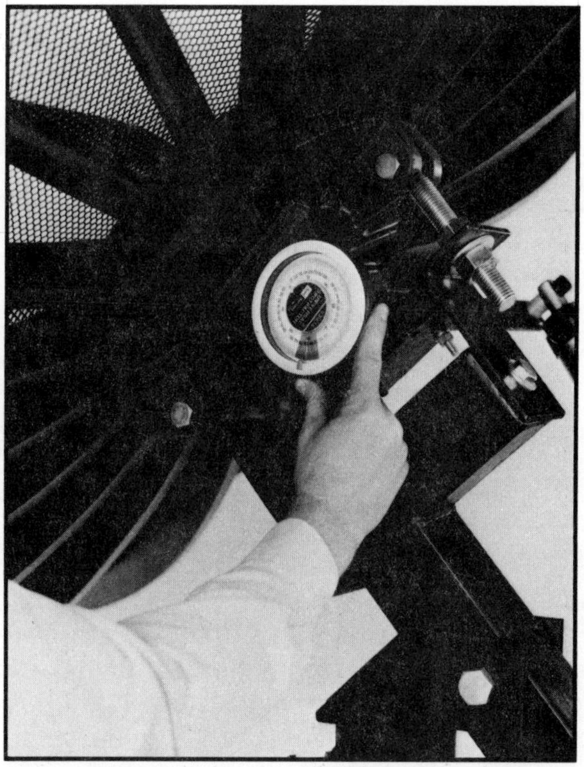

Figure 8-79. Setting the Declination Angle. *The declination offset angle is set by placing an inclinometer on a flat rear surface of the dish that is parallel to the line between the rims and adjusting the declination bolts until the reading equals the site latitude plus declination. (Courtesy of Brent Gale)*

Declination offset angle, like elevation, is measured with an inclinometer. The difference between two readings, one on the main part of the mount, the axis bar, and one on a flat spot on the back of the dish, should equal the declination offset angle. The easiest way to set declination offset is with an inclinometer placed on a back surface that is parallel with the face of the dish or via a flat board placed in a vertical direction spanning the dish rims (see Figures 8-79 and 8-80). This reading should be set equal to the sum of site latitude plus offset angle. For example, in the American city of Denver at latitude 40° north, the polar axis angle is set equal to 40° and the declination offset angle is 6.226° so the face of the dish should have a total elevation of 46.26°.

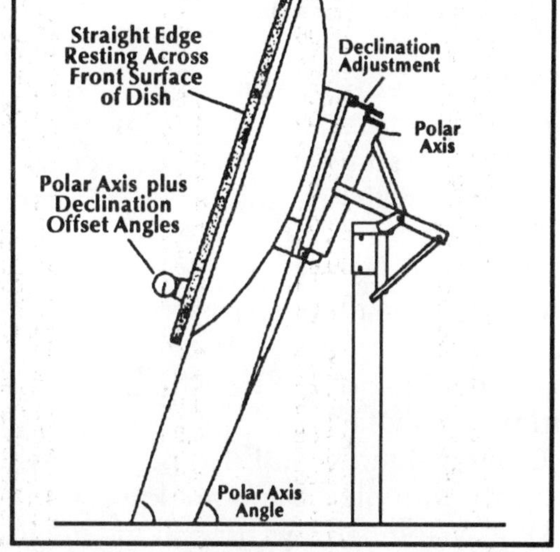

Figure 8-80. Using Dish Face to Set Declination Angle. *If a flat surface cannot be found on the rear of the dish or on the hardware that supports the declination adjustment, the declination angle can be set by placing an inclinometer on a flat board that rests the face of the dish.*

Figure 8-81. Arc-Set Alignment Tool. *This installation tool is set by "capturing" three critical angles which are copied from a properly aligned, local "mother" dish. It can also be used to troubleshoot installed dishes located within a 50 mile radius of the "mother" dish. It can free an installer from the need to fine tune the north/south position with a compass because one of the plumb adjustments allows for setting north/south orientation. (Courtesy of Gourmet Entertaining)*

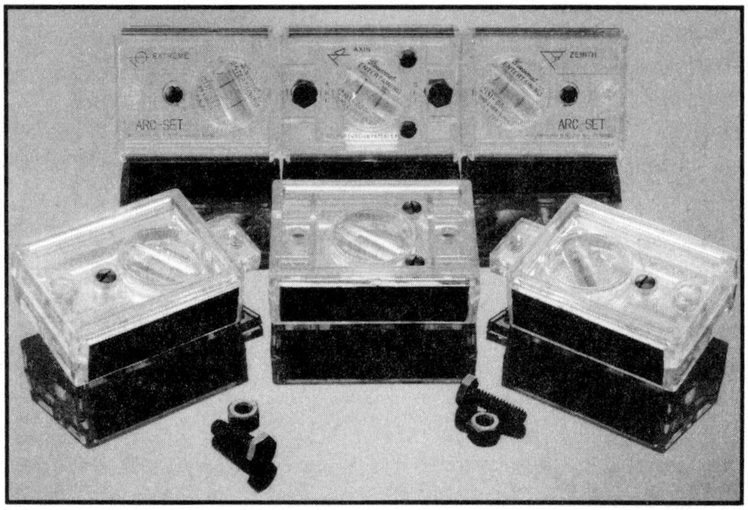

An Alternative Method to Set Angles

There are some alternatives to the method suggested above for setting the tracking angles. For example, the arc set kit consists of three levels that are set on a "mother dish" and then used to set angles on any other dish within a radius of approximately 50 miles (see Figure 8-81).

Aligning Onto the Arc for Analog Broadcasts

The process of aiming a tracking dish onto the arc to receive analog broadcasts is described here. At present just a handful of satellites worldwide relay MPEG-2 DVB satellite transmissions so tracking is necessary only for conventional broadcasts.

All connections and leads should be checked one final time before powering on equipment. In some instances, especially when using smaller dishes, the initial adjustments may be perfect and clear pictures immediately appear. This is less likely with larger diameter, more narrow beamwidth reflectors.

For dual-band, C/Ku systems, the arc can be more easily targeted by first aligning onto the C-band signals. In those rare cases when tuning to both digital and analog signals via a tracking system, always tune in the analog before the digital broadcast.

A good starting point is to target a satellite that has many active transponders, preferably one at the highest point on the arc that relays analog broadcasts. For example, Galaxy 6, Anik E1 and Galaxy 1 have numerous active transponders. Then attempt to find a picture by tuning through all active transponders. As soon as anything other than white noise appears, the process can begin. All that is needed to start the alignment fine adjustments is just a semblance of a picture. Tune to an active channel on one of the satellites which is broadcasting at the time of the installation. Then aim the dish as closely as possible to this satellite and begin the hunt.

Once a rough picture has been detected, try to improve it by moving the dish ever so slightly east and west and by adjusting the skew and video fine tune. If a side lobe of an off-target downlink antenna has been received by mistake, picture quality generally will not improve beyond a point. It may then be necessary to move the dish until the picture disappears and then reappears more clearly. If possible, the receiver's automatic frequency control (AFC) should be set off during this procedure. Also make sure that when a mechanical polarizer is used, the probe is free to move so that the skew can be adjusted at will.

Once the clearest possible picture has been obtained by using the standard mechanical adjustments, gently push the dish up and down from its front bottom rim. Take care to not bend the reflector. If the picture improves, adjust the elevation in that direction until the best picture is received or until the reading on the signal strength meter is maximized.

Adjust the feed one last time. These methods are discussed in the following section. Once the feed has been peaked, it should remain in this optimal position so that the dish can be aligned as closely to the arc as possible.

Next, target a second active satellite near one end of the arc. Set the receiver to an active transponder on this spacecraft. Once a picture is detected, the skew and video fine tune as well as the polar axis angle will likely have to be adjusted. Note the movement needed to improve pictures. If it is necessary to raise or lower the dish, the north/south axis usually needs some minor correction. For example, if pictures had been peaked on Satcom F1, a far westerly satellite, but the reflector had to be slightly lowered after being swept east to Galaxy 6, the dish and mount would have to be rotated slightly west. The geometry underlying these adjustments is outlined in the Figure 8-82.

North/south re-adjustments should be made in very small increments, on the order of 1/32nd to 1/16th of an inch (1 or 2 mm) as measured by the movement of the mount relative to the support post. This fine tuning can most easily be done by partially loosening the mount-to-pole bolts and by gently pushing on the outer rim of the dish where leverage is greatest. Pushing from the mount is more difficult and can cause jerky, unpredictable movements. It is also wise to use a piece of chalk to mark the starting point so that if too much movement occurs, the process need not be started over from scratch. A dish can often be heavy and should not be allowed to pull the mount forward on its supporting pole. The bolts should be only slightly loosened. Even so, the fine adjustments are often made with the mount tilted slightly forward so that when the bolts are tightened, the mount will pull back and the dish will be pulled slightly off target. Be aware that this can happen and that in most cases a slight decrease in elevation can correct the error. Do not make more than one adjustment at a time; it generally simply causes confusion.

After the dish is tracking the arc reasonably closely, the initial fine tuning can begin. Accurate readings are now taken from the signal strength meter. It is important during these steps that the scale chosen on the meter be left unchanged so that comparisons are possible between these readings before and after changing any settings. For example, assume that the meter reads 6.23 on Satcom F1. Then the dish was moved to the Galaxy 6 and fine adjustments in north/south position and polar axis angle were made. When the dish was returned to the same channel on Satcom F1, the new reading should have been compared with 6.23. After readjusting to obtain the best signal, this reading should have been at least as high as the old one.

When the dish is accurately aimed, maximum signal strength readings should be observed on all satellites in the arc. Generally this can be accomplished by adjusting just the north/south position and the polar axis angle. However, the declination offset angle may occasionally need some small readjustment. The accuracy of this iterative fine tuning procedure is better if satellites at far ends of the arc are used. If a stronger signal can be obtained on any satellite by pushing either up or down or sideways on the dish, the arc is still not perfectly targeted. For further clarity, this procedure is outlined step by step in Table 8-7.

Six common tracking problems are illustrated in the Figure 8-82 for locations north of the equator. South of the equator reverse the sense of the north/south movements described in this illustration. The best way to visualize these is to picture the arc of satellites as one half circle and the tracking movement of the dish as another. Both of these circles must be aligned for perfect tracking. It is that simple. It is important to understand that none of these procedures will work if the pole is not perfectly plumb or if the mount is not sitting vertically on the pole. If problems are encountered, check these mechanical settings. Be perfectly sure that the dish does not have the opportunity to be driven into a table, toolbox or even an installer's vehicle by the actuator while it is being repeatedly scanned across the arc.

This fine tuning process is completed by repeating the necessary adjustments at both ends of the arc. Satellites at the center of the arc should also be targeted to ascertain that all the fine adjustments are correct. Tighten all bolts and nuts and then recheck to make sure that doing so has not slightly altered the elevation alignment.

During this procedure, familiarity with the programming relayed by each satellite is invaluable. By far the easiest way to learn this information is by having experience with a working satellite TV system. It becomes quite clear what to expect in a relatively short period of time. Become familiar with those satellites and transponders having test patterns with the satellite name and transponder number displayed. These are invaluable landmarks in the search.

Aiming a Polar Mount – An Alternative

Gourmet...Entertaining has developed a patent pending method of aiming a polar mounted dish that does not require use of a compass, just an inclinometer to measure angles, and that can be used even when the pole or foundation supporting the mount is not perfectly plumb.

The procedure is as follows: After carefully adjusting the axial inclination and declination angles, move the dish with the actuator to the side that has the lowest visible satellite, furthest east or west, until the dish aims at the elevation angle for this satellite. Then loosen the mount's attachment to its support and rotate the entire dish/mount in azimuth to peak the signal level from that satellite. This action brings the axis of the mount into a true north/south and sets the polar mount accurately onto the arc.

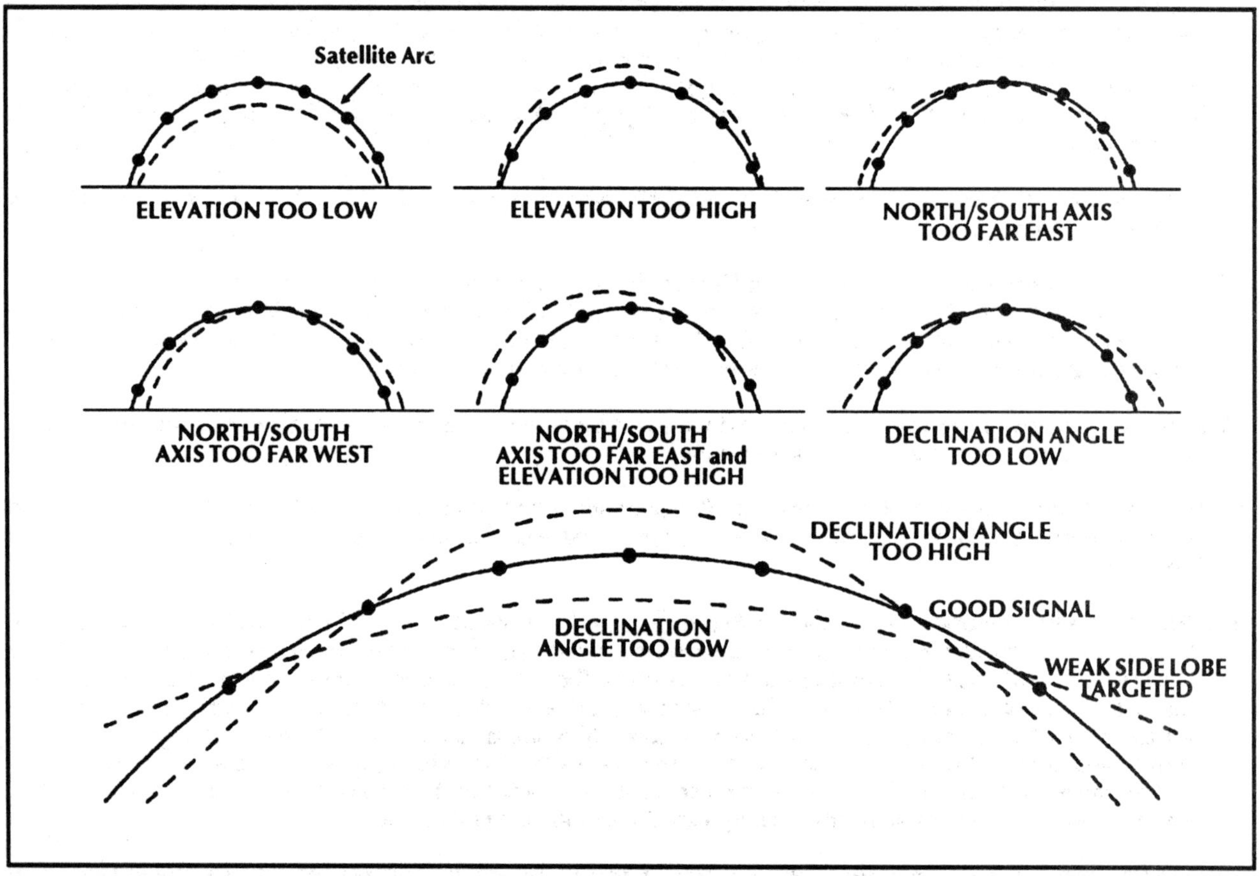

Figure 8-82. Common Dish Tracking Problems and Solutions. *Most tracking problems are associated with an incorrect north/south orientation. However, if the declination angle has not been correctly set, tracking will also be incorrect. Lining up a dish with the geosynchronous arc of satellites is simply a matter of lining up two half circles, the one in which the satellites are located and the one scanned by the polar mount. The north/south axis corrections in this figure must be reversed when aiming a dish in the southern hemisphere.*

TABLE 8-7. THE BASIC DISH ALIGNMENT PROCEDURE – ANALOGUE SIGNALS

1. Carefully check that the feed system is centered and located at the correct focal distance. It is important that the feed be peaked before tracking is started.

2. Set the two tracking angles, polar axis and declination offset angles, and align the dish polar axis in a north/south direction as well as possible. It is easiest to adjust the polar axis and declination offset angles on a satellite high in the arc, i.e. at the center of the arc, and to adjust the north/south orientation during the following step on a low elevation satellite at either end of the arc. The simple rule is elevation adjustment on a high satellite; north/south adjustment on a low satellite.

3. Check all electrical connections and turn the power on.

4. Aim the dish at a preset position or slowly track across the arc to find even the faintest picture. Attempt to target a satellite located as close to due south as possible so that the polar axis angle can be accurately adjusted.

5. Once even a semblance of a picture has been detected, move the dish slightly east and west and adjust the polarity and video fine tune controls on the satellite receiver to get the best possible reception. Then make any adjustments necessary in the polar axis angle to further improve reception.

6. Re-peak the feed system to maximize the signal strength reading. Once this has been accomplished, do not change the feed position again.

7. Move to another satellite, preferably at one far end of the arc, and attempt to tune in a second channel. During this and procedure number 3, having a receiver with a scan setting may be very useful. Otherwise, set the receiver tuner, both frequency and polarization, on a channel known to be active on the satellite being targeted. If signals from this transponder cannot be located on this satellite, move to another one closer to the first successful target. If three or four satellites in a cluster but no others farther away in the arc can be found, the north/south alignment is probably substantially off position and needs correction.

8. Tune in the second satellite as well as possible with the video and polarity fine tune controls and east/west scanning.

9. Next, determine if the dish has to be pulled up or pushed down to improve reception. If neither, move to another satellite at the other end of the arc and try again. In those rare cases when all satellites are targeted accurately, jump to step number 12. If adjustment is necessary, consult the common tracking error figure to determine which direction to rotate the north/south axis.

10. Make a small movement of the north/south axis in the correct direction. Remember to mark the original position in case the mount moves too much.

11. Return to the original satellite and repeat this procedure to further zero in on the arc. It makes sense to do this for one satellite at the far eastern and far western end as well as one in the center of the arc for best results.

12. Next, if possible, choose one satellite at the middle and one at each end of the arc, all three of which have the same polarization format and a common active transponder. Adjust the skew and video fine tune controls one last time to maximize the reading on the signal strength meter on one satellite. Do not touch these controls until this procedure is completed. Repeat steps 4 through 9 again to get maximum readings on all three satellites. At this point, some adjustment to the declination offset angle could possibly be necessary. It will probably be necessary to move back and forth between the three transponders on the three satellites until all three readings are consistently at their maximum value. The accuracy of this procedure can be improved by using signal measuring instruments.

13. As a final test, tune to the lowest power transponders in the arc. If excellent pictures can be obtained on these as well as on satellites all across the arc, the adjustments are optimized and the arc is correctly targeted.

14. Tighten all bolts carefully, watching to be sure that the signal strength meter maintains its maximum readings.

In cases where the supporting pole or foundation is not plumb, the elevation angle will change as the dish is rotated when attempting to peak the signal. The arc can still be tracked. During this process use the actuator to find the correct elevation angle that will properly target the satellite. Then reset the axis inclination because it will have changed from its original setting when the dish/mount was being rotated. When the axial adjustment is large, reception of the satellite signal will be not perfect because the elevation angle has been altered. A final step of moving the dish to the correct elevation angle and peaking with a another rotation should complete the alignment.

To confirm that the arc of satellites is precisely tracked in both cases, move the dish to the opposite end of the visible arc. When the far satellite has been targeted, gently move the dish up and down and tug it left and right. The signal strength should decrease. If not, the angles are not precisely set or the feed is not properly centered and aligned at the focal point.

Aligning a Polar Mount Without a Compass

After some experience in the field, many installers have developed useful shortcuts to align a polar mounted dish onto the arc of satellites. One such method that allows a complete installation to be accomplished using only an inclinometer is described below.

First, two satellites that can be used as reference points are located. A satellite east and another west of due south are chosen. Note that due south always corresponds to a satellite located at an orbital slot equal to the longitude of the receive site. So, for example, in Denver, Colorado, due south would correspond to an orbital slot of 105°W. The closest satellite to this point is Anik E2 at 107.3°W.

The dish is aimed in a fashion similar to aligning an az-el mount. The elevation angle of the first reference satellite is set. Then the dish is rotated around the azimuth axis, i.e. the mount is rotated on the supporting pole, until a signal from this satellite is detected. The most common technique is to slowly rotate the dish and watch the TV screen for even a flicker of a picture. This method can be preferable to using a spectrum analyzer because it allows visual confirmation of picture content to verify that the satellite observed is the actual target. When the signal is at a maximum, this first point is noted on the pole with a marker or tape.

Next, a second satellite is located by the same procedure. If the two satellites are at equal distances from due south they would have equal elevation angles. This would simplify the aiming procedure but is not crucial. Once the elevation is set and the second satellite is found, its position is also marked on the pole.

The calculation of the heading for south is now simple. The circumference of the base pole is next measured. This distance divided by 360° gives the linear distance per degree. Since the azimuth bearings to the reference satellites are known, the linear distance of due south from each of the reference points can easily be calculated. Once this is done, the mount axis can be rotated to a true south heading as measured from the previously marked bearings from either of the two reference satellites.

This technique is simple to implement. It is also quite well suited to cold weather installations since most of the effort and time spent in tweaking a polar mount is focused on finding a due south heading. In contrast, adjusting the elevation and declination angles can be accurately and rapidly accomplished using an inclinometer.

The azimuth and elevation bearings to the two reference satellites should be calculated before traveling to the installation site. This can be done using the equations in Appendix B or via a computer program such as the one listed in the reference materials.

An Example

The example of this simple technique in Table 8-9 uses the satellites Telstar 303 at 123°W and Galaxy 3 at 93.5°W. The same procedure would obviously be applicable in any location on the globe. Table 8-8 is a useful guide in determining the information required in this procedure.

TABLE 8-8. INFORMATION REQUIRED to FIND TRUE SOUTH BEARING WITHOUT a COMPASS

	Satellite 1 (East of south)	Satellite 2 (West of south)
Name		
Azimuth		
Elevation		
Channel & Polarity		

A1 = Angular distance of satellite 1 from true south
= 180° – Satellite 1 azimuth

A2 = Angular distance of satellite 2 from true south
= Satellite 2 azimuth – 180°

Cp = Pole outside circumference
= 3.142 x outside diameter

D_{deg} = Angular distance per degree
= Pole outside circumference/360

D_{deg} * A1 = Distance of true south from satellite 1

D_{deg} * A2 = Distance of true south from satellite 2

Outside Diameter Inches	(mm)	Circumference (mm)	mm per degree
3.5	88.9	279.3	0.7758
4.0	101.6	319.2	0.8866
4.5	114.3	359.1	0.9975

TABLE 8-9. FINDING TRUE SOUTH WITHOUT a COMPASS – An EXAMPLE

	SATELLITE 1	SATELLITE 2
	Galaxy 3 (93.5°W)	Telstar 303 (123°W)
Azimuth	151.0°	195.4°
Elevation	42.2°	40.1°

Site Location: Latitude – 40.0° N
Longitude – 105° W
Mount base pole outside diameter: 88.9 mm (3.5")

The data above can be used to calculate the following:

Angular distance from satellite 1 to true south = 180 – 151.0 = 29.0° from west
Angular distance from satellite 2 to true south = 195.4 – 180 = 15.4° from east
Mount pole outside circumference: 3.1415 x 88.9 = 279.3 mm
Distance per degree: 279.3/360 = 0.78 mm per degree
Distance on pole of true south from satellite 1 = 0.78 x 29.0 = 22.6 mm to west
Distance on pole of true south from satellite 2 = 0.78 x 15.4 = 12.8 mm to east

Final Feed Centering Adjustments

Once one satellite or all satellites in the arc have been tracked as closely as possible, the feed should be adjusted one final time, if possible. In some brands of offset fed dishes the feed is set firmly at one position and cannot be adjusted. This position is likely to be the optimal one.

Picture and sound quality are best when the feed is precisely at the "phase center" where all the signals are focused. Moving it away from its correct position will degrade picture quality by lowering gain and increasing the amount of noise detected. If the reflector is warped, the optimal reception point may be centimeters away from the recommended focal point. This problem must be corrected, preferably early in the installation.

Moving the feed assembly closer to the dish decreases the signal power relative to the amount of noise detected via the side lobes. This shows up as a decrease in C/N ratio and ultimately in the S/N ratio. Moving the feed away from the dish past the prime focus also decreases the S/N ratio as the signal strength decreases and as more noise from the surrounding ground is detected. If both the TV picture and signal strength meter are carefully monitored at the same time, knowing these effects, the feed can be located precisely at the phase center. The payoff in improved picture quality is more than worth the trouble.

If the feed is improperly aligned, there can be substantial degradation in system performance. For example, if 2 dB of signal is lost, this is equivalent to reducing the performance of a 2 meter dish to one with a 1.5 diameter. In addition to centering the feed and correctly setting the focal length, the body of the feed should not be twisted relative to the face of the reflector. If this does occur, additional gain would be lost because areas beyond the reflective surface would be more fully illuminated.

Alignment with Test Instruments

While a reception system can be aligned by simply watching the picture on a television set, aiming can be improved by using a more sensitive test instrument such as a carrier level meter. Many digital-ready satellite receivers now feature built-in signal level meters displayed on-screen for ease of use.

There are at least two reasons to use a portable signal level meter. First, smaller dishes are sometimes installed above ground where they can be reached either from a position on a ladder or by standing on a roof. So lugging a satellite receiver and portable TV monitor up to the dish may prove impracticable. Second, none but the most trained technician would notice a change in analog picture quality on a monitor as the C/N ratio increases from, for example, 12 to 15 dB. Nobody could see a difference with an MPEG-2 DVB signal. But this extra margin of 3 dB may be crucial during a rain storm when the C/N ratio may be reduced.

Carrier level meters fall into a number of rather broad classes: simple wide band detectors, narrow band carrier monitors, spectrum analyzers and various combinations of these. Use of such tools allows for more efficient use of the available equipment so that even finer adjustments can be made to the feed position, north/south orientation, polar axis angle and declination offset angle.

Signal power is generally measured at the LNB output. It is not realistic to expect to measure power levels at the LNB input since instruments to do so are very expensive.

RF Heads – Simple Wideband Detectors

An "RF head", a device first developed to measure power levels from C-band downconverters, is neither a voltmeter, designed to detect much lower frequency signals, nor a field strength meter, which monitors AM signals on a cable TV system. A "carrier level detector" monitors the full range of LNB output frequencies up to 1700 MHz and can be built from simple passive circuit components. These instruments are not particularly sensitive to the small changes in signal level that occurs when a dish is aligned close to its optimum position.

Narrow Band Carrier Meters

Narrow band carrier level meters monitor just one video channel at a time and thus eliminate much of the broadband noise. These are, in essence, the front-end of a satellite receiver that can be tuned to a particular channel and are generally more complex and costly than simple RF heads. In addition, they can be bulky and therefore less convenient for use on ladders. Like RF heads, they are usually powered from either internal batteries or via the RF coax from a receiver.

Spectrum Analyzers

A spectrum analyzer displays the relative levels of signals being detected plotted across the entire satellite frequency band. It is used in, for example, commercial installations where alignment procedure generally requires the use of excellent electronic test gear. This instrument while quite accurate tends to be bulky and relatively expensive compared to simpler carrier level detectors. Nevertheless, relatively high quality but reasonably priced analyzers having fully adjustable frequency ranges and signal sensitivities are available today.

An analyzer can display the entire frequency band or can be used to zero in on the structure of a message transmitted on a single carrier. With top quality devices, even the weakest microwave signals can be observed. An analyzer can be used to determine if an LNB is working properly. When the LNB is connected, its noise floor should increase. It can even be used to troubleshoot hand-held remotes which also relay RF signals.

A spectrum analyzer can be connected via a 2-way power passing splitter to the LNB output. Its internal battery then provides power to the LNB. The second port of the splitter can be connected to a carrier level detector that offers both an audible tone and digital read-out of signal strength. Alternatively the second port can be connected directly to a satellite receiver, which would power the LNB.

More Practical Meters

More practical carrier level detectors designed to monitor signal power levels at the LNB output have been introduced (see Figure 8-83). In general these have sensitivities adjustable over a range that could perhaps span 5 dBmv to -80 dBm. Many have an audible tone output that increases in frequency as signal strengths rise. Most now can be directly connected into between the output of the LNB and the input of the satellite receiver. In this configuration they draw power from the receiver. Some brands can be powered with a portable power supply so a dish could be aimed before running the RF cable to the receiver.

Figure 8-83. *The input to this meter is a coax line from the LNB; the output is to the satellite receiver. It has a range of 10 MHz to 1.9 GHz and features an audio tone that indicates when the meter reading is high. (Courtesy of Natropolis International)*

FINE TUNING the SATELLITE RECEIVER

Once the dish has been accurately aimed, the receiver can be moved indoors to its final installation location, if this has not already been done, and final fine tuning can be completed. For tracking dishes, make sure that each active satellite in the entire arc can be selected via actuator controls. Next receiver IF gain should be adjusted to receive the optimal picture across all transponders.

In those installations where linear polarities are being received via mechanical polarizers, some receivers have back panel or internal potentiometers to adjust the skew control so that the knob is near the center of its range of movement. Also, make sure the probe is not sitting at the end of its range of movement by observing its position or by feeling the motor for vibration when the skew knob is set at its end positions. Some brands of actuators have timers which shut the motor off in 5 seconds so this test requires two people, one at the receiver and one at the dish. Both the IF gain and skew fine tuning adjustment are usually accessed from the rear or bottom panel.

At this point, if pictures are not as perfect as expected, check that all connections to the receiver are secure. A customer's TV set may sometimes require fine tuning. Its channel frequency should precisely match the satellite receiver's modulator output. This channel is rarely used for over-the-air programming but may have been used if the customer was a cable TV subscriber or used a VCR. In addition, to guarantee that the customer's TV set is in good health, tune to a strong over-the-air broadcast or use a VCR input.

PROGRAMMING the RECEIVER and ACTUATOR

Each brand of programmable receiver and actuator is different and the manufacturer's instructions should be carefully followed in programming. Standard 9-volt batteries must be added to some remote controls.

It is good policy to ensure that a customer either fully understands any programming procedures for a system, i.e. the system should be "user friendly." This latter approach might mean attaching the dealer's telephone number to a satellite receiver and charging a fee for an extended warranty. Both are strategies designed to accomplish the same objective, that of making a system "user friendly."

Installers should also be familiar and comfortable with any components that are installed. They may require service in the future. For those who install numerous systems using a common receiver type, it is wise practice to maintain a consistency in programming this particular component. For example, if possible, ensure the preset channels are identical in all receivers or that satellites are stored with similar abbreviations or mnemonics on all actuator controllers. If such continuity is maintained, a set of simple instructions for in-field work can be created.

A copy of the chart for each receiver along with a detailed setup procedure could then be issued to each installer (see Table 8-10a and 8-10b as examples). The installer's copy should have a diagram of the remote control and also be sufficiently detailed to enable the technician to talk the user over the telephone.

Many receivers have a hardware switch on the back panel or via the remote control that enables the programming and re-programming procedure. If this switch is set in the locked position, it prevents the receiver from being inadvertently re-programmed by the remote control. This switch should always be in the locked position after a receiver is programmed.

WATERPROOFING

Water in the form of vapor, rain, ice or snow is the chief enemy of a TVRO. It can corrode connectors, accumulate in feeds or weigh down dishes. Once the system has been completely tested, is fine tuned and is working perfectly, the next step is waterproofing. This step is a crucial one.

TABLE 8-10a. SAMPLE CUSTOMER CHART

Satellite Name:
Satellite Abbreviation:

Channel Number		Service
Channel 01	....	....
Channel 02	....	....
| |	|	|
Channel nn	....	....

Satellite Name:
Satellite Abbreviation:

Channel Number		Service
Channel 01	....	....
Channel 02	....	....
| |	|	|
Channel nn	....	....

All connectors and junctions where water could enter should be sealed. Some types of caulks are made with ammonia or lactic acid which can corrode metals; others harden, become brittle and crack. A sealant that stays soft is recommended.

The feed/LNB at the dish focus should be protected by a cover, a small expense and excellent insurance. Drip loops should be used whenever applicable.

Be sure that water cannot accumulate anywhere on the mechanical system. There have been cases where water has collected at the bottom of mounts which were open at the top but closed at the bottom: the water froze, split the tubing and weakened the mount. The cable trench, if any, can be covered up and the sod replaced after waterproofing is completed. The installation is now finished.

TABLE 8-10b. SAMPLE INSTALLER CHART

Satellite Receiver Type:
Positioner Type:
Satellite:
Satellite Abbreviation:

Channel	Frequency	Service	Bandwidth	Audio
Channel 01................	....	....	....	....
Channel 02................	....	....	....	....
| |	| |		| |	
Channel nn................	....	....	....	....

Detailed receiver setup procedure:
Battery type in remote control:
Detailed positioner setup procedure:
Include parameters used in programming a channel into memory

CONNECTING ACCESSORIES

Every installation has slightly different requirements for in-house equipment. This may include extra TVs, VCRs, projection TVs integrated into a surround sound home entertainment center or even advanced computers. These systems may range from a very simple two-way splitter to a full-blown distribution system. More elaborate systems are described in more detail in Chapter 10.

One receiver can power two or three nearby television sets without amplification or any number of TVs with insertion of the necessary line amplifiers. The simplest way to accomplish this is via splitters that divide the video signal into 2, 3, 4 or even more equal ones. The placement and type of splitters used depends upon the location of each television. For example, assume that there is one set adjacent to the satellite receiver and three at the far end of the house. A 2-way splitter would be used at the receiver output to drive the nearby television and a trunk line having a 10 to 20 dB line amplifier would be connected to a location near the other sets. There, a 3-way splitter would divide the signal. This would certainly be a much better strategy than using a 4-way splitter at the receiver and running three long cables.

The type of cable selected depends upon the distance traveled and number of televisions connected. For short distances, RG-59 is adequate. RG-6 can be used for lines up to 150 feet. RG-11 can be used for longer trunk lines. However, for example, if 30 televisions all in close proximity to the receiver were driven by this receiver, it would still be necessary to use lower loss RG-6 or RG-11 since the original signal would be split or tapped so many times. The rule is common sense and minimizing cable runs. When necessary, line amplifiers can be inserted via F-connectors.

Hooking up Stereos and VCRs

All satellite receivers process the incoming signal and produce a composite baseband signal, the raw video, audio and data messages as originally created at the studio. They all feature some standard outputs as summarized below.

The RF output feeds an AM (amplitude modulated) signal via an F-connector into a TV set. This AM signal, produced by a modulator inside the satellite receiver, converts the baseband signal into an AM form that is acceptable by conventional televisions. In North America, users have a choice between VHF channels 3 or 4. In Africa and Europe, the baseband signal is modulated onto one of the channels in the UHF range from 21 through 69. (Satellite receivers and IRD modulators are usually set to UHF channels 38 in Africa and 36 in Europe). All satellite programs can be selected by setting the TV set to this channel and tuning via the satellite receiver. Microcomputers, video recorders and video games also use the same channels. Sometimes, if necessary, a modulator may be tuned to another channel so that it does not interfere with these other additional video sources.

The baseband signal can also be fed directly into a TV monitor or VCR without being modulated. The baseband outputs are usually labeled video out, left audio out and right audio out and connected via phono jacks.

A TV monitor differs from conventional televisions that have built- in AM demodulation circuits designed for detecting over-the-air AM broadcasts. When a signal is fed into a TV, the baseband signal must first be amplitude modulated in the satellite receiver. It is then demodulated back to baseband in the television. Higher quality reception can generally be attained when using a monitor, VCR or laser disk players since these components have composite video and audio outputs, the raw video and audio signals. These are fed directly into a monitor so additional modulation and demodulation steps are eliminated. Like a monitor, a VCR also accepts the composite baseband signal. It has its own built-in modulator for output to a television set or, if it has the ability to tune individual TV channels, can feed a baseband signal directly to a TV monitor. VCRs can record higher quality videotapes from the satellite baseband signals than from over-the-air AM broadcasts. Whenever possible, highest quality reception can be obtained by using composite audio/video signals and avoiding converting to radio frequency (RF) signals.

Table 8-11 summarizes the various methods to hook up televisions, VCRs and monitors.

The output ports on the different types of satellite receivers are usually but not always labeled in the same fashion. Thus, for example, the F-connector RF input from the LNB might be labeled *Satellite In* or *RF In.* Each input could be labeled *V* and *H* for vertical and horizontal. The off-air antenna input might be marked *In From Antenna.*

Table 8-11. HOOKING UP TELEVISIONS, VCRs and TV MONITORS

SATELLITE RECEIVER to TELEVISION SET

Satellite Receiver Outputs	Stereo In/Out	VCR In /Out	Television Inputs	Monitor Inputs
RF Out	—	—	In	—

SATELLITE RECEIVER to VCR and TELEVISION SET, #1

Satellite Receiver Outputs	Stereo In/Out	VCR In /Out	Television Inputs	Monitor Inputs
RF Out	—	In From Ant/To TV	In	

SATELLITE RECEIVER to VCR and TELEVISION SET, #2

Satellite Receiver Outputs	Stereo In/Out	VCR In /Out	Television Inputs	Monitor Inputs
RF Out		Video In	—	—
Left Audio	—	L Audio In	—	—
Right Audio	—	R Audio In	—	—
	—	To TV	In	—

SATELLITE RECEIVER to TV MONITOR

Satellite Receiver Outputs	Stereo In/Out	VCR In /Out	Television Inputs	Monitor Inputs
Video*	—	—	—	Video In
S-Video*	—	—	—	S-Video In
Left Audio	—	—	—	Left Audio In
Right Audio	—	—	—	Right Audio In

SATELLITE RECEIVER to VCR and TV MONITOR

Satellite Receiver Outputs	Stereo In/Out	VCR In /Out	Television Inputs	Monitor Inputs
Video*	—	Video In/Out	—	Video In
S-Video*	—	—	—	S-Video In
Left Audio	—	L Audio In/Out	—	Left Audio In
Right Audio	—	R Audio In/Out	—	Right Audio In

SATELLITE RECEIVER to VCR, STEREO and TV MONITOR

Satellite Receiver Outputs	Stereo In/Out	VCR In /Out	Television Inputs	Monitor Inputs
Video*	—	Video In/Out	—	Video In
S-Video*	—	—	—	S-Video In
Left Audio	L Audio In/Out	L Audio In/Out	—	Left Audio In
Right Audio	R Audio In/Out	R Audio In/Out	—	Right Audio In

TOOLS REQUIRED for SATELLITE TV INSTALLATIONS

The old boy scout adage "be prepared" applies to a satellite TV installer. Having to make an extra round trip of 60 km for an F-connector coupling or a 75 to 300 ohm transformer is certainly a waste of valuable time and money.

Essential installation tools include a compass and inclinometer for a site survey, a crimp tool for fitting F-connectors, and other standard items including screwdrivers, a drili, spanners, a soldering iron and power extension cords. These should be supplemented by an assortment of connectors. A complete list of tools is presented in Table 8-12.

Tools of the trade can be carried in a small pickup or a thoroughly equipped truck. The type of equipment needed depends upon what type of dishes are being installed, the climate and how a dealer wants to organize his installations. For example, when installing a roof- or wall-mounted dish a sturdy, reliable ladder is required. Or when installing a larger ground-mounted dish, one crew can be used to plant the pole, assemble the dish and run the cable and then another may do all the electrical and alignment work. Alternatively, a single crew can complete the job in one visit.

Heavy Machinery

Heavy machinery includes vehicles needed to transport poles and dishes and all equipment required for excavation and concrete work. Although some 3 meter mesh dishes can be packaged to fit into a small car, spun or stamped reflectors come in one piece and require at least a small pickup truck for transportation. At the very least, a small pickup truck or a single or tandem axle trailer which can be pulled behind a car should be owned by any serious installer. Excavating tools, when necessary, include a wheelbarrow or a portable cement mixer, a pickax, breaker bar, hoe, rake, hole digger and shovel. Others may be required in different situations.

Major Tools

This category includes all those tools required for every installation. This includes an inclinometer and compass, extension cord, wire strippers and carpenter's level.

Occasionally Needed Tools

Tools on this list are needed occasionally in special situations. For example, if no power is available at the site, a portable generator may be required. For long pole or roof mounts, an extension ladder may be needed.

Connectors, Cables and Other Accessories

This category includes all those small items that are most often required for an installation. These should always be on hand for both installing and troubleshooting. The lack of a small but essential item could be a source of great frustration.

The installer's equipment checklist in Table 8-12 has been provided for convenience. Modifications to this list can be made to suit one's tastes or needs. However, using a checklist before leaving the shop to install or troubleshoot a system is a recommended practice.

TABLE 8-12. INSTALLER'S EQUIPMENT CHECKLIST

Heavy Machinery
___Pickup Truck or Trailer
___Wheelbarrow or Cement Mixer
___Post Hole Digger or Power Auger
___Shovel
___Pickax
___Rake and Hoe
___Bucket and Hose
___Breaker Bar
___Trencher

Major Tools
___Inclinometer
___Compass
___Signal Strength Meter
___150-foot Extension Cord with 3 or 4-way Adapter
___Extra TV or Monitor
___Multimeter (volts, ohms, etc.)
___Hex Crimp Tool for RG-6 and RG-59 Connectors
___Wire Strippers/Exacto Knife
___Tape Measure (at least 100 feet)o Ultrasonic "Tape Measure"
___Soldering Gun (20 and 40 watt) and Solder or Gas Powered Iron
___Carpenter's Level
___Flashlight and/or Fluorescent Light
___1/2" (12 mm) Heavy Duty Drill, Hammer - Operation Switchable
___3/8" (10 mm) Variable Speed Drill and Bits
___1 kilo Hammer
___Hack Saw and Blades
___Large Diagonal Pliers
___Large Long Nose Pliers
___Small Long Nose Pliers
___Vise-Grips
___Assorted Screw Drivers
___3/8 to 1-1/8" (10 to 30 mm) Set of Oped Ended Spanners
___Set of Socket Spanners
___Small Set of Metric or SAE Sockets
___Large and Small Crescent Wrenches 12 to 15 inches (30 to 40 cm)
___Awl or Center Punch (for Aligning Holes in Dish Petals)
___Set of Allen Wrenches
___A Short Ladder
___String and Tape
___4 to 5 cm Electrical Conduit, Solvent and Sweeps
___Metal File and Assorted Sandpaper
___5/8 by 18" (16 mm by 45 cm) Wood Drill Bit
___5/8 by 12 to 18" (16 mm by 30 to 45 cm) Masonry Bit

Often-Needed Tools
___Extension Ladder
___Generator
___Fish Tape (for feeding cable inside walls)

Connectors, Cables and Other Accessories

___F-Connectors
___F-to-F Bullets (F-81 Barrels)
___Right Angle and Straight F-connector adapters
___RCA Jacks, MolyPlugs, Lugs, Scotch Locks and Assorted Adapters
___Video Connectors Kit with Cables
___Rolls of Direct Burial, RG-6 and RG-59 Cable
___75 to 300 Ohm Transformers
___A/B Switches
___2-, 3-, and 4-Way Splitters
___75-ohm Terminators
___Spare Fuses and Voltage Regulators
___Electrical Tape
___Box of Assorted Wire Nuts
___Box of Romex Staples
___Can of Touch Up Spray Paint
___Coaxial Sealant
___Non-Conducting and Non-Corrosive Sealant
___A Spare Polarotor Servo Motor
___Surge Protector with Multiple Outlets
___10 and 20 dB Amplifiers
___Ground Rods and Heavy Gauge
___Grounding Wire
___T-25 Arrow Roundtop Staple Gun

COLD WEATHER INSTALLATIONS

Doing an installation in very cold weather can be uncomfortable. However, there are many factors to consider other than discomfort. Things that are taken for granted do not necessarily happen as expected. For example, electrical tape which normally sticks perfectly well becomes brittle and useless. It is necessary to buy a good brand and keep it in a warm place or a pocket until it is needed. Low power soldering guns cannot produce enough heat to melt solder. A very large gun is needed at -30 °C. In these cases, crimp-on connectors, not solder, may be chosen.

When it is sub-freezing, fiberglass becomes brittle and can shatter (see Figures 8-84 and 8-85). Many fiberglass dishes have been damaged in these conditions by over-tightening the bolts that hold the reflector to the mount. The answer to this problem is, if possible, to assemble dishes indoors, out of the cold weather. If this option is not possible, let the unassembled pieces sit in the sun under a plastic covering, until they are warm and be very careful.

Figure 8-84. Assembling a Fiberglass Dish. *Here the shipping crate was taken apart and laid down on the snow to keep water from freezing between the dish panels as bolts were tightened. It was important to be sure the panels were lined up without any gaps between them. (Courtesy of Brent Gale)*

Concrete will not set properly in very cold weather unless the right precautions are taken (see Figures 8-86, 8-87 and 8-88). Two percent calcium chloride must be added to prevent freezing and the surface must be covered with a blanket, straw or another insulating material until it sets. Other binding materials dependent upon chemical reactions, like Dish Set™, set more slowly when cold. A simple practice to aid the hardening of either this material or a fast setting cement is to place the pole indoors or under the exhaust of a running vehicle to warm it up just prior to planting it.

Some brands of silicone sealants will not harden. Make sure that a silicone designed specifically for cold weather is used. Other products behave in surprising ways when cold. For example, one of the authors purchased two thousand 12 inch wire ties one summer expecting that they would have a long service life. When winter came, the ties were so brittle that they snapped when used. When warmed up they were fine, but at below freezing temperatures they were useless. Another brand manu-

Figure 8-85. Penetrating the Frost Line. *This photo shows the use of a digging bar to penetrate the frost line so that a power auger could subsequently be used. Snow cover does have the advantage of acting as an insulator and keeping the ground from freezing. In contrast, those areas of the country that have extreme cold but little or no snow can have frost penetrations of over 3 feet. In such cases, a barrel filled with burning wood or charcoal can be placed over the site of the pole; this thaws the ground in a matter of hours. (Courtesy of Brent Gale)*

factured specifically for cold weather had to be ordered. Similarly, most non-colored nylon wire ties will fail after only one year's exposure to sunlight. Black, blue, brown or specially UV-stabilized brands will, however, last years longer.

Tools can become very brittle in the cold, especially at temperatures below -25°C. For example, ends of screwdrivers snap off and crescent wrenches break. A solution is to use oversized wrenches which have greater leverage and need less torque, such as 50 cm instead of 30 cm crescents.

Figure 8-86. Assembling a Dish Directly on the Mount. *In this case, the dish had to be assembled directly on the mount. This was somewhat more difficult than a typical installation.(Courtesy of Brent Gale)*

Plastic components are especially susceptible to damage in such extreme conditions. If TVs are treated roughly in the cold, their plastic containers easily break. Knobs also snap off receivers and TVs. In addition, some brands of satellite receivers do not work properly below a certain temperature. A recommended method is to keep the receiver and television in a warm car or truck during testing. Many types of cable will not unbend when cold and should be kept warm until needed. Some cables are made especially for such use. Center conductors of some coaxial cables will snap in two when cold. Therefore, care should be taken when uncoiling coax in cold weather.

Cold metal can be dangerous. For example, one dealer was working on a system when the chill factor was -50 °C. He went inside to check a picture and ran back to raise the elevation adjustment without his gloves. A crescent wrench which had been outside froze to his hand and could only be removed by running water over it. It is just as easy to pull skin off a hand by tightly grasping a very cold mount. Always wear a good pair of gloves.

Figure 8-87. Use of a Power Auger in Cold Weather. *This two-man power auger is a very useful tool that allows rapid excavation of holes in cold climates. (Courtesy of Brent Gale)*

Figure 8-88. Use of Leveling Stakes. *This photo was taken shortly after the concrete was poured. The mount was installed on the pole only after curing was complete. (Courtesy of Brent Gale)*

Many problems can be traced to freezing of water which enters components either by condensation or leakage. Water left inside a waveguide can easily freeze and crack the housing. Even small amounts of condensation can cause problems in polarizer motors in extremely cold climates. Ferrite devices having no moving parts are recommended in such situations instead of mechanically rotated probes. There have been cases of water collecting in mounts or actuator motors and cracking the surrounding metal. Using drain holes, effective sealants, LNB/feed covers and other protective measures is especially recommended in such climates (see Figure 8-89).

Figure 8-89. Weather Sealing. *A thorough weather sealing including the use of drip loops and coax seal is especially important in cold climates. (Courtesy of Brent Gale)*

A properly equipped dealer should be able to install in any weather, be it rain, snow or blizzard. For example, in very cold weather prefab concrete bases poured in a shop can be trucked to an installation site. But there are times when it is simply too cold or wet or snowy to warrant the effort.

CUSTOMER RELATIONS

Customer relations are common sense behavior, although common sense must sometimes be learned, often at a high price. The most important piece of common sense dictates that a dealer or installer be punctual and reliable.

Some customers have an amazing amount of technical know-how and can teach most any dealer some interesting tricks and methods. Often a dealer who knows how to spot such people will make a sale where others have failed.

Some customers may want to participate in an installation by simply helping to assemble the dish, plant the pole or run cable lines, or by actually completing the whole job except for the final wiring and fine tuning. Ninety percent of the time they can do an excellent job with the proper guidance and may become a source of many referrals. Other customers are pleased to do the entire installation knowing that a competent dealer is ready waiting in the wings when needed. Such installations are often examples of fine workmanship.

Whether or not a customer wants to participate, every installation should be be treated as if it were the dealer's own. It should be a source of pride and referrals and should last for years. Before leaving a site, a professional will take one last look around the premises in order to see if anything else can be done more perfectly.

Contracts and Warranties

Never make a sale or do an installation without a signed contract. Customers should be informed of all warranties and special conditions of this legal document. Each section of the contract should be explained so that there are no hidden expectations or unwanted surprises. Contracts protect both the dealer and the customer. Whenever a situation deteriorates to the point that a lawyer must be called in, both parties usually have already lost.

DOCUMENTATION

It is important that records be kept of each installation in case any problems arise in the future. Documentation for both the customer's and installer's records should include the following:

- Date of Installation
- Original Site Map
- Dish Type
- Mount Type
- Pad or Pole Support
- LNB Temperature and Make
- Receiver Make
- Actuator Make
- Type and Length of Cable
- Number of TV Sets
- Serial Numbers of All Components
- Any Special Installation Details

During the installation, a table outlining the measurements taken during final adjustments should be created. This will prove very useful if performance comparisons are required at a later time (see Table 8-13). Remember, however, that as satellites age their transponder power will decrease and these numbers will slowly fall.

Signal strengths can be measured using equipment such as the Bulz-I-Meter or other signal strength meters.

Business Documents

A variety of proposal, quotation and other business documents is used in the everyday running of a satellite TV dealership. Documents kindly provided by Ron Long and presented on the next four pages provide excellent samples.

TABLE 8-13. SIGNAL STRENGTH COMPARISON TABLE

Date: March 10, 1996
Location: R.F. Jones, St. Laurent, Quebec.

Satellite	Transponder Number	Signal Strength
Satcom SN2	9	5.0
Galaxy 7	6	5.9
Anik E2	13	5.5
Satcom C4	8	6.4

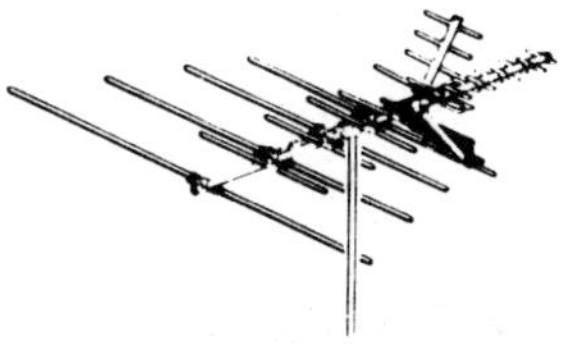

LONGFELLOWS
ANTENNA SERVICE

532 Dawson Drive, Suite 4-N
Camarillo, California 93010
(805) 484-2432

№ 1103

DATE ____________

NAME
ADDRESS
CITY — PHONE
NATURE OF SERVICE REQUEST

QTY	DESCRIPTION	PRICE	AMOUNT
	Antenna		
	Antenna		
	Antenna		
	Amplifier		
	Amplifier		
	Coupler		
	Coupler		
	Cable		
	Connectors		
	Base plate		
	Outlets		
	Outlets		
	Wall plates		
	Wire		
	Matching xfmr.		
	Eyebolts		
	Screw eyes		
	Seperators		
	Guy—Wire		
	Mast		
	Misc. Items		

Service Details:		
	TOTAL PARTS	
	TAX	
	TRIP CHARGE	
	TECHNICAL CHARGE	
	TOTAL	

CUSTOMER SIGNATURE ______________________ DATE ____________

MATRIX PRINTING

LONGFELLOWS
ANTENNA SERVICE

532 Dawson Drive, Suite 4-N
Camarillo, California 93010
(805) 484-2432

№ 1068

DATE

NAME

ADDRESS

CITY PHONE

NATURE OF SERVICE REQUEST

QTY	DESCRIPTION	PRICE	AMOUNT
	Dish		
	Feed		
	LNB C		
	LNB C		
	LNB KU		
	Actuator ACME/BALL		
	Receiver		
	Receiver		
	Satellite Cable		
	Coaxial Cable		
	2-way Splitters		
	Coax Connectors F/BNC		
	Concrete		
	Ground Pipe		
	PVC Pipe and Accessories		
	Satellite TV Week Subscription		
	Misc.		

Service Details:		
	TOTAL PARTS	
	TAX	
	TRIP CHARGE	
	TECHNICAL CHARGE	
	TOTAL	

CUSTOMER SIGNATURE ______________________ DATE ______________

MATRIX PRINTING CENTER '89

QUOTATION

LONGFELLOWS
ANTENNA SERVICE
532 Dawson Drive, Suite 4-N
Camarillo, California 93010
(805) 388-5453

INQUIRY NO. ______________

DATE ______________

TERMS ______________

DELIVERY ______________

PRICES QUOTED ARE F.O.B.:

____

TO ______________

____

____

____

WE ARE PLEASED TO QUOTE ON YOUR INQUIRY AS FOLLOWS:

QUANTITY	DESCRIPTION	PRICE	AMOUNT

QUOTED BY:

MATRIX PRINTING

LONGFELLOWS
ANTENNA SERVICE
532 Dawson Drive, Suite 4-N
Camarillo, California 93010
(805) 388-5453

No.________________
Date________________
____

PROPOSAL

Proposal Submitted To:

Name________________
Street________________
City________________
State________________
Phone________________

Work To Be Performed At:

____
Street________________
City________________State________
Date of Plans________________
____

We hereby propose to furnish the materials and perform the labor necessary for the completion of

All material is guaranteed to be as specified, and the above work to be performed in accordance with any drawings or specifications required for above work and completed in a substantial workmanlike manner for the sum of

Dollars ($).

with payments to be made as follows:________________

Any alteration or deviation from above specifications involving extra costs, will be executed only upon written orders, and will become an extra charge over and above the estimate. We are not responsible for damages due to acts of God.

Respectfully submitted________________

Per________________

Note—This proposal may be withdrawn by us if not accepted within _______ days.

ACCEPTANCE OF PROPOSAL

The above prices, specifications and conditions are satisfactory and are hereby accepted. You are authorized to do the work as specified. Payment will be made as outlined above.

Signature________________

Date________________ Signature________________

MATRIX PRINTING

CHAPTER 9

UPGRADING SATELLITE TV SYSTEMS

In some instances a service call can evolve into a potential system upgrade. For example, the actuator tube may be bent and parts for this particular unit may not be available, the receiver or LNB may have given up the ghost, or the signal quality may be poor to terrible because coax or connectors need replacement. Occasionally, a customer may order a new receiver and an older LNA/downconverter might have to be replaced. Or someone might have purchased a home with an existing satellite system that is not functioning. A person might have purchased a home with a functioning satellite system and then requested help in learning to operate it only to realize that an upgrade was necessary. Thus it is also not uncommon for a casual conversation during a simple troubleshooting visit to turn into a request for an upgrade.

However the opportunity arises, some system upgrades can tax the ingenuity of even a seasoned professional. During the relatively short evolution of the satellite industry, numerous engineering approaches have come and gone. For example, it is difficult if not impossible to obtain parts for some potentiometer and optical sensors used as actuator counting mechanisms. The challenge is to salvage as much of the old equipment as is reasonably possible when upgrading the system to be compatible with modern technologies.

Service calls and upgrades also offer a dealer a potential profitable sale of other home entertainment components ranging from surround sound receivers to full-blown home entertainment systems. Owners of satellite systems tend to be on the cutting edge of market innovation and are often interested in the new and unusual.

COMPONENT EVALUATION

LNAs and LNBs

Although a few LNA/downconverter systems are still functioning, many of the original manufacturers are out of business and, in any case, parts are rarely available. Furthermore, the 80 to 120K noise temperatures typical of older models are inadequate in the age of the 15 to 20K LNB.

Whenever a new satellite receiver is purchased, an existing LNA and downconverter must be replaced with a new LNB. This may be true even if an LNB was included in the previous system because some older receivers were mated with units having uncommon output frequency ranges such as 430 to 930 MHz, 900 to 1400 MHz or 930 to 1430 MHz.

Feeds

Earlier satellite television systems incorporated feeds and polarizers that used techniques some of which have been phased out. Some products included feeds with ferrite polarizers or low noise feeds (LNFs) that were LNAs with built-in polarizers. Many older feeds accepted from 3 to as high as 15 Vdc for driving dc polarizer motors. Polarizer voltages for three wire servomotors ranged from 3 to 8 Vdc. There were dc motors requiring both two and three-wire hook-ups. Some motors had just a limited 100° range of rotation while others could rotate a full 180°. The autotune feature on some modern IRDs could burn out the motor of some of these limited range polarizers by running them up against a limit. Trying to match any of these product features to current equipment could be as difficult as trying to find parts for antique automobiles.

Modern feeds include mechanical, voltage-switched and some ferrite polarizers. In North America, most systems use a mechanical polarizer with a servomotor that rotates 270° and that can accept approximately 8 Vdc without motor damage.

Actuators

Pioneer actuators were in many respects crudely designed and manufactured. A variety of unusual count mechanisms was employed. Although the potentiometer and optical counting mechanisms are now historical curiosities, there are still a few such units in the field. Some of the pioneer actuators did not even have counters or limits but were moved by simply depressing a "west" or "east" movement button on the controller. Some actuators had circuit boards attached to the mount that was specially designed for a particular dish. Today, such boards are normally located in the satellite receiver or actuator controller. In these cases the only possible upgrade is to replace both the dish and actuator.

Actuator motors have been powered by 12, 24 and 36 Vdc. The latter voltage is the standard today. Actuators used on older systems often are not capable of moving much past satellites as high as 30 degrees above the horizon.

Cable and Connectors

Some coax manufactured and installed during the infancy of the satellite industry was substandard. It was generally not necessary to sweep test this cable because 70 MHz was often the highest frequency relayed. Transmissions at 950 to 1450 MHz suffer much higher losses than at 70 MHz. The extremely long runs of RG-59 cable installed in many older single and dual-downconversion systems are simply inadequate to relay signals from a modern LNB.

A few systems that use RG-58, RG-62 or RG-92 coax are still on-line. These cables have characteristic impedances different from the 75 ohms required with satellite receiving equipment. While this mismatch could be tolerated at 70 MHz, at which frequency almost any type of cable will function properly, such cable is a disaster at 950 to 1450 MHz.

Some coax used for pioneer TVROs that was white or beige in color would not tolerate ultraviolet radiation. The dielectric around the center conductor would shrink, causing the characteristic impedance to change. The end result is a deterioration in picture quality. This cable damage shows up as an outer jacket that is tacky or sticky to the touch and has blisters. Moisture could enter at the blisters further altering the characteristic impedance.

Some earlier all-in-one cables were enclosed in a single round jacket. In order to separate the actuator, polarizer and LNA/downconverter or LNB runs, the outer jacket had to be removed. This exposed the wires to the degrading effects of ultraviolet rays and moisture. Most installers did not even attempt to tape these awkward configurations to reduce damage.

Cable that was installed years ago without a conduit could easily have been damaged by the action of rodents or water. Similarly, connectors may have degraded to the point that substantial signal losses occur.

Ribbon cables manufactured in the past used the second extra coax line to transmit LNB power and the IF signal on separate lines. Additional wires may also be unused if an actuator was not installed. These extra lines may have frayed or degraded over time.

In these situations as well as others where seemingly excellent cables were installed, it is crucial to be certain that no damage has occurred. If any damage is detected, all cables should be replaced.

Satellite Receivers

The satellite receiver must be properly mated with the LNB, actuator and polarizer. Voltages, size and number of wires and electronic control circuitry must be compatible.

If the existing receiver has a TI filter attached, it is a wise policy to replace the existing feed with one that has good TI rejection and the receiver with one having built-in filters or, at least, an IF loop. In the latter case, select a receiver that has an IF loop frequency for which filters are available. Notch filters with 70 MHz IF are readily available.

General Considerations

Any new component must be compatible with any other with which it interacts. Unfortunately, many upgrades require replacing almost everything other than the dish.

In an upgrade, there are many concerns similar to those that arise when troubleshooting any newer system. For example, the dish might need an alignment, connectors may have to be replaced or a mount may have to be stabilized. In some cases, installing an entirely new system may be the simplest solution.

UPGRADING to KU-BAND OPERATION

Dishes, feeds and other components designed for Ku-band reception are subject to more rigorous performance criteria than are their C-band counterparts, in large part because the C-band signal wavelength is approximately one third of that at Ku-band. Therefore, when considering a Ku-band retrofit to an existing C-band TVRO, each component should be examined to determine if compatibility problems might arise. For example, in some cases it may be necessary to purchase a second dish or receiver, or to alter the structure of a mount or actuator drive.

Dishes

In most cases, C-band systems were installed long before the possibility of retrofitting for Ku-band reception was considered. Unfortunately, many existing dishes do not have efficiencies high enough for effective operation at higher frequencies, either because they were not initially designed to have the necessary surface accuracy or because they have suffered the rigors of time and have not aged well.

Dish efficiency is three times as sensitive to surface imperfections or roughness at Ku-band than at C-band. For example, signal loss is 2 dB, about 58%, if ripples of about 1 inch occur with a regularity across the surface of a Ku-band reflector. While the average deviations from a perfect parabola should be less than 1/6th inch (4 mm) at C-band for good performance, these should be one third of this value or 1.3 mm in the Ku-band.

One piece spun or hydroformed dishes can perform well in both frequency bands. However, as the number of panels increases, deviations from the designed parabolic shape are more likely to occur, even if the surface of each panel is perfectly smooth. Expanded wire mesh reflectors often depart quite dramatically from the ideal shape. Imperfections are sometimes hidden. Poor quality fiberglass reflectors may be perfectly smooth to the touch but have embedded metal mesh that may have partially delaminated from their rigid backing and result in poor performance. Dishes often have more accurate surfaces at their centers than in regions closer to their outer rim. While this inaccuracy may have negligible effects at C-band, performance at Ku- band might suffer.

Expanded and perforated metal dishes can have unacceptable losses at higher frequencies if the surface holes are larger than perhaps 1/5th of the Ku-band wavelength, i.e. 1/5th of an inch. If so, not only does reflection efficiency decrease but higher amounts of ground noise originating behind the dish can leak into the feed. The net effect is a decrease in gain and an increase in noise and side lobe performance.

In all cases, the larger the dish the higher the chance that it can be used to receive Ku-band signals. For example, a 12-foot dish has four times the gain of a comparable 6 footer because gain is proportional to reflector surface area. If the 12-foot dish had just a 30% efficiency, it would still perform slightly better than a 6-foot unit having a 70% efficiency.

Mounts and Actuators

Since dish beamwidths are more narrow at Ku-band than at lower frequencies, the reflective surface as well as the feed must be rigidly secured. This is a critical. Even slight wobbling of the support in the wind or play in an actuator arm may throw a reflector far enough off target to ruin reception. Furthermore, a dish that seems quite solidly mounted may still not perform up to par if it weighs enough to cause sagging as it tracks between satellites. For example, a heavy fiberglass dish which may be accurately aimed at the center of the arc could twist the support enough when moved to either end to cause inaccuracies. This could develop into even more of a problem under wind or snow loading. In a few instances, a pole might even have to be replaced with a three-legged tower for the required additional support.

Some brands of actuators are not geared finely enough to allow the reflector to be accurately and securely aimed at a satellite. Those having too few sensor counts per length of motion would probably be unable to aim the dish at a Ku-band satellite within the necessary fractions of a degree. The actuator arm or both the arm and controller may have to be replaced. Many relatively new linear actuators and nearly all horizon-to-horizon actuators are quite stable and have little gear play or backlash.

Feeds

Two central issues must be addressed when examining whether or not a feed/LNB can be used to receive a Ku-band broadcast: which frequency band can it manage and how does is select polarity.

Figure 9-1. C/Ku Dual-Band Feed. *This dual-band feed is designed to receive both C- and Ku-band broadcasts. The Ku-band LNB has an input waveguide that is about 1/3rd of the diameter of its C-band counterpart. The same servo motor is used to adjust polarity in both bands. (Courtesy of California Amplifier)*

A C-band feed cannot properly detect Ku-band signals and vice versa. Ku-band and C-band feeds, both prime focus and offset, are similar except the Ku-band unit is about a third the size to account for the difference in wavelengths between these frequency bands. If a C-band system is being transformed into one to receive Ku-band signals, the feed must be changed. Alternatively, a dual-band C/Ku feed that simultaneously detects both C- and Ku-bands can be installed (see Figure 9-1).

A feed/LNB assembly must be securely supported, especially when used at Ku-band. Any swaying can easily move the throat of the feed away from the "sweet spot" at the dish phase center. While the area that the beam must hit is the size of a tennis ball for C-band signals, this shrinks to the size of a large marble for Ku-band. Since two LNBs and the necessary feed can weigh enough to simulate the "bowling ball on a bamboo stick" phenomena, guy wires attached by S-hooks or other non-obtrusive supports may be required.

Polarity Selection

Satellites broadcast either linearly or circularly polarized signals. Feeds that detect either or both senses of polarity are available and in some cases feeds designed for one purpose can be modified to detect both senses. For example, scalar feeds which select between horizontal and vertical polarities, can be modified to detect circularly polarized signals by inserting a teflon dielectric slab, albeit at a loss of signal power. In this case a better choice is to install a new feed designed to detect both circularly and linearly polarized transmissions with little to no power loss on either type (see Figure 9-2).

Since feeds can select polarity via mechanical, ferrite or pin diode methods, the correct feed must be chosen for the intended satellite receiver. For example, if a mechanical polarizer has been used to detect analog C-band North America broadcasts and a new dual-band feed is being installed to receive both C-band as well as perhaps DirecTV/USSB digital signals, voltage-switching would be used on the Ku-band LNB.

LNBs

LNBs are designed to downconvert signals that fall within a particular frequency range. The most obvious distinction is between C- and Ku-band frequency ranges. However, an LNB intended for Ku-band broadcasts in one area may not function in another. For example, in African MultiChoice Ku-band transmissions are relayed at 12.5 to 12.75 GHz so an LNB designed for the American 11.2 to 11.7 GHz range would be useless.

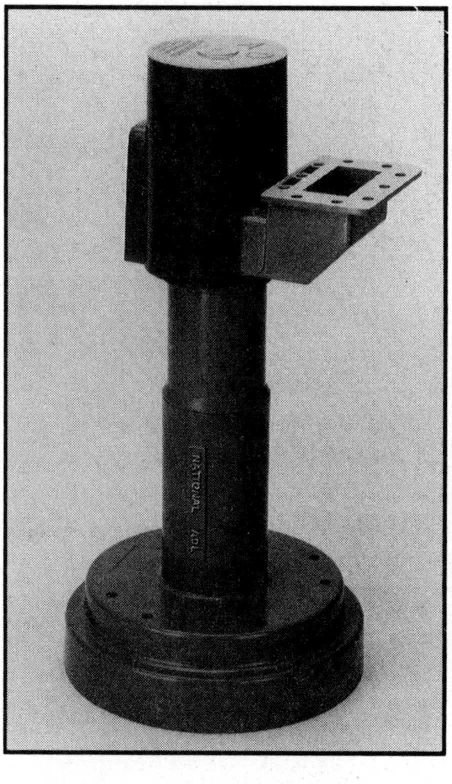

Figure 9-2. C-Band Dual-Polarity Feed. *The RP1/CP400 feed uses a die-cast quarter wave plate instead of the common teflon insert to create a feed capable of discriminating between both RHCP and LHCP as well as horizontal and vertical polarity signals. The more than 1 dB increase in C/N ratio compared to retrofitting a feed designed just for receiving only linear polarity signals makes a substantial difference in performance. This device operates over an f/D range of 0.335 to 0.425 and employs mechanical polarity selection. (Courtesy of ADL)*

To further complicate issues, LNBs are designed to handle a particular signal bandwidth. For example, LNBs intended for North America and Asia are typically built to downconvert a 500 MHz band of frequencies to the 950 to 1450 MHz range, while European and African LNBs manage a 750 MHz bandwidth and downconvert to the 950 to 1700 MHz range. While these bands overlap, an American LNB would miss the upper portion of the European channel range on some satellites.

LNBs within the Ku-band can have different input ranges. For example, while the standard 11.2 to 11.7 GHz frequency is now common on Ku-band broadcasts in the United States, DirecTV/USSB digital signals are relayed in the 12.2 to 12.7 GHz band.

Satellite Receivers

Once an LNB (or LNBs in the case of dual-band transmissions) has been selected, the next choice is the correct satellite receiver or receivers, if necessary. The fact that both C- and Ku-band LNBs can downconvert incoming signals to the same general range does make life a little simpler. However, satellite receivers designed for digital broadcasts cannot handle analog transmissions, and vice versa.

Receivers that process both C- and Ku-band analog broadcasts must have the capability to invert the video signal. This stems from the fact that C-band downconverters generally use high-side injection while Ku-band systems use low-side injection. The signals are inverted relative to each other. For example, if the local oscillator (LO) produced a mixing frequency of 5.15 GHz from which the 3.7 to 4.2 GHz C-band were subtracted, the resulting frequencies would range from 1450 to 950 MHz. If a Ku-band receiver LO injected 10.75 GHz which was subtracted from 11.7 to 12.2 GHz, the resulting band would span 950 to 1450 MHz. The same band of frequencies is also produced in this case, but it is reversed so the channel order as well as the video signal are effectively flipped. If the same channel format had been used on Ku-band broadcasts, C-band channel 1 would become Ku-band channel 23, channel 2 would become channel 24, and so on. This confusion is further complicated because different channel formats are used on different Ku-band satellites.

Dual-band analog receivers must also have a fine tuning capability. While C-band channels on many satellites are evenly spaced at 20 MHz intervals across the allocated bandwidth, no standard has evolved for the frequency allocation of Ku-band channels. Their center frequencies can range as far as 10 MHz away from those for the preset C-band channels. Since each Ku-band spacecraft of interest could conceivably have a different frequency plan, receivers that have a programmable memory capability are a great convenience. Every channel, both C-band and Ku- band, can then be programmed into memory for easy selection. Some Ku- band satellites also operate with more than the C-band standard of 24 channels. For example, Canada's Anik C3, E1 and E2 satellites carry 32 transponders. So, a fully versatile satellite receiver is required.

Digitally capable satellite receivers also process the incoming signal in a similar fashion with similar constraints to analog units. However, following tuning and limiting the signal enters a modulation/processing stage which is totally different in both varieties of receivers.

Decoders

Numerous scrambling and encryption methods are used to control access to satellite television broadcasts. Fortunately, for North American C- band broadcasts the VideoCrypt II+ is the accepted standard. But some business television is relayed using either the B-MAC or Leitch encryption system. Furthermore, the new Ku-band digital small-dish programmers use a variety of methods including Videocrypt, Digicipher and perhaps one or two others in the near future.

INSTALLATION PROCEDURES

Retrofitting an existing C-band TVRO to either C/Ku-band or operation must begin with an inspection of the existing system. Then a plan of action can be formulated and the necessary equipment can be selected.

Mechanical Requirements

In those retrofits in which a Ku-band LNB and a new feed are to be installed, the feed support must be strong enough to securely support the extra weight. Dual-band feeds are certainly heavier than, for example, a C-band scalar feed. This stability consideration may be important even when a feed is being replaced with a slightly heavier one capable of managing both linear and circular polarity signals.

Where an actuator driven C-band system is being upgraded to C/Ku-band operation, the arm may have to be replaced with a more finely geared or stable unit. Any points on the supporting structure or mount where sagging or rocking can occur must also be stabilized. Some dishes simply flop around too much and may not be of sufficient quality for retrofit. The option might be to purchase a second smaller, more accurate dish exclusively for Ku-band reception.

Wiring

When wiring a dual-frequency band installation, the signal from the Ku-band LNB may be channeled via a separate length of RG-6 coaxial cable to the Ku-band input port on a dual-band receiver. Most current receivers have dual-band capability.

Tracking the Arc

Once the necessary mechanical connections have been completed, the easiest approach is to first track the full arc of C-band satellites. Locating active transponders on Ku-band spacecraft is relatively more difficult because the larger diameter C-band dishes have more narrow beamwidths at Ku-band. Then when the C-band arc is properly tracked, final adjustments to both the feed position and mount must always be completed when aiming at Ku-band satellites. This ensures that both Ku- and C-band broadcasts are automatically targeted.

The Effect of a Dual-Band Retrofit on Signal Reception

Unless proper care is taken in upgrading a system for dual-frequency or dual-polarity reception, the end result can be a deterioration in performance on one or both signals. If the existing system had been installed with an adequate signal margin, the visible effect could be minimal. However, if a marginal system is used or if terrestrial interference (TI) exists at levels just below detection, a poorly conceived retrofit could have unacceptable results.

DIGITAL and ANALOG RECEPTION

Most current home satellite systems are designed to receive analog C-band broadcasts. However MPEG-2 digital video broadcasts (DVB) are now beginning to make some major inroads around the world. For example, Multi-Choice of South Africa launched a digital service via PanAmSat-4 to provide compressed digital signals at C-band throughout Africa and at Ku-band in southern Africa. While other broadcasters such as DirecTV, USSB and PrimeStar in the United States also already offer a compressed digital service at Ku-band, similar activity at C-band is still in its infancy. How will this evolution to new technology affect analog C-band home satellite transmissions in North America? Will upgrades to digital be economically possible and even desired?

C-band home TVROs in North America have access to possibly the widest selection of programming of any broadcast medium including cable TV, the new Ku-band digital services and certainly terrestrial methods. The central advantage is the universality of the VideoCipher II+ encryption service. The new direct broadcast service (DBS) providers such as DirecTV, USSB, PrimeStar Partners, EchoStar and AlphaStar will likely use four distinct and incompatible encryption systems.

It now seems clear the parent of VideoCipher, General Instrument Corporation, also the developer of the DigiCipher I system now used by PrimeStar, will introduce a backwards compatible DigiCipher II module for the C-band home satellite industry. The good news is that this module will be capable of deciphering analog VideoCipher II+ as well as digital DigiCipher II signals. The bad news is that consumers in North America who decide to view whatever C-band digital transmissions are available will have to purchase a new upgrade module. However, it will likely be at least a few years before the number of programmers switching to DigiCipher II make this purchase necessary.

Owners of C-band systems in North America also have another option to upgrade to digital broadcasts. An existing large-dish system can be retrofitted with a dual-band feed that accepts signals in the C-band as well as in the DBS range, 12.2 to 12.7 GHz. One coax run would then feed the C-band IRD and another a digital-ready IRD. The digital system could be used to receive premium programming; the C-band systems for all the "wild" feeds and other special interest programming not carried on digital.

CHAPTER 10

TELEVISION DISTRIBUTION METHODS

TELEVISION DISTRIBUTION SYSTEMS

Satellite TV signals can be combined with terrestrial broadcasts, VCRs, video game equipment, closed circuit television security systems or any other video and/or audio sources and then distributed to any number of customers in a home, a condo or apartment building, a hotel, mobile home park or a city. There is really no limit to the design possibilities. In all cases, one or more satellite receive dishes as well as other program sources are shared among users tied into the distribution network.

In this chapter the basic components required to assemble networks ranging from the simplest in-home distribution system to a full-scale SMATV (satellite master antenna TV) system are first examined. Then a few basic systems are presented. In the next section intermediate frequency (IF) distribution systems are outlined. In such networks each user who is connected to one or more dishes actually has an on-premise satellite receiver.

In the final section the design of "headends" and larger- scale distribution networks are studied. In such systems all satellite receivers are located at a central point, the headend. A user connected to an SMATV network tunes into one channel at a time in one of two ways: either via a standard cable converter that is modulated onto one channel (typically channel 3 or 4 in North America or a UHF channel elsewhere) or, for systems with fewer channels, via the tuner on a standard television set. Our book *Wireless Cable and SMATV* explores this subject in much greater detail.

COMPONENTS of DISTRIBUTION SYSTEMS

Even the most complex channel combination and distribution systems are assembled from similar, simple-to-understand basic components. When the different pieces of this jigsaw puzzle are independently studied, creating a virtually unlimited number of different designs becomes quite straightforward. The headend, where signals are received, processed and combined, and the distribution system includes components such as receive dishes, feeds, LNBs, coaxial cables, satellite receivers, off-air antennas, VCRs, combiners, modulators, amplifiers, A/B switches, splitters, line amplifiers, attenuation pads, terminators, dc power blocks, coaxial relays, taps, and various types of signal processing equipments to create any possible design.

A/B Switches and Combiners

An A/B switch is used to select between either one of two input signals (see Figure 10-1). A high quality A/B switch has at least 40 to 50 dB isolation between its two ports. This means that if a 10 dBmV signal is present on the input port that is not in use, it would be attenuated by 40 or 50 dB when "viewed" from the port that is in use. Thus its level at the in-use output terminal would range from -30 to -40 dBmV (10 dBmV less 40 or 50 dB). Most modern satellite receivers have built-in electronic A/B switches that select between satellite and over-the-air broadcasts. When power is turned off the conventional broadcast is automatically chosen.

A signal combiner does what the name implies. It combines satellite and over-the-air signals and can be used in place of an A/B switch, although with extra signal losses (see Figure 10-2). When these signals are combined it is important to balance their power levels before they enter the combiner. This is especially important if the combiner is of poor quality. If two adjacent channels are being combined, the lower sideband of the higher frequency channel should be eliminated with a bandpass filter. This method is discussed below.

Cables, Connectors and Splitters

Coaxial cables, connectors and splitters are used to carry signals to any chosen destination. Each component attenuates power in a characteristic way that depends upon the frequency of the signal. This behavior must be accounted for when designing a distribution system.

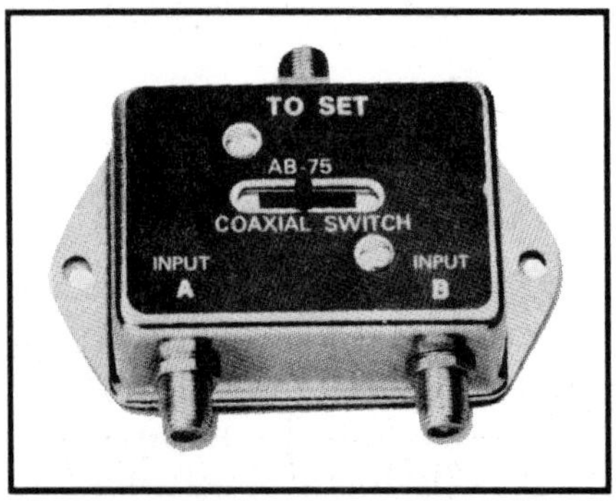

Figure 10-1. A Coaxial A/B Switch. *A/B switches come in a variety of models with isolation between ports ranging from 30 to 90 dB. They have a variety of uses such as switching between a satellite signal and a terrestrial over-the-air broadcast. A/B switches are built into many satellite receivers (Courtesy of Pico Products)*

Signal attenuations in standard 75 ohm cables such as RG-6, RG-11 and RG-59 have been outlined in Chapter 5. A ballpark figure for RG-59, a standard cable used in video distribution networks, is 2.1 dB per 100 feet (6.6 dB per 100 meter) at 50 MHz and 4.0 dB per 100 feet (13.1 dB per 100 meter) at 200 MHz.

Splitters do exactly what their name implies, they split or divide a signal into two or more branches (see Figure 10-3). In certain situations, these devices may also be used in reverse to combine signals, although this is not a recommended practice. Each type of splitter is rated to handle a specified frequency range. For example, when dividing the 950 to 1450 MHz output of an LNB, a splitter rated up to 1450 MHz (1700 MHz in Europe and Africa) must be used or losses will be high. Some

Figure 10-2. Signal Combiner. *This combiner consists of a high quality, 6-stage bandpass/bandstop filter with adjustable balancing attenuators. It allows 3, 4 or 6 signals from a videotape recorder, TVRO or video game to be combined on a distribution system carrying adjacent channels. Properly balancing signal levels in an MATV system eliminates the need for an A/B switch.*

brands also have built-in bandpass filters so that frequencies below and sometimes above the designated range are sharply attenuated. Limiting the bandwidth in this manner can also protect cables from ingress interference. Typical insertion losses for two, three, four, eight and sixteen-way splitters are listed in Table 10-1.

A two-way splitter cuts the signal a little more than in half. This is because the power reductions include an "insertion loss" as well as a 3 dB reduction, a halving of power. A 3-way splitter has one 3.5 dB and two 7 dB output ports; a 4-way splitter has four -7 dB ports. For example, if a signal of 3 dBmV enters a 4-way splitter, each output leg would produce a signal of 3 dBmV reduced by 7 dB which equals 1964 dBmV. This voltage level would not be enough to produce a quality picture because it is well below acceptable input levels for a television. Noise would begin to overpower the signal. If this signal were also relayed down 200 feet of RG-59 cable, additional losses of 4.2 dB at 200 MHz would result in an even weaker signal of -8.2 dBmV at the TV input.

F-connectors and at times even splitters are occasionally mounted onto wall plates so that components can be hooked into a distribution system in an attractive fashion.

Line Amplifiers

Line amplifiers that are inserted via F-connectors directly into a coax line are available for boosting signals spanning either a single channel or a range of channels. Less-expensive, simpler units are powered from the dc voltage relayed down the coax (see Figure 10-4). Most of

TABLE 10-1 SPLITTER LOSSES PER LEG

Type of Splitter	Loss/Leg (dB)
2-way	3.5
3-way	3.5 and 7
4-way	7
8-way	10.5
16-way	14

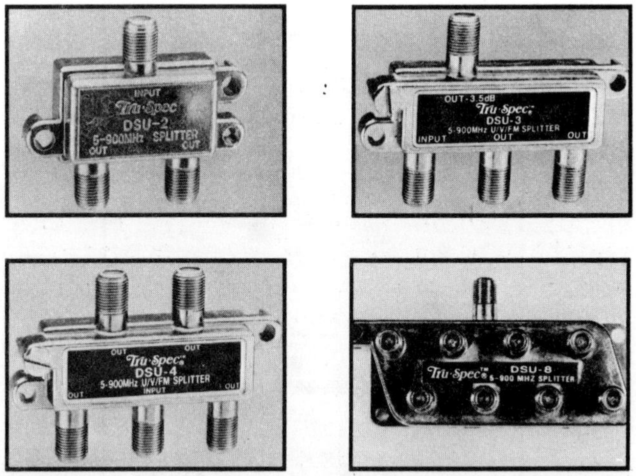

Figure 10-3. Splitters. *2, 3, 4 and 8-way splitters are most commonly employed when dividing signals in distribution system applications. The usable frequency range of this brand ranges from 5 to 900 MHz which extends past the UHF channel frequency band. Others rated up to 2000 MHz are available for splitting output signals from LNBs.*

these inexpensive amplifiers which operate in the LNB output range are installed between the LNB output and receiver input. Similar amplifiers operating in the 5 to 950 MHz range in an over-the-air distribution system are installed between the pre-amp output and the channel processor. More advanced MATV (master antenna TV) commercial amplifiers draw current from a regular wall outlet and convert it to dc power via an internal power supply (see Figure 10-5). This is similar to the way preamps designed for distributing over-the-air broadcasts over cable systems operate.

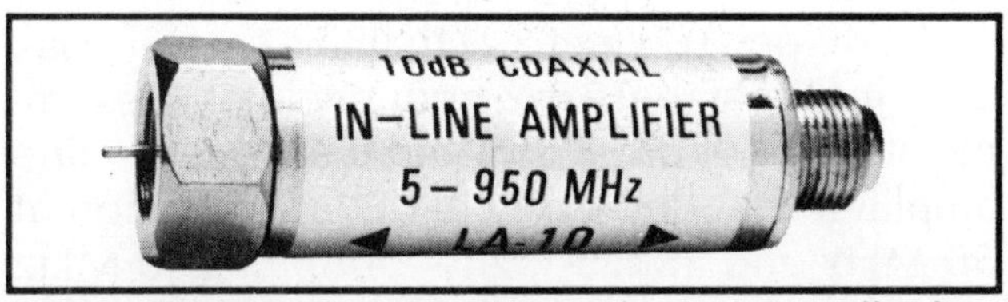

Figure 10-4. Bullet Amplifier. *The LA-10 Bullet line amplifier is rated for frequencies from 5 to 950 MHz and is normally used in SMATV distribution systems or in conjunction with an LNB having a 400 to 950 MHz output. It can be inserted as a line amplifier between a pre-amp output and a channel processor input in long cable runs, but must receive line power to operate. (Courtesy of Pico Products)*

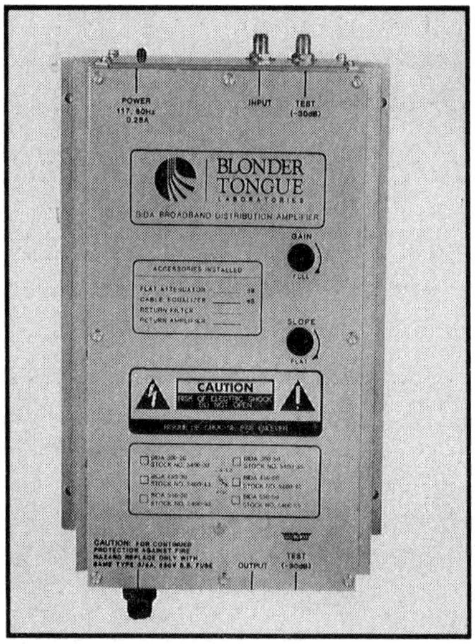

Figure 10-14. CATV Amplifier. *The BIDA series of broadband amplifiers are available in gains of 30 or 50 dB. They have built-in variable gain and slope control of 10 dB. Input and output test ports allow testing without interrupting service. These units are also field-upgradable to two-way capability. Attenuator and cable equalization options are available. (Courtesy of Blonder-Tongue)*

Amplifier Tilt

Line amplifiers can have either fixed or adjustable gains. Some types designed for longer cable runs have gains that increase with frequency to compensate for the increased attenuation of signals at these higher frequencies. The variation in attenuation with frequency is called tilt. For example, if RG-59 is used a tilt of about 2 dB per 100 feet between 50 and 200 MHz (North American channels 2 and 13) is necessary for cable attenuation compensation. This is based on the difference between the 4 and 2 dB per 100 feet loss that this type of coax contributes at these two frequencies. For example, if a cable run of 200 feet were used, a line amplifier that delivered 20 dB amplification at 50 MHz and increasing to 24 dB at 200 MHz would properly compensate for the 4 dB difference in cable losses between these frequencies. As a result the output at the end of the span of cable would be flat across the entire range of frequencies. This equalizing of signal levels reduces the possibility of adjacent channels interfering with each other and generating intermodulation distortion.

Attenuation Pads

An attenuation pad, more simply known as a pad, is used to reduce the strength of an overly powerful signal (see Figure 10-6). Pads are small, inexpensive devices that insert directly into a coax line via F-connector fittings. They are available with either fixed or variable rated attenuations and occasionally with built-in tilt compensation. Most pads are not designed to pass dc power so they cannot be used in the line between receivers and LNBs.

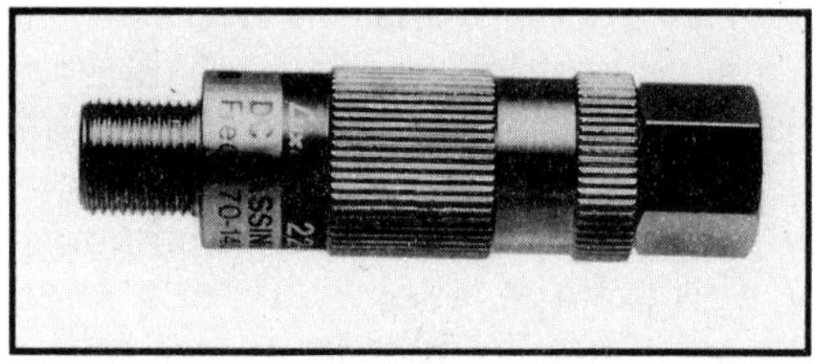

Figure 10-6. Attenuation Pad. *This pad has one male and one female port for insertion into a standard RG-59 or RG-6 coaxial line. It provides 10 dB of signal attenuation and passes dc power.*

Terminators

Any output port on a video or audio distribution network must end in an appropriate device such as a television set, a satellite receiver or a terminator (see Figure 10-7). If not, interference could leak into the unconnected port or the resulting impedance mismatch could cause signals to be reflected back into the system as "ghosts." A terminator is simply an encased resistor which screws onto an F-connector fitting. It has a matching impedance, 75-ohms in the case of RG-59 or RG-6. So the cable does not "see" an opened end or discontinuity at the unused port but behaves as if the cable effectively had no end.

dc Power Blocks

A dc power block is a simple device that allows higher frequency TV signals to pass unattenuated, but blocks the passage of any direct current (see Figure 10-8). The purpose of a dc block is to isolate various components in a distribution system from dc voltages. This device is nothing more than a simple circuit element called a capacitor encased with a male and female F-connector fitting at either end.

Input Selectors

Input selectors allow the user to select among two or more coaxial cable inputs, often using the polarizer voltage control output of a satellite receiver for the voltage source (see Figures 8-9 and 8-10). These devices are electrically operated A/B switches that pass signals in only one direction at a time while isolating the unused port by at least 40 dB from the output port. They are quite similar to the voltage-controlled devices used to switch some types of polarizers.

Input selectors are commonly used in smaller scale distribution networks where distances are short enough to enable the transmission of the block IFs via cable, as described below. In dual-band systems, selectors are often used to switch between two input cables. Input selectors are also occasionally used in conjunction with dc power inserters.

Figure 10-7. Terminator. *Terminators are installed on an unused splitter port or at the end of a distribution line in order to provide a 75 ohm impedance match. Doing so eliminates both signal reflections and ingress interference. (Courtesy of Pico Products)*

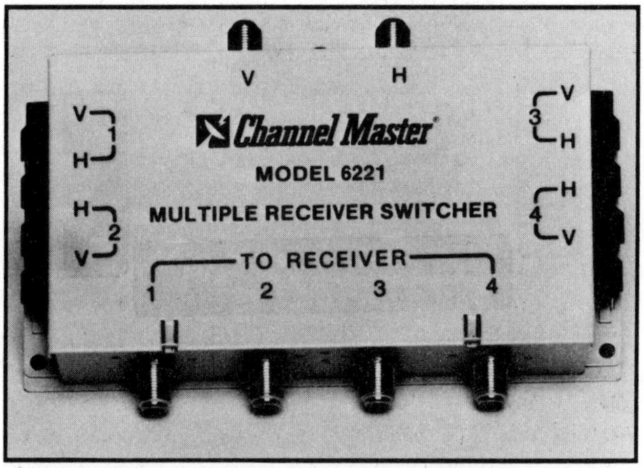

Figure 10-9. Channel Master Multiple Receiver Switcher. *This switcher/divider is designed to take a horizontal and vertical LNB input, divide this signal and then select a path to any of four satellite receivers giving each independent polarity control. (Courtesy of Channel Master)*

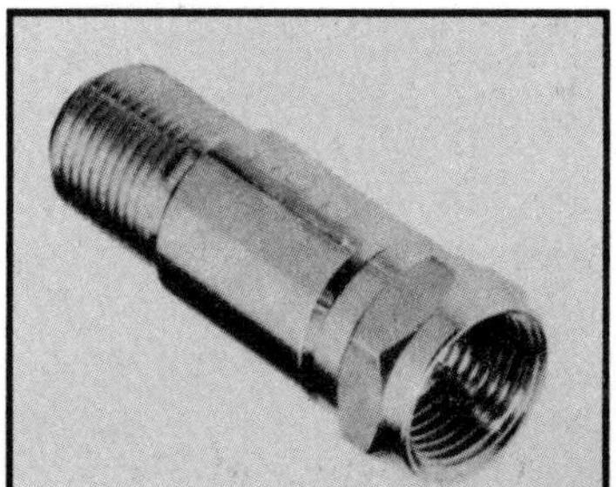

Figure 10-8. Voltage Blocking Coupler. *VBCs are used to block any dc or low frequency ac power which might be present on a coaxial cable. They can be used to protect equipment such as televisions or some types of satellite block distribution system relays from damaging power surges. (Courtesy of Pico Products)*

Figure 10-10. Global Magic Switch. *This 4 x 4 switch allows up to four dishes to be connected to four or more separate receivers. It is driven by dc pulses generated by most receivers. For more details see Figure 10-. (Courtesy of Global Communications)*

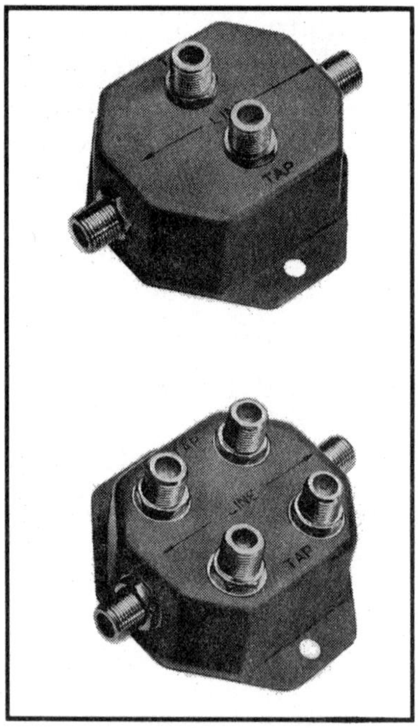

Figure 10-11. UHF/VHF/FM Multiport Tap. *This tap operates in the 5 to 890 MHz frequency range and is ac/dc isolated on its output ports. It is available in 12, 17 and 23 dB tap values with insertion losses at VHF frequencies of 1.5, 1.3 and 1.0 dB, respectively. (Courtesy of Qintar, Inc.)*

Taps

A tap, also known as a directional coupler or directional tap, extracts a specified portion of an incoming signal while allowing most power to pass through to its output (see Figure 10-11). For example, feeding a 30 dBmv signal into a 24 dB tap siphons off 6 dBmv and passes the remaining portion, less a small insertion loss of about 0.5 dB, through its output port. As seen in Table 10-2, the lower the rated tap value, the more signal is extracted and the higher the insertion loss. The actual insertion losses for any particular brand of tap can be obtained from the manufacturer.

To illustrate, a 40 dBmV signal is reduced to 39.5 dBmV after passing through a 24 dB tap (with an insertion loss of 0.5 dB) but to 37.8 dBmV after passing through a 6 dB tap (with an insertion loss of 2.2 dB). The former tap extracts 16 dBmV of signal; the latter 34 dBmV.

TABLE 10-2. TAPS and TYPICAL INSERTION LOSSES

Rated Tap Value (dB)	Insertion Loss (dB)
30	0.5
27	0.5
24	0.5
20	0.5
16	0.8
12	1.0
9	1.5
6	2.2

A tap differs from a splitter which divides a signal equally into two or more output legs (a 3-way splitter, the only exception, has two equal output ports and one with twice the power). But both types of signal allocation devices are used to accomplish the same function. Taps are usually installed to pull signals off a main feeder line since the throughput losses are much lower than splitters. Following this point, splitters are used for distribution on local cable branches to individual TV sets.

Modulators, Signal Processors and Bandpass Filters

The purpose of a modulator is to "rebroadcast" and convert a television signal from one frequency range to another. For example, those modulators that are built into a satellite receiver take the composite baseband audio and video satellite signals and translate them onto a standard terrestrial AM broadcast format. A modulator-like device known as a channel converter shifts the frequency range of one particular channel onto that of another. For example, if all the VHF channels were being used, a converter might accept VHF channel 7 and translate it onto UHF channel 14. The frequency selection circuits in such converters are usually crystal controlled to prevent frequency drift.

The baseband output signal from a satellite receiver can be modulated onto any chosen VHF or UHF television channel. In those cases where some channels are occupied by local off-air broadcasts, satellite signals can be modulated onto any other desired frequency. The FCC allocation of all 58 television frequency slots in the United States is listed in Table 10-3.

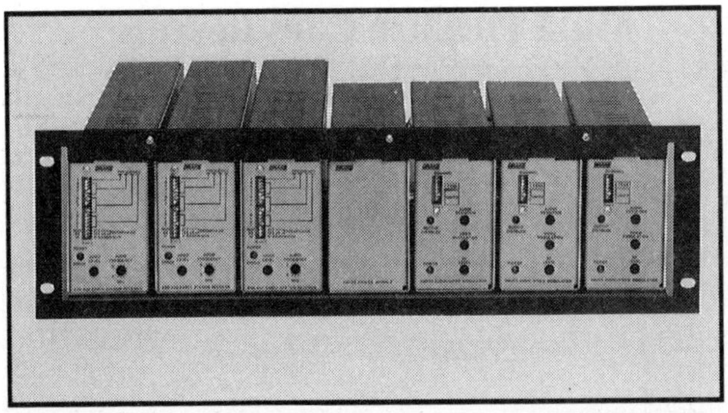

Figure 10-12. Satellite Receiver/Modulator Combination. *The three frequency agile modulators in the right of the case span PAL VHF channel 2 to 13 and are designed for the international market. This headend features a synthesized video carrier and selectable audio intercarrier of 5.5 MHz or 6.0 MHz. The output level is adjustable across a 14 dB range from the front panel (Courtesy of R.L. Drake).*

Modulators fall into three broad classes. "Home style" brands have evolved along with home VCR technology. In North America and Asia these usually modulate signals onto channel 3 or 4 in the VHF range; in Africa and Europe onto one channel in the UHF 21 through 69 range. They have low output voltages, typically on the order of 0 to 3 dBmv but always lower than 10 dBmv. This variety does not have built-in bandpass filters to eliminate the lower video sideband. Commercial grade modulators have much higher output powers, with gains typically greater than 30 dB (see Figure 10-12). A distinction is drawn between those brands of commercial modulators with and without built-in bandpass or SAW filters (surface acoustical wave filters which have a very sharp frequency cutoff). Those not incorporating filters can generate a signal that may interfere with an upper or lower adjacent channel when adjacent channels are being distributed.

Less expensive crystal stabilized modulators typically are permanently set on one output channel. Many crystal stabilized modulators used in home receiver modulators in North America can usually be switched between

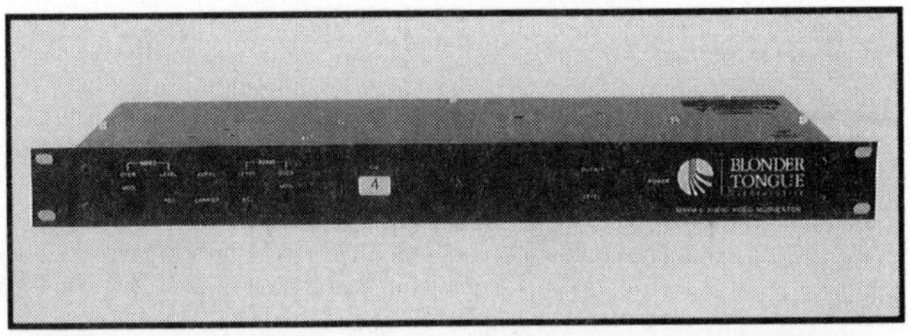

Figure 10-13. MAVM-s Audio/Video Modulator. *This modulator uses a crystal-referenced phase-lock-loop oscillator to provide an extremely stable output. An external IF loop allows signal scrambling systems to be interfaced. SAW filters are used to preserve vestigial sidebands. (Courtesy of Blonder-Tongue Laboratories)*

channels 3 and 4, although a different pair of output channels is occasionally chosen. Those modulators which are LC (an abbreviation for inductor/capacitor) driven can be tuned by a small set screw across a range of channels. When this adjustment is made a non- metallic screwdriver must be used because metal affects the circuit elements and can make fine tuning onto the selected channel difficult. The set usually needs to be fine tuned after this to match the modulator and TV frequencies.

Commercial modulators have adjustable audio and video output levels as well as RF output controls (see Figure 10-13). The audio and video settings control the "percentage of modulation" or how much of these signals are added to the carrier. Higher video outputs make for a brighter TV picture but too high a level causes a buzz in the audio and washed out pictures and could damage the front-end of a TV receiver. The RF adjustment sets the output voltage level.

TABLE 10-3. VHF and UHF TELEVISION CHANNELS

Designation	Channels	Number of Channels	Frequency Range (MHz)
Lo-VHF	2 to 6	5	54 - 88
FM Radio			88 - 108
Aeronautica	I		108 - 120
Midband	A to I	9	120 - 174
Hi-VHF	7 to 13	7	174 - 216
Superband	J to W	14	216 - 300
Hyperband	AA to WW	23	300 - 438

Channel processors also have outputs that can be combined with signals from modulators. However, processors take already modulated signals, usually from terrestrial broadcasts, and "clean up" the signal. They do this by demodulating down to baseband and amplifying and filtering both the input and output signal to eliminate the unwanted sideband. Then the output is remodulated onto the same channel as the input signal.

Multi-Purpose Components

Some manufacturers have introduced multipurpose components in an effort to simplify installing common system designs. For example, a 4-way coaxial relay might be integrated with four dc power blocks and an output line amplifier all in one box. Examples of such components are shown in Figures 10-9 and 10-10.

BASIC DISTRIBUTION SYSTEMS

The function of a distribution system is to take the signal from a satellite receiver or the combined signal from the headend and deliver a high quality signal to each device it feeds.

The design of a distribution system is conceptually simple. Once the signal voltage at the output of the satellite receiver or the SMATV headend "launch" amplifier is known, it is a matter of subtraction to calculate the losses at each tap, splitter or cable run to guarantee that an adequate signal reaches each set. If losses are too great, distribution amplifiers are inserted at critical points. The noise added by amps must be low enough to be consistent with the objective of preserving channel S/N ratios.

A distribution network can be compared to water distribution systems. With insufficient pressure in the system, water will just trickle out of shower heads; with too much pressure, pipes can burst and water knocks develop. When receiving television broadcasts, snow can appear if the signal is too weak and picture flutter or jitter may be seen if the signal is overdriven.

Two types of distribution systems can be designed using either taps or splitters. Splitters are generally used in smaller systems. Designs ranging from larger SMATV systems to citywide cable networks use a combination of taps feeding the signal to a customer and then use splitters to feed additional television receivers. A main trunk line is never tapped but uses only special hybrid splitters/amplifiers so that the distribution system is well protected from ingress interference. Then signals are tapped to individual customers from feeder lines. For example, if a hotel having two 5-room buildings were being cabled, a 2-way splitter might be used to feed each building. Then a network of either taps or splitters could be used in each building to feed all five rooms.

Examples of Distribution Systems

Two simple distribution systems, one using taps and one using splitters are shown below in Figure 10-14. This diagram demonstrates how power levels can be traced from their source in the headend to their final destination.

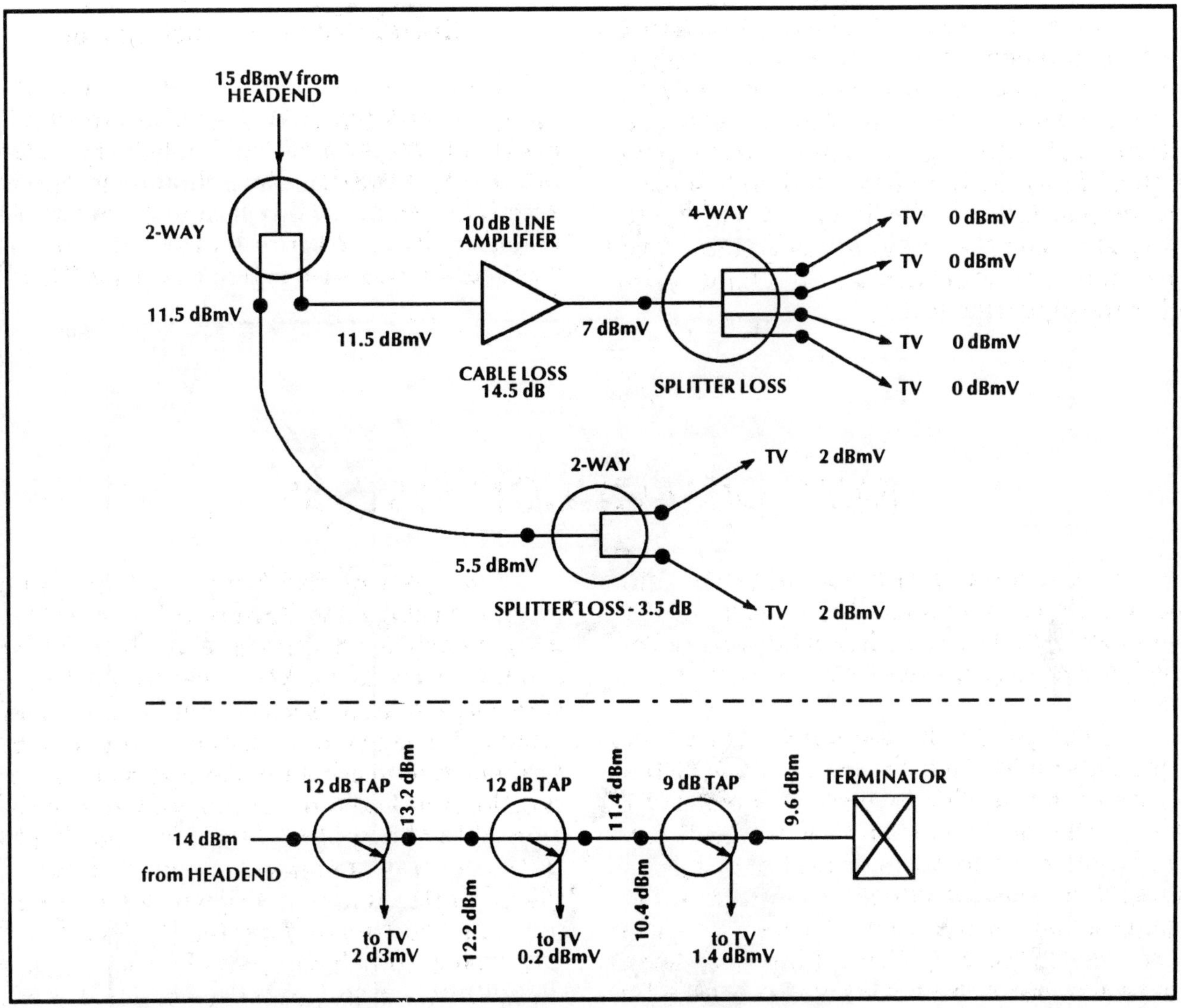

Figure 10-14. Sample Distribution Systems. *These two examples illustrate how power levels are calculated for both tap and splitter distribution networks.*

Television Signal Requirements

The ultimate target of broadcast signals that are processed by a headend and distributed by a cable network is the television set and audio system. Televisions function best with an input signal level of 0 to +3 dBmV, although the optimal level can vary among various brands. The designation dBmV stands for decibels relative to 1 mV. For example, 0 dBmV equals 1 millivolt and 3 dBmV equals 2 mv. When signal levels exceed approximately +3 dBmV, some older model TVs can be overdriven and pictures will be distorted. However, most sets now have automatic gain control circuits that compensate for overly strong signals. This allows inputs exceeding 3 dBmV to be managed, to a point. But the 0 to 3 dBmV is a good target range which should be achieved for best results at TV inputs throughout the distribution system.

It is much easier to manage signals that are too powerful rather than too weak. If the input C/N ratio has been allowed to drop too low or is "in the mud," no level of amplification can improve the situation because, at this point, as much noise as signal is amplified. However, when signal power levels exceed the upper limits, attenuation pads can always be used. Therefore, too much power is never really a problem but allowing signal strength to deteriorate can be a cause for concern.

BASIC TVRO SYSTEMS

Basic Off-Air/Satellite TV System

In the most basic satellite TV/terrestrial system, an LNB relays the signal directly to a satellite receiver. The terrestrial off-air antenna with optional pre-amp and power supply feeds a signal into the rear panel of the receiver. Modern satellite receivers have an internal switch that is used to select between off-air and satellite television signals (see Figure 10-15).

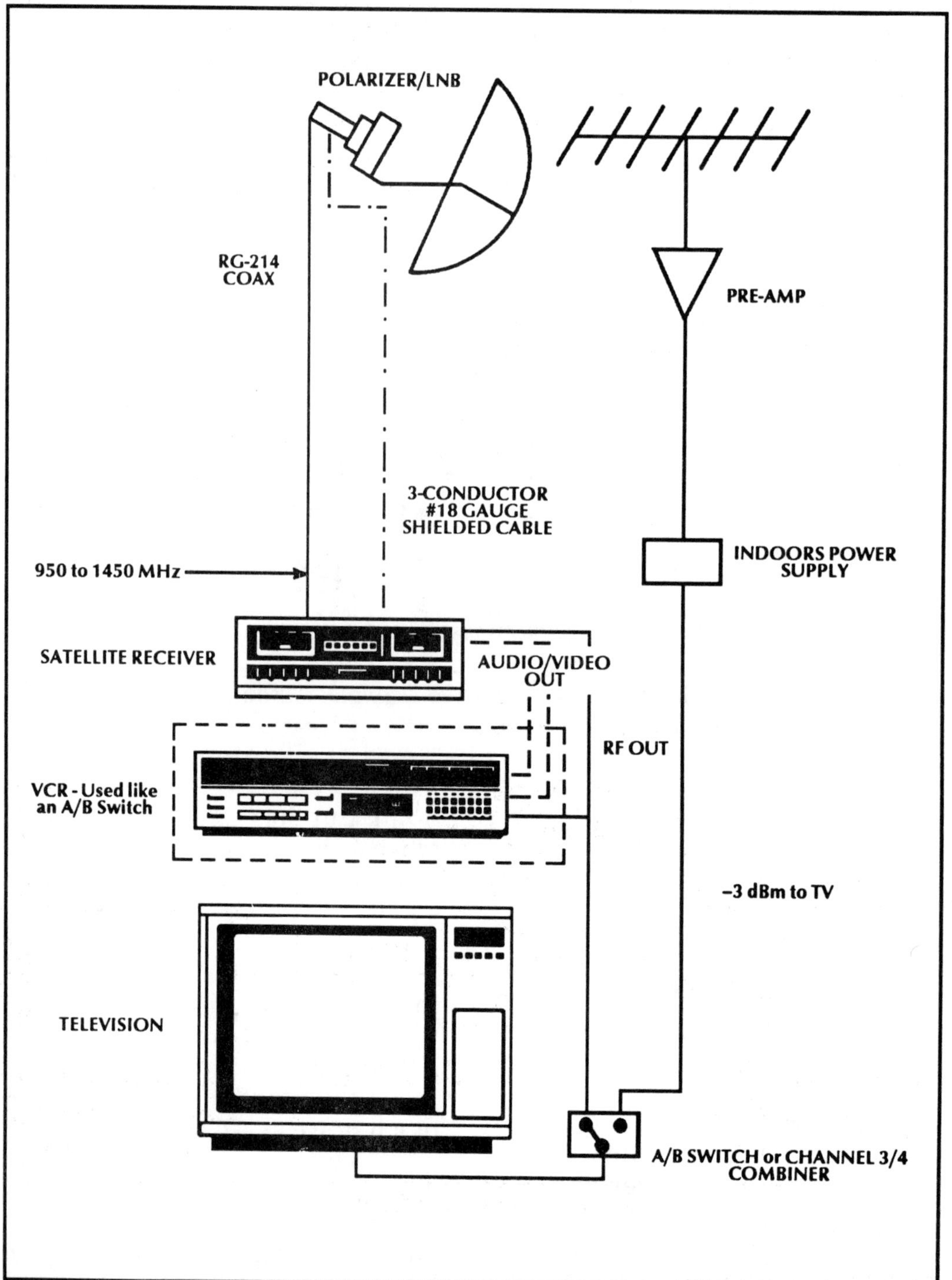

Figure 10-15. A Basic Off-Air / Satellite TV System.

Single Receiver, Two TVs and Extra Remote Control Headend

This design is similar to the basic system except that an extra TV has been added (see Figure 10-16). The main television is located near the satellite receiver which is assumed to have an infrared, line-of-sight remote control. The satellite receiver can also be controlled from the second infrared transmitter, which is located in another room some distance away but usually in the same building, by use of an extra remote control which communicates via the coax to the receiver. Using such a "remote-remote" can be very convenient for controlling the receiver from a bedroom or family room. This system can be controlled from up to four rooms.

Some brands of receivers feature a UHF radio frequency, hand-held remote. The satellite receiver can then be controlled from any location within a radius of about about 200 feet (60 meters) without the need for extra wires. Even though this remote operates on UHF frequencies in the vicinity of automatic garage door opener range, it is rare that these two different systems interfere with each other. When extra range is needed on such a UHF remote, additional antennas can be installed by running coax from the antenna output port to remote locations via standard cable and splitters.

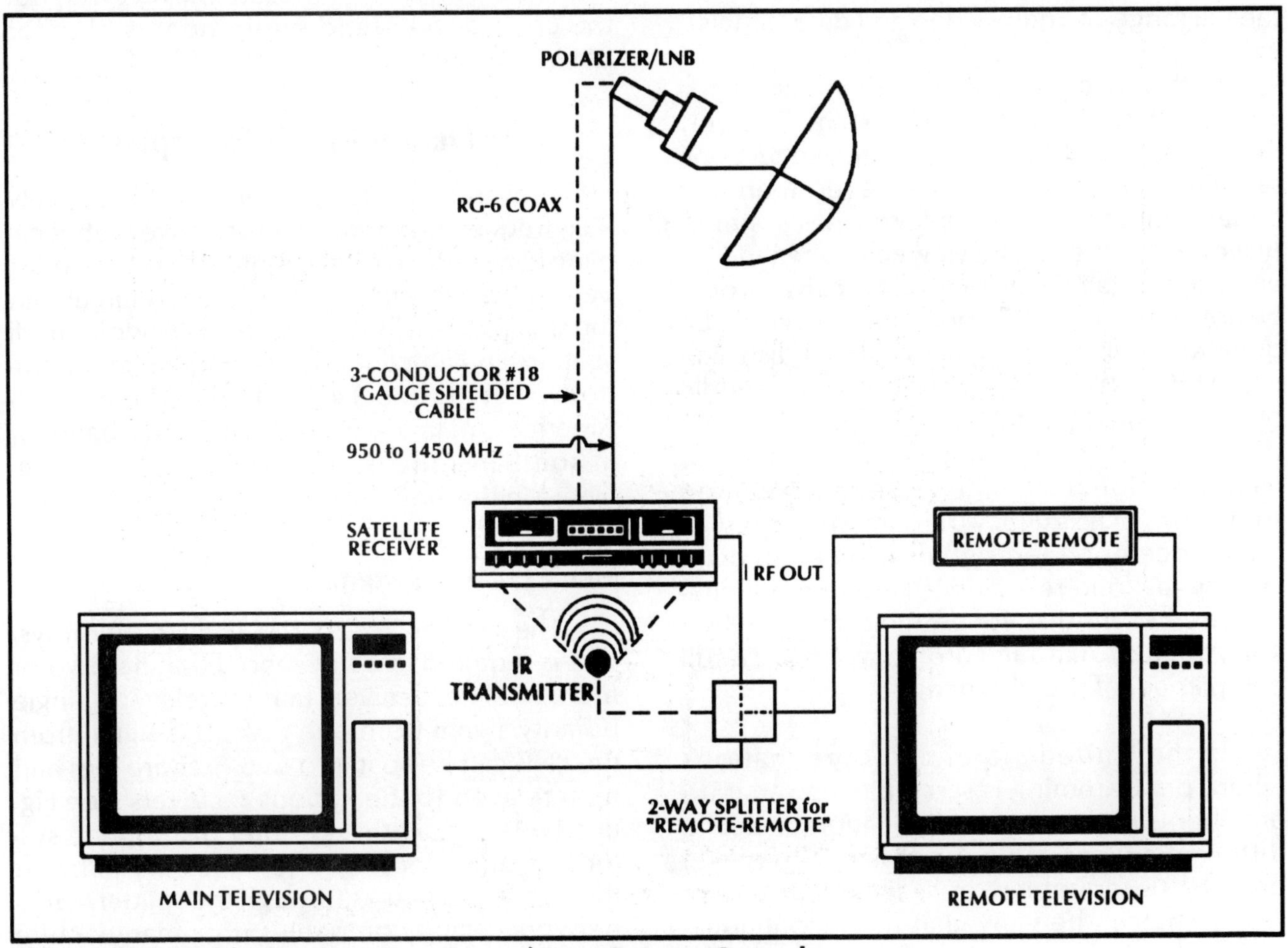

Figure 10-16. Single Receiver, Two TVs and Extra Remote Control

IF DISTRIBUTION of SATELLITE SIGNALS

In a conventional SMATV system, all incoming signals are converted to baseband, processed and remodulated onto frequencies into the VHF, midband or UHF ranges. These are then combined and distributed. The SMATV operator chooses programs from a variety of sources to be distributed. An alternative method that is much more attractive in some instances is IF distribution of satellite signals.

With IF distribution the modulated satellite signals are received and then directly distributed from the output of one or more LNBs. There is no intermediate processing other than amplification, if necessary. The entire LNB output band is thus received by subscribers, each of whom has a satellite receiver that can tune to the full range of audio, video and data signals.

Each subscriber connected into an IF network effectively has a TVRO system. The only difference is that the dish is shared so just one satellite can be targeted. However, there are certain limitations in multiple receiver systems. Just one satellite can be viewed at one time for each dish installed and only one receiver would be able to control dish movement. It would be quite annoying to have an actuated dish that might be moved by one customer in the middle of a movie being watched by another. An IF distribution net requires an additional dish for each additional satellite to be received and some form of frequency translation to avoid interference between signals from each dish. Having just one receive dish is a minor limitation for systems that, for example, receive PAS-4 in Africa or Astra in Europe, and have a full compliment of programming.

In the United States and other nations where programming is scrambled, a separate decoder is required at each location to view a different subscription program. Otherwise, only the unscrambled, non- subscription programming can be viewed at the second location. The user must pay a fee for both descramblers which each have different authorization numbers. Programmers have no method of knowing if more than one decoder is in the same residence.

There are some economic and performance tradeoffs between IF and SMATV distribution systems. Subscribers connected into an IF network can tune into all programs carried by the satellites that feed the system, whether they are audio, video or data. An SMATV customer views only the programming selected by the operator. While an IF system has a comparatively inexpensive headend, SMATV is much more economical for larger scale networks. Another way of stating the same fact is that the cost for subscriber equipment is quite high for an IF network, primarily due to the need for each household to own a satellite receiver, but the cost for headend equipment is high for SMATV systems.

IF Distribution – Examples

IF distribution systems operate at relatively high frequencies and therefore have higher cabling losses. RG-6 or even lower loss coax must be installed to minimize attenuation in all but the smallest distribution systems. In addition, all splitters and amplifiers must be rated for operation in the 950 to 1450 MHz band in North America and in the 950 to 1750 MHz band or, in some instances, 950 to 2050 MHz, elsewhere in the world.

Single Polarity System

The simplest form of an IF distribution system is required in a household that has two or more satellite receivers that are relayed a single polarity signal from one dish. The signal from the LNB can be split into two or more legs and then relayed to the various receivers (see Figure 10-17). As with any other distribution system, an amplifier might be necessary to boost the signal as it passes through the splitter/cable network. Some companies now manufacture "lossless" splitters that not only divide but also amplify the signal. These multifunction components are ideal for small scale IF distribution networks.

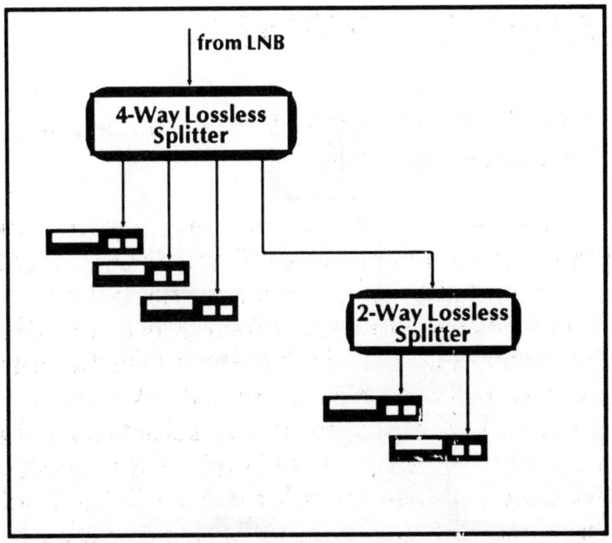

Figure 10-17. One Polarity IF Distribution Net. *Channels of a single polarity are distributed at IF in this illustration. The LNB output is split by a "lossless" 4-way splitter, one of whose legs also feeds a lossless 2-way splitter.*

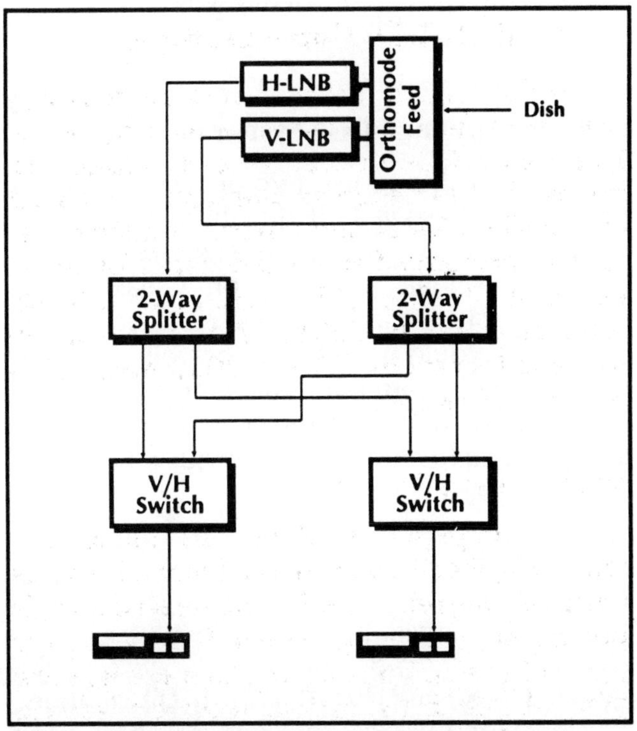

Figure 10-18. Dual-Polarity IF Distribution. *Channels of both horizontal and vertical polarities are distributed here. Two LNBs on an orthomode feed send signals to 2-way splitters. These each route both outputs to V/H switches controlled by the 13/17 Vdc receiver polarity switching voltage. Both satellite receivers can then choose either polarity independently of the other.*

Dual-Polarity Single-Dish Systems

An IF system designed to distribute signals of two separate polarities requires the use of two LNBs, one for each polarity. Although two receivers in one household can receive both polarities from just one LNB, conflicts would likely arise between which receiver controls polarity selection.

The basic dual-polarity system is diagrammed in Figure 10-18. Two LNBs installed on a single orthomode feed or a single "dual-output" LNB that has two LNBs and an orthomode feed in one unit can be installed to feed the system. Each signal, assumed here to be vertical and horizontal polarities, is split into two branches and then routed into V/H switches. The 13/17 Vdc polarity switching voltage relayed along the coax from each receiver controls a switch that selects either the vertical or horizontal signal. The combination of two 2-way splitters and V/H switches is now available in self-contained units, switching matrices, from various manufacturers. Some of these are also lossless, like the splitters mentioned above.

The V/H switches used must have interport isolations of at least 20 dB and preferably 25 dB to prevent signals from one channel interfering with the other. In addition, the cable runs from the LNB, often side-by-side, to the switches should be well shielded to prevent cross- talk. Double-braided cables that are a minimum of 90% shielded should be used.

Dual-polarity IF distribution systems can feed any number of televisions by simply cascading 2 by 2 or 2 by 4 switching matrices. A 2 by 2 matrix has two inputs while a 2 by 4 matrix has two inputs and four outputs. Each output is connected to a satellite receiver whose switching voltage control with polarity is received (see Figure 10- 19). Such matrices that also accept an off-air antenna input can also be factored into such designs (see Figure 10-20).

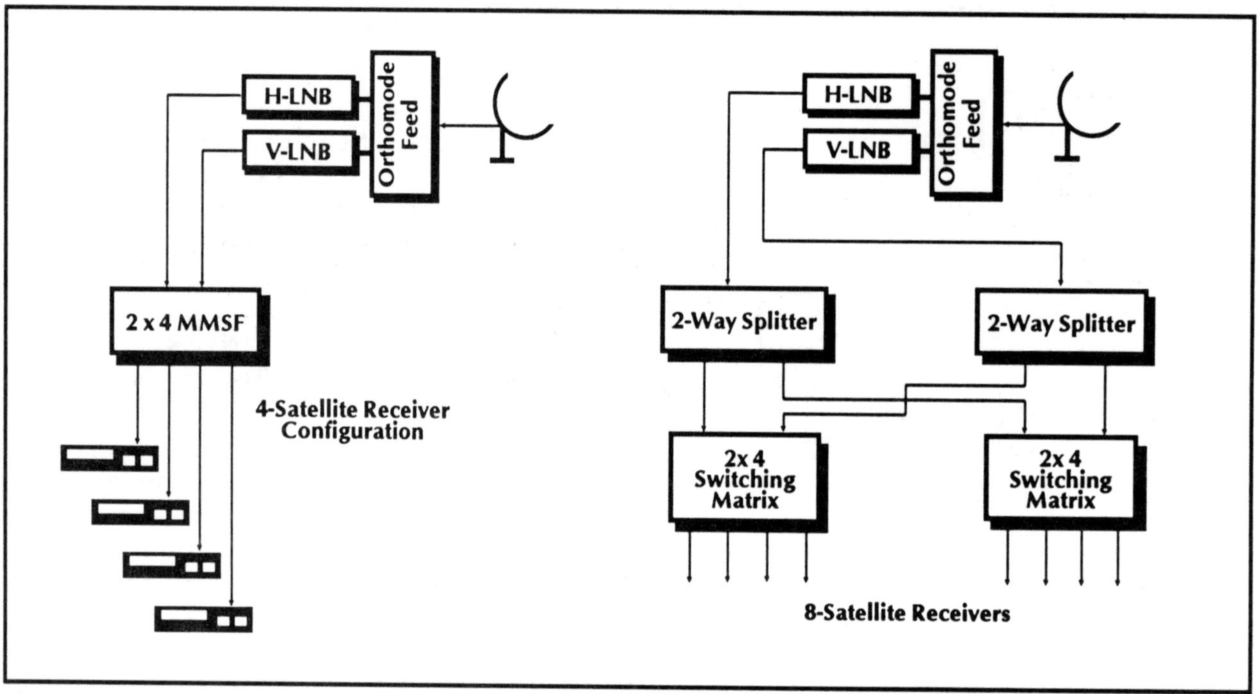

Figure 10-19. 4 and 8-Port IF Distribution Configurations. *Switching matrices, 2 by 2 or 2 by 4, can be cascaded to create distribution systems that feed any number of ports. The MMSF matrix switches from Global Communications are assumed in this diagram. Each receiver can independently switch between V and H.*

Full-scale IF distribution systems can be created in the same fashion that SMATV nets are designed via use of switching matrices in the form of conventional taps (see Figure 10-21).

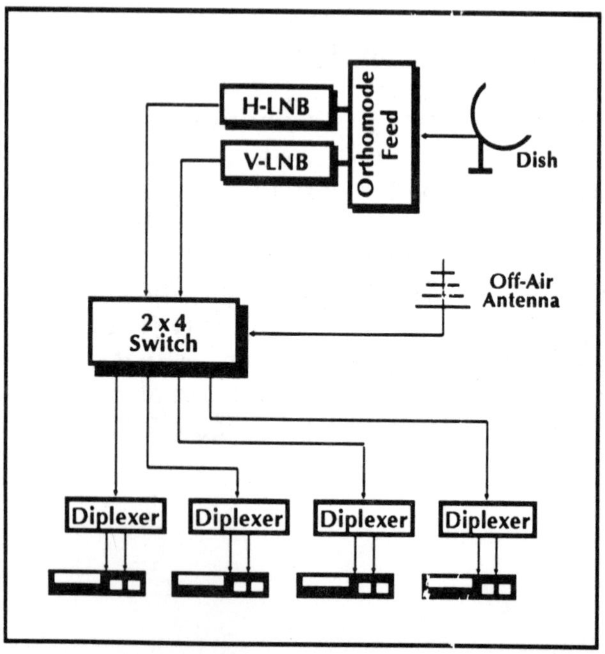

Figure 10-20. Off-Air and Dual-Polarity IF Distribution. *Both horizontal and vertical satellite as well as off-air signals are distributed here. Two LNBs on an orthomode feed both feed a 2 x 4 switching matrix. Off-air signals are also fed into this switch. A diplexer, basically a bandpass filter, that separates out the satellite and off-air signals, is used before the input to each satellite receiver. The receiver acts to switch between H and V satellite signals via its 13/17 Vdc voltage and also selects between off-air or satellite via an internal A/B switch.*

Dual-Polarity Dual-Dish Systems

It is possible to use two or more dishes as program sources in an IF distribution system. A system in which one dish receives vertical and horizontal signals, perhaps from Galaxy 6 and Anik E1, is shown in Figure 10-22. Each switching matrix would have to be controlled by four voltage levels from each receiver. Some satellite receivers are configured to deliver 0, 3, 6 and 9 dVc to accomplish this requirement.

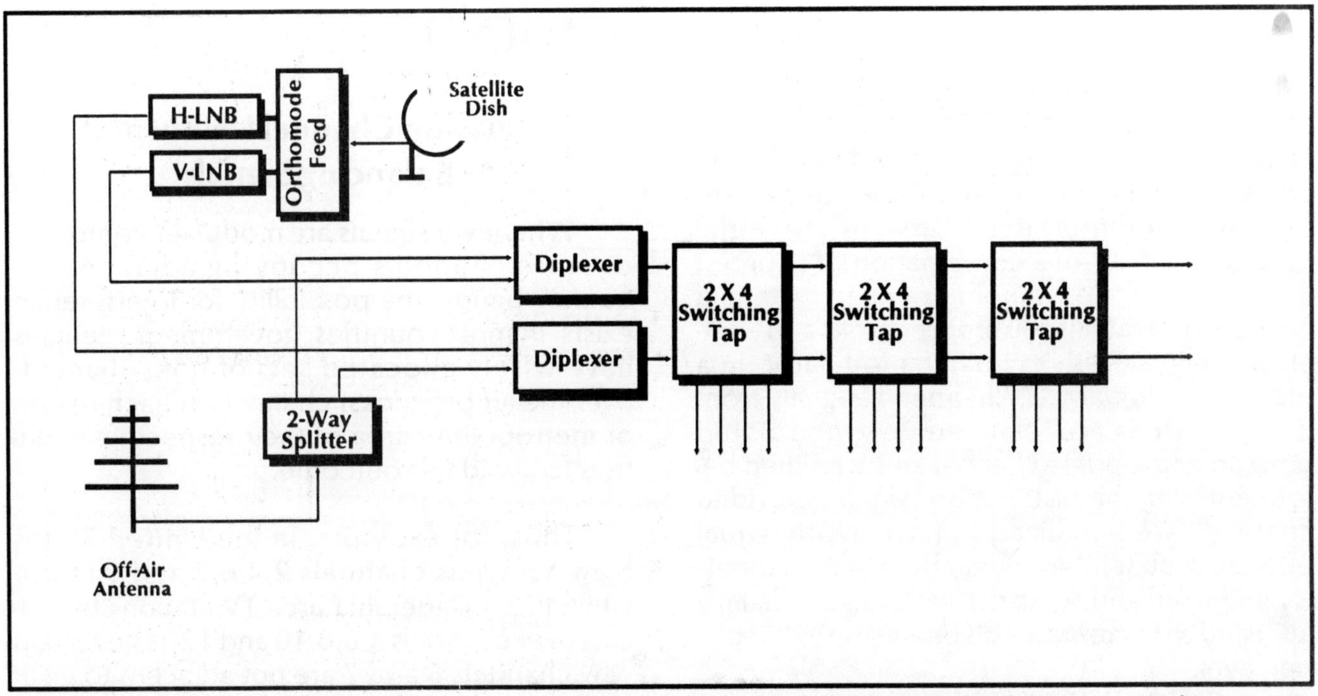

Figure 10-21. IF Tap Distribution System. *Both terrestrial and satellite signals feed this tap switching matrix distribution system. Each matrix, independently switched by each satellite receiver, feeds four customers.*

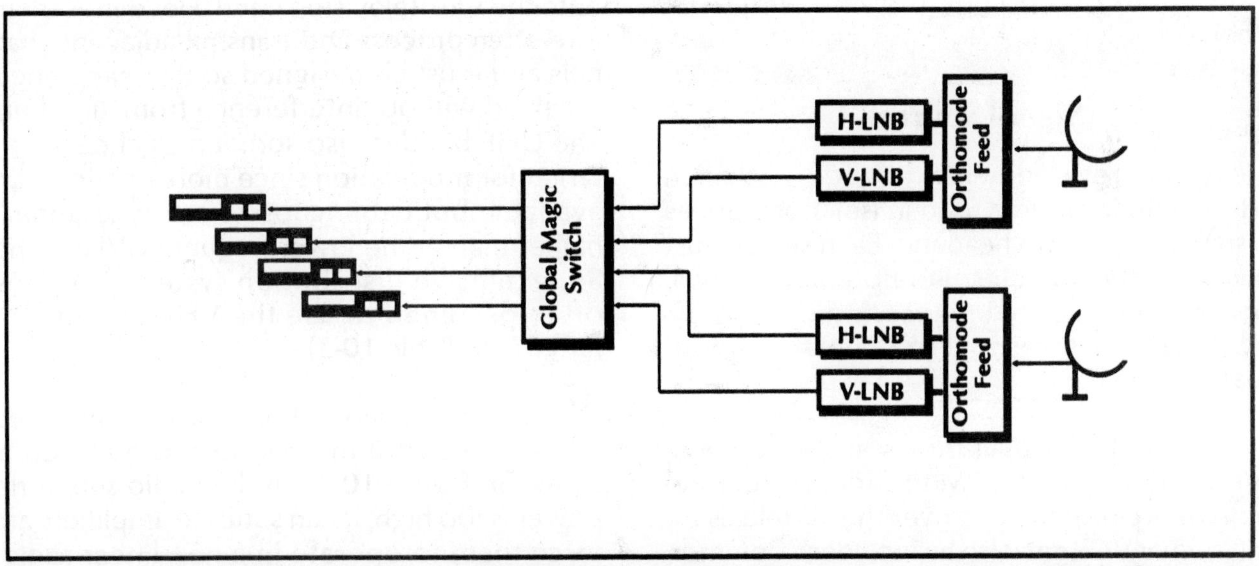

Figure 10-22. Dual-Dish, Dual-Band IF Distribution System. *This system receives signals from two dishes and channels them into a Global Communications "Magic" switch. Each satellite receiver independently switches between the four sources via four voltage levels, 0, 3, 6 and 9 Vdc (see also Figure 10-10).*

The SMATV HEADEND

Video and audio broadcasts from a variety of sources are captured, processed and combined at the headend. Signals may originate from one or more satellite and/or terrestrial broadcasts or from a combination of sources. In its simplest form, a home satellite system, a headend is a satellite dish, a feed/LNB and a satellite receiver with its built-in modulator. In a more complex SMATV headend, signals from many sources are "cleaned up", modulated onto an appropriate channel and amplified before entering the distribution system. All video channels are equalized to have nearly equal powers so that interference between channels is minimized and audio carrier levels are usually adjusted to be about 15 dB below the video carrier level.

The overall objective in designing an SMATV headend is to provide each television and audio receiver on the network with "clean" signals and, if adjacent channels are present, with signals that do not interfere with each other.

Signal Sources

All the requirements that apply to each video or audio source used in isolation applies to signals entering a headend. Each signal must have an adequate signal-to-noise ratio. Thus, for example, satellite broadcasts must be sufficiently above threshold to maintain picture quality.

Other input signals from sources such as a home video camera, FM broadcasts, condominium security system, over-the-air television antenna or VCR can also be included. Commercial-quality systems usually process and amplify each component signal before combining all onto a distribution network. Home multiplexer units which add either composite baseband or terrestrial over-the-air inputs are now available to create a "professional" headend in a low-cost system.

Television Channel Layout and Balancing Signals

Whenever signals are modulated onto two or more channels occupying adjacent frequency bands, the possibility for interference exists. In most countries, government agencies have wisely allocated sets of non-adjacent, over-the-air broadcast channels in each region or metropolitan area of their respective countries to avoid this outcome.

Thus, for example, in the United States New York has channels 2,4,6,7 and 9 in use while the Philadelphia area TV stations broadcast over channels 3,6,8,10 and 12. (The reason why channels 6 and 7 are not adjacent to each other is explained below).

A modern SMATV system may be designed to relay, for example, five over-the-air channels along with six satellite TV broadcasts onto the VHF television band. Headends therefore often process and transmit adjacent channels and must be designed so that each one is received without interference from all others. The UHF band is also sometimes chosen as a target for modulation since more channels are available but cabling losses are substantially higher than in the lower frequency VHF band. So, extensive distribution systems are most often designed to use the VHF or midband range (see Table 10-3).

Each NTSC video broadcast has its video, color and audio information organized as shown in Figure 10-23. If the audio subcarrier power is too high, it can saturate amplifiers and force them to operate in a non-linear region where signals interact to create unwanted intermodulation products. The result is that the audio can "bleed" into the upper adjacent channel video and cause a cross-hatch or pattern over its picture. If the video bandwidth has not been filtered so that all signal energy below the lower channel edge has not been eliminated, it can interfere with the lower adjacent channel and cause scratching sounds and "cross-talk" on its audio output. In light of these considera-

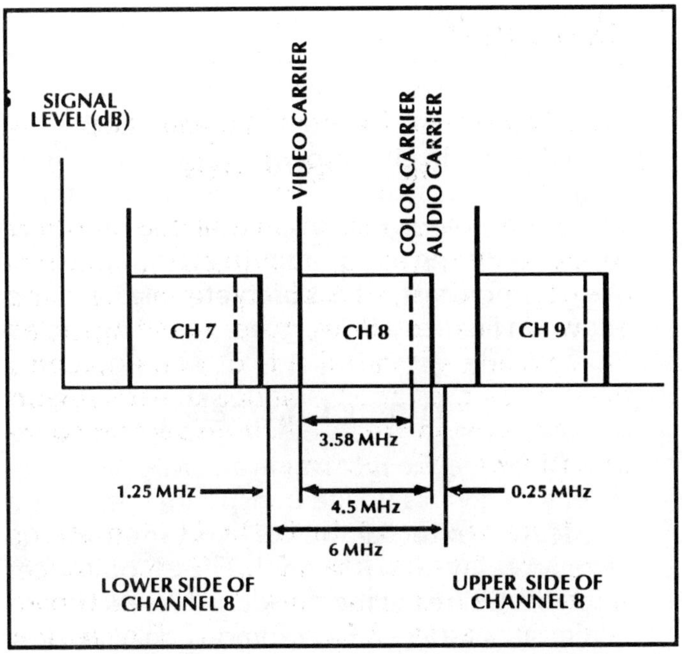

Figure 10-23. Layout of NTSC Television Channels. *Each NTSC television channel is organized into a video subcarrier centered 1.25 MHz above the lower side of the channel, a color signal centered 3.58 MHz above the video carrier and an audio subcarrier at 0.25 MHz below the upper edge of the channel.*

tions, whenever adjacent channel modulation is used, two rules must be followed: the signal must be restricted by a bandpass or SAW filter to protect the lower adjacent channel and the audio level must be maintained at 15 dB below the video carrier. Since most acceptable quality modulators properly bandpass video signals so the most important task is to ensure that the audio carrier level is properly set.

Table 10-4 outlines channel frequency allocations in North American. This is important information for the headend designer and shows which channels should be considered adjacent in the American VHF band. Since there is a large space between the VHF low band (2 through 6) and VHF upper band channels (7 through 13), channel 6 does not have an upper adjacent and channel 7 does not have a lower adjacent. Fortunately for system designers there is also a frequency space between channels 4 and 5, a 4 MHz guard band which is occupied by marine radio communications.

This table has valuable information for satellite TV technicians. For example, assume that a small motel wants to send two satellite channels to each room along with its off-air broadcasts. If channels 2,4,7 and 9 are occupied, satellite TV could be modulated onto channels 11 and 13 with no trouble. But what if an off-air station is already relaying a broadcast on channel

TABLE 10-4. AMERICAN VHF TV CHANNEL ALLOCATIONS

Channel Number	Frequency Range (MHz)	Upper Adjacent	Lower Adjacent
2	54-60	3	none
3	60-66	4	2
4	66-72	none	3
5	76-82	6	none
6	82-88	none	5
FM Band	88-108		
7	174-180	8	none
8	180-186	9	7
9	186-192	10	8
10	192-198	11	9
11	198-204	12	10
12	204-210	13	11
13	210-216	none	12

11? A good choice would be to modulate the satellite signal onto channel 6 which has no upper adjacent channel as well as onto the free channel 13. This sensible choice would save time and money in designing a headend. Less expensive modulators and signal balancing techniques could be used since non-adjacent channels were chosen instead of channels 5 and 6.

Efforts should be made to equalize signal levels being fed into a distribution system. However, precise balancing is not nearly as important when non-adjacent channels are used. Therefore, systems should be designed, if possible, with non-adjacent channels as targets for modulation. When adjacent channels are used, however, high quality processors and modulators that "clean up" the audio and video signals and which provide filtering and amplification are necessary.

Headends for more complex systems are usually housed in a temperature controlled location. Any substantial temperature or humidity variations can affect system performance. For example, modulator gains can increase by over 10 dB when temperatures drop from those of a hot summer day to a cold winter night. A high humidity may cause electrical components to degrade and have a shorter than normal life span.

Combining and Distributing the Signal

Signals from each modulator are combined once power levels have been equalized. This requires appropriate adjustment of modulator output controls, bandpass filtering, and occasionally the use of signal processors. Power levels can be calculated to a fair degree of accuracy before installation. For example, if a modulator with 40 dB gain is drawing a -3 dBmV baseband signal from a satellite receiver, it will input 37 dBmV to the distribution system. If an adjacent channel is being relayed at 32 dBmV, balancing could be achieved by either inserting a 5 dB pad in the output line or by readjusting the modulator RF output level.

Output signals can be combined at the headend using either splitters or taps in reverse. Thus, for example, a 3-way splitter could take three separate signals and combine them into one stream. Or a string of taps in series could feed modulator outputs onto one trunk line. The insertion losses are the same when using both devices in either forward or reverse directions. A similar but more expensive component, a combiner, may also be a good choice to meet this design requirement. An example of all the components for a complete headend are shown in Figures 10-24.

Tuning Channels

Each satellite receiver in an SMATV system is set to a given satellite transponder and has a dedicated external modulator tuned to any chosen television channel. The signals after being combined at the "headend" are relayed to all the subscribers on a common cable. In a system that relays a relatively limited number of channels that can fit into the VHF band, customers can tune to desired channels using their conventional TV sets. Customers connected to an SMATV system that relays more channels or a cable network tune channels via a converter that sits on top of the TV set. Channels are typically transmitted at mid or super-band and the converter modulates all programs onto either channel 3 or 4 (or UHF channels in other parts of the world).

Figure 10-24. SMATV Headend. *These commercial frequency agile SMATV satellite receivers accept an input of 950 to 1750 MHz at a -60 to -20 dBm level from an LNB. The measured threshold is a C/N of 9 dB. The power supply and four receivers are resting on their one-piece mounting plate in this photo. Each unit has a built-in modulator. (Courtesy of Televés of Spain)*

Example of SMATV Headends

Three examples of SMATV headends are illustrated below:

4 Channel Headend

This SMATV headend may be located at a hotel in an area where no off-air TV is available. Four satellite TV channels, two horizontally and two vertically polarized, are being distributed (see Figure 10-25). Note that the even and odd channels from the block downconversion receivers are combined in reverse 2-way splitters before being added together on one line. The system is designed this way to protect adjacent channels from interfering with each other. Note that since channels 4 and 5 are not adjacent in frequency, lower cost modulators without bandpass filters could be used, but this is not advised.

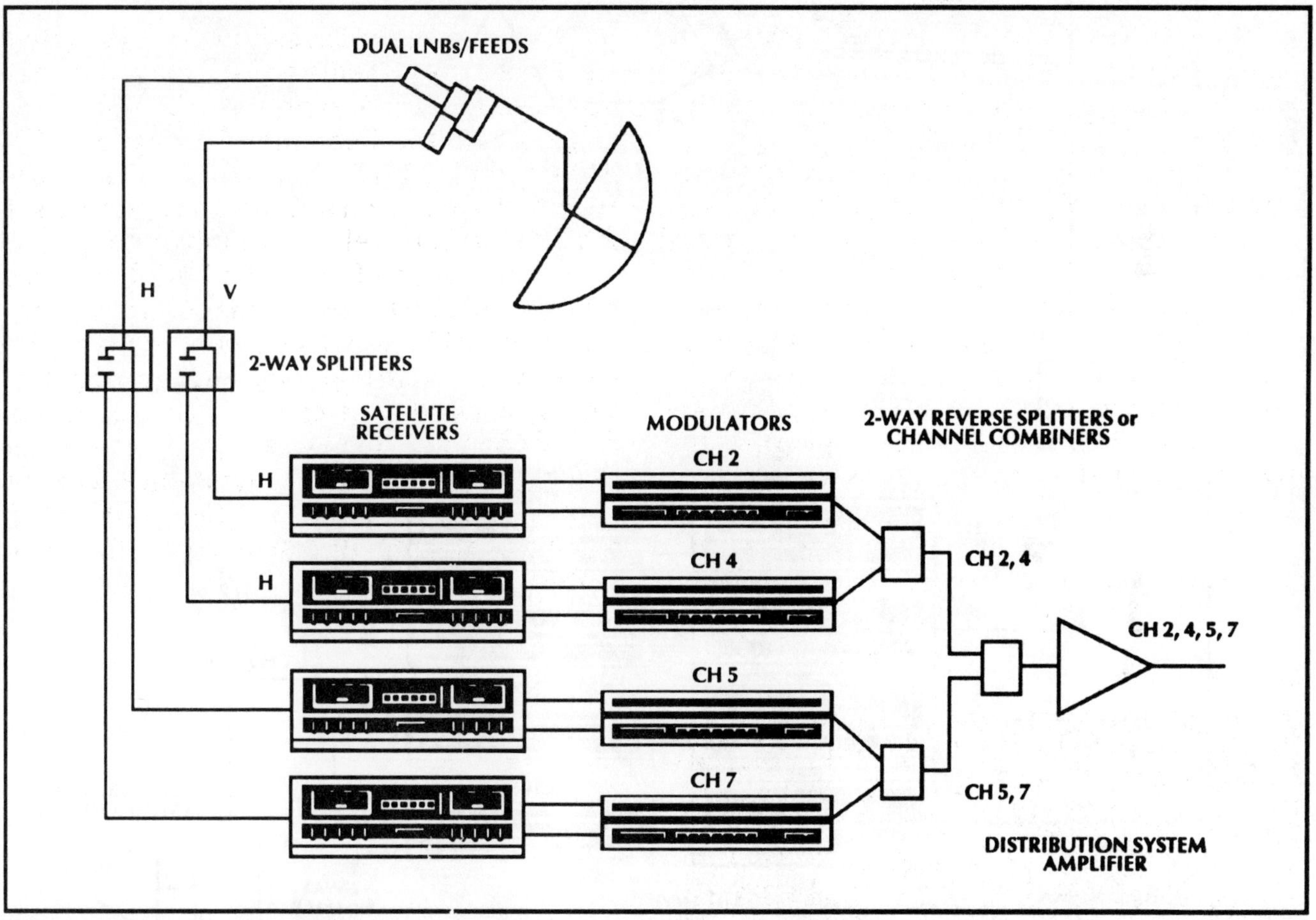

Figure 10-25. 4-Channel SMATV Headend

7 Channel Headend With Off-Air TV

This 7 channel headend combines four satellite TV channels with 3 off-air stations into one distribution network (see Figure 10-26). In this case, care has to be taken in protecting channels 8 and 9 from interfering with each other by properly balancing signal levels and by using a bandpass filter either as an extra component or built into the channel 9 modulator.

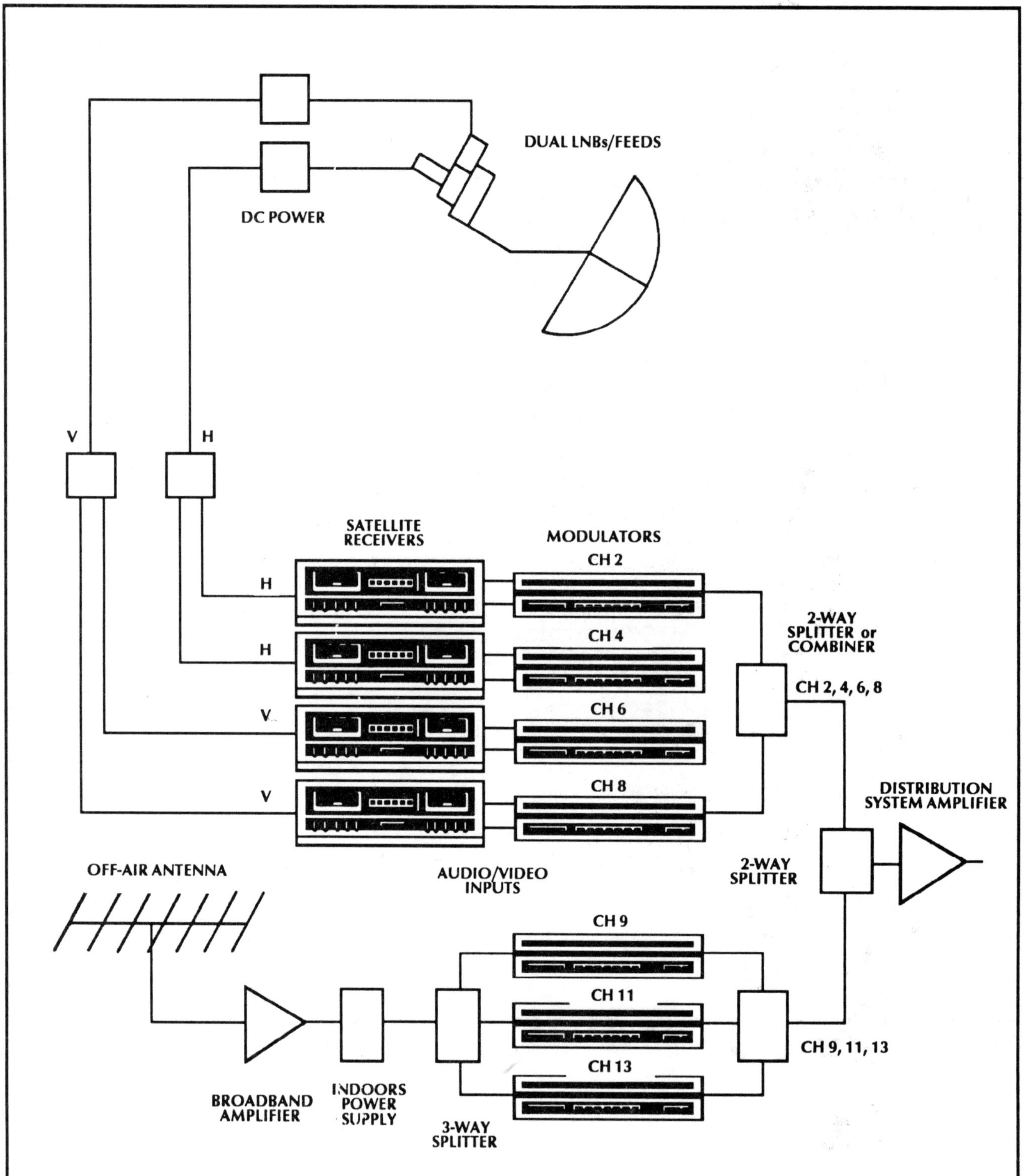

Figure 10-26. 7-Channel SMATV Headend

12 Channel Headend

This 12 channel headend combines all the available VHF channels (see Figure 10-27). Three 4-way splitters used in reverse, sometimes also labeled as combiners, feed their outputs into a 3-way combiner and subsequently into the distribution network.

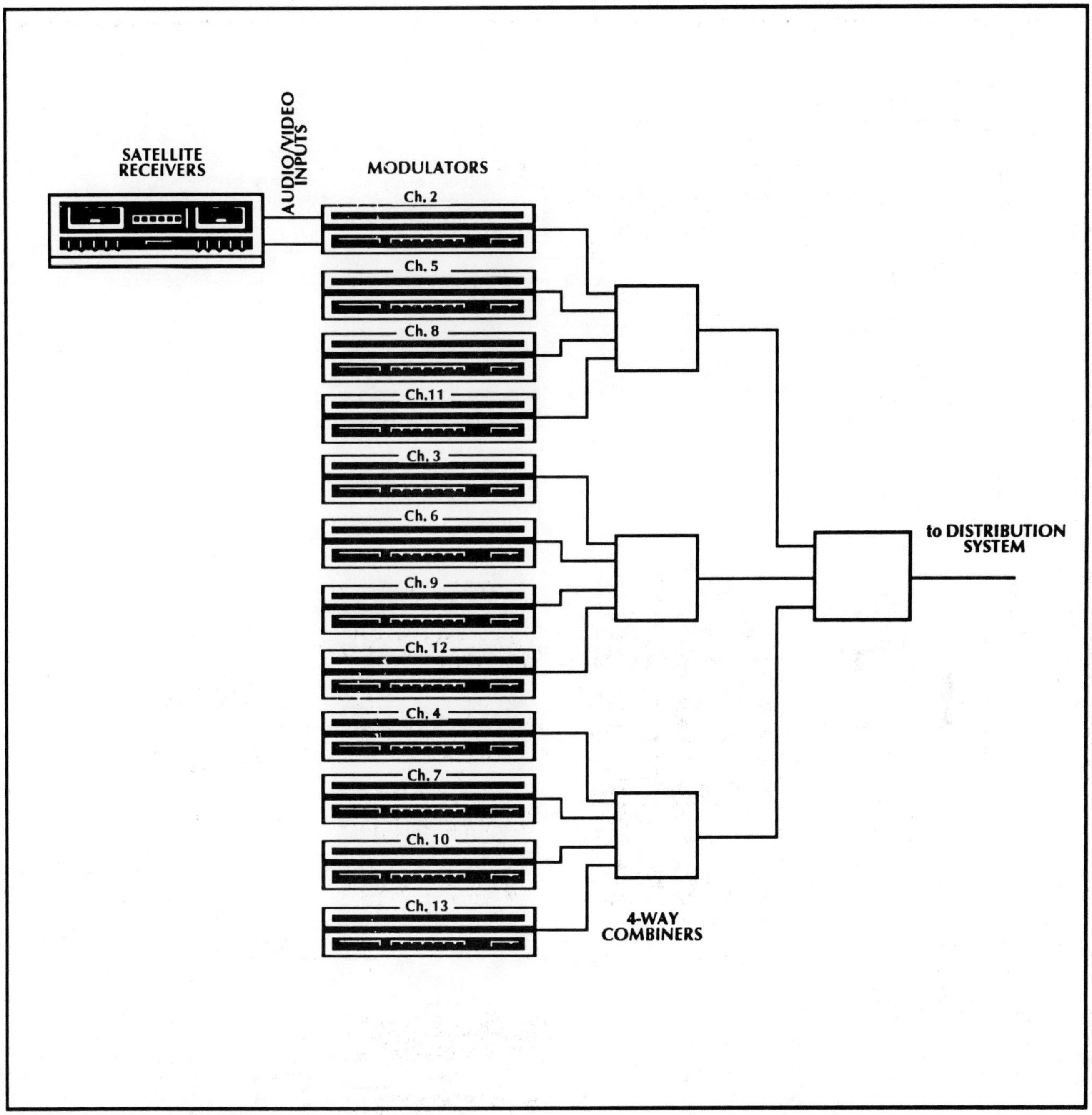

Figure 10-27. 12-Channel SMATV Headend

CHAPTER 11

TROUBLESHOOTING & REPAIR

INTRODUCTION

Every TVRO will eventually have operational problems of one sort or another. When this might occur depends to a large extent upon whether equipment was properly selected and installed. Poor weatherproofing is the most common installation error. Although it is clear that no man-made device will last forever, all problems can be solved by thoughtful troubleshooting. The purpose of this chapter is to provide a consistent method to understand and solve any difficulties that might arise. Finding out what is wrong, the diagnostic procedure, can in itself be an interesting and rewarding exercise if a correct approach is followed.

Troubleshooting can be relatively easy when only one component of a system goes bad, often a short time after an installation has been completed. Then diagnosis and repair can generally be quite simple, especially if the service call is made by the same person who installed the system. The installation documentation should facilitate locating the fault. Unfortunately, if another person installed the system, diagnosis may be more difficult. Some outrageous mistakes have been discovered when servicing home satellite systems which were poorly installed.

However, most problems often arise over a longer period of time when several unrelated, minor faults that result in poor quality pictures simultaneously develop. For example, connectors can corrode, concrete may settle and throw a pole slightly off from a vertical orientation, a center conductor in a coax may slightly contract and not make perfect contact, the dish may have sagged slightly or rotated from its proper alignment, or the feed may have moved noticeably over time. Eventually a customer may telephone to complain about poor reception. As a result, a check of dish tracking, feed alignment and all connections would be required.

Troubleshooting during an installation can be much more difficult, especially if proper procedures have not been followed from the start. These "proper procedures" are very important in order to avoid future headaches. To illustrate, assume that the components had not been tested before installation and the LNB was faulty and that a cable which had already been buried also had an internal short. The problem with the LNB would have been easily located if it had been tested before installation. But once the installation is completed, trying to separate the two problems without going back to square zero could be difficult.

As the number of installed systems continues to grow and as equipment in the field ages, the troubleshooter's role can only become more important and more profitable. The hallmark of a successful troubleshooter has been and will be an approach that combines technical knowledge with a systematic and methodical process for diagnosing, isolating and repairing TVROs.

A SYSTEMATIC APPROACH to TROUBLESHOOTING

Diagnosing an operational problem in a satellite TV system should follow a logical path to a rapid solution. The strategy is to take the simplest route having the highest probability of solving the problem before proceeding with more complicated steps. However, even these more complicated steps can be approached with a similar perspective of first tackling the most quickly solved problems.

The first step in a systematic approach to troubleshooting and repair is an in-depth interview, preferably over the telephone to save time and expense. Of course, this interview can be successful only if the underlying theory is clearly understood. This first step is aided by a technique known as the subsystems approach that can be quite useful in isolating problems in any satellite system. If necessary, the next step would be a visit to the installation site for a visual inspection. If a simple inspection does not isolate the problem, a series of technical troubleshooting procedures would follow to eliminate the fault.

None of these procedures is magical or complicated. All that is required is knowledge of how a satellite TV system works and a clear, level-headed approach. No problem is too difficult to solve and, with experience, solutions come more quickly.

The Subsystems Approach

Every satellite system can be broken down into three subsystems: the mechanical, consisting of the mount, dish, feed support and feed, the electromechanical consisting of the actuator and the polarizer, and the RF consisting of the LNB, satellite receiver and TV (see Table 11-1). Each can be diagnosed somewhat independently even though symptoms may be similar in nature. The only reason for defining and isolating these three subsystems is to simplify troubleshooting and repair procedures.

TABLE 11-1. SATELLITE TV SUBSYSTEMS

Mechanical	Electromechanical	RF
Mount	Actuator	LNB
Dish	Polarizer	Receiver
Feed Support		TV Set
Feed		

The Diagnostic Interview – Identifying the Problem

A telephone conversation between the customer and an experienced technician can often solve problems and eliminate the need for most service calls. This is especially true during the initial few weeks or months of ownership when customers are not familiar with many aspects of their systems. Troubles can stem from simple factors such as a television set that is incorrectly tuned to the modulated channel or from a customer's inexperience with a hand-held remote or operation of the parental lockout feature on a receiver.

In particular, a new owner generally has expectations about how the pictures should appear and would certainly notice any problems or irregularities. Complaints about poor picture quality are most typical.

The interview can also identify symptoms of a problem that may require a service call. Such information can prove extremely useful during the initial diagnosis as well as in assembling the proper equipment to isolate and repair the system. For example, if the customer had been out tilling his garden when the system failed, certainly bring some extra cable and connectors. There is nothing worse than driving out to a site and then having to return to the shop for one small forgotten component such as an extra fuse or voltage regulator. If adequate documentation about an installation was on file, questions can be guided by knowledge of this particular system. The documentation should have included information about dish size and make, location of cable runs, type of electronics installed and satellites targeted.

A satellite reception system can have three general categories of symptoms: completely inoperative system, intermittent where the operation alternates between normal and bad; and unusual operation where out-of-the-ordinary behavior is exhibited. The questions can be guided with these categories in mind.

In summary, the interview is an important first step before anything else is done. Some sample questions are:

- When did the trouble first start?
- Did it occur gradually or suddenly?
- Describe the problems.
- Are these problems constant or do they occur at regular or irregular periods during the day or night?
- Was the equipment recently moved or has any component been replaced?
- Which channels and how many channels are coming in clearly?
- Can all satellites be received?
- In a tracking system, does the dish move easily from satellite to satellite and does it make any unusual noises in doing so?
- Have you possibly cut or disconnected any cables by mistake?
- Does picture quality vary from channel to channel or from satellite to satellite?
- Is the audio or video most affected?
- Has there been any recent construction in front of or near the dish?
- Did the problem first occur after a storm and did it persist when the weather cleared up?
- Do your problems occur more during rainstorms than on clear days or more during hot than during cold weather?
- Is it possible that a fuse has been blown in the receiver? Do the lights come on? Could a circuit breaker in the house that provides power to the receiver have been tripped?
- Have there been any children playing around the dish?
- Did you recently dig a garden or do any other excavation work?
- When the wind blows can you hear any part of the dish rattling?
- Can your television set receive over-the-air channels?
- Did these problems appear in the spring?

A competent troubleshooter is like a doctor. The patient often knows what is wrong but simply does not have the vocabulary to express it. A well conducted interview gathers useful and important information. With a little experience, a complete list of questions can be created so that such an interview will proceed smoothly.

Many problems can be quickly solved by this technique. For example, a customer may complain that pictures are perfect except during heavy rain or snow storms. This might be caused by rain fading because the system was under-designed. Or a wet connector or leaky gasket between the feed and LNB may be at fault. Also, for example, if every transponder went bad after a severe lightening storm, the cause could likely be an LNB with a blown stage or simply a blown fuse or circuit breaker.

Then there are occasional satellite problems or sun outages which happen infrequently enough that people forget that they are even possible. Sun outages, as described below occur twice a year during the spring and fall equinoxes. Some technicians are simply not aware of and could be baffled by these occurrences.

VISUAL INSPECTION

Once the customer interview and a thoughtful evaluation of the assembled facts are completed, a visual inspection may be required. This is especially important if the system was installed by another person.

The visual inspection can be guided by symptoms of the problem. For example in a tracking system, if the customer reports that he can receive television from only one satellite, the first step would be to check the actuator and its control box to see if the dish is stuck in one position. If all channels are received with crystal clarity on some satellites but poorly on others, the first visual check would be to determine if the reflector is warped, if the feed is moving off the focal point as satellites are being tracked or if the north/south alignment has been disturbed.

Many symptoms can stem from the same problem. It is easier to initially conduct a visual inspection when a not-so-obvious symptom is encountered than to jump into a more technical troubleshooting approach or to immediately start exchanging components (see Figure 11-2). It is important not to create additional problems. But symptoms will usually give an indication of where to start the visual inspection. For example, if picture quality were poor, it would certainly make more sense to first try moving the dish east and west with the actuator or by simply pushing it gently up and down than to start measuring voltages. A second sensible step would be to check for loose or corroded connectors and a third one would be to check that the feed is properly centered and the dish is correctly aligned. Perhaps the mount has rotated in a strong wind because the bolts on the collar were not tightly secured or pinned. Sight with a compass to be sure that the polar mount is still properly oriented.

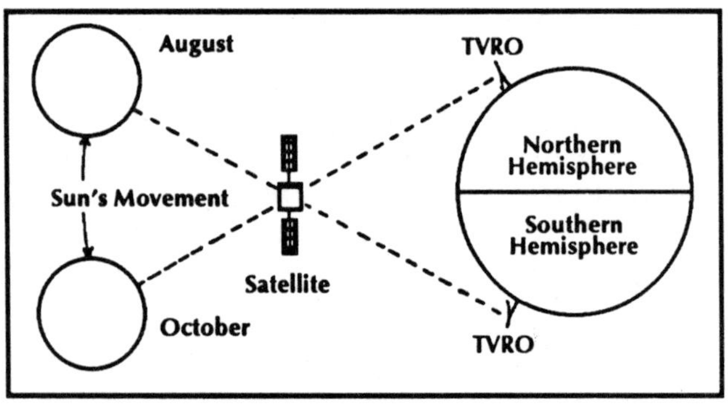

Figure 11-1. Sun Outages. *The timing of solar outages is directly related to the site latitude. They occur when the sun lines up directly along the boresight of a dish. Solar outages always take place within 3 1/2 weeks of the equinoxes on March 21 and September 21, when the sun crosses the equator during its annual north/south journey. Solar outage in February, March and April begins at the northernmost latitudes and moves southward until satellite systems on the equator have outages at equinox time. Then TVROs in the southern hemisphere begin having outages until those at southernmost latitudes experience them 3 1/2 weeks after the equinox. In August, September and October, the entire pattern is reversed because the sun is moving in the opposite direction. Outage season lasts for a little over a week in any given location. During this period, there is an outage daily on each satellite. It first appears as small amounts of video noise, rapidly becomes annoying interference and, on the days at the center of the outage season, peaks out as a total loss of incoming signal. Dishes should be aimed away from this satellite during outages in order to avoid heat build-up on the feedhorn and LNB. Spun aluminum and shiny surfaced fiberglass dishes can raise the temperature at the focal point sufficiently to cause severe damage and possibly destroy components.*

Figure 11-2. Dish with an Obstructed View. *This dish was installed in the middle of dense foliage. Although it may not have been completely screened during the winter when the surrounding trees had no leaves, chances are excellent that few satellites in the arc can be clearly received in the summer.*

The basic questions below show the thought process underlying a simple visual inspection. Many more questions that may arise are presented in the discussion of technical details in the following sections.

RF Subsystem

In general, the visual inspection should begin at the system control center, the satellite receiver/television set. Turn on the TV and what do you see?

1. Does the television work properly on over-the-air channels?

2. Is the picture completely snowy or infected with sparklies? If there is a weak picture in the background perhaps the dish is aimed onto a side lobe of the satellite.

3. Is the receiver turned on and are all the indicator lights working? Maybe a rear panel or internal fuse has blown or a 120 or 220 Vdc circuit breaker in the house has been tripped.

4. Is the television tuned to the correct channel that matches the modulator output?

5. Have any cables behind the receiver been disturbed? Perhaps the coax was improperly crimped and the coax is disconnected.

6. Are any other wires loose or disconnected? If this is a new installation make sure that all the wires are well attached to the proper place before powering on.

7. Is there a high signal strength meter indication but no picture? This would suggest that the modulator may be bad, the TV set is not tuned to the right channel or TI is present.

8. Is the receiver shutting off because it is overheating? Most receivers have ventilation slots that allow cooling by natural air flow. If heat is not carried away, the voltage regulator that generates substantial amounts of heat might be internally compensating and shutting down.

9. Are all the cable runs intact? Check all the cable routes from indoors to the dish. Perhaps some over-enthusiastic gardener has severed the cable with a weed-eater, shovel or lawnmower. Maybe a dog or rodent has chewed through the cable.

10. Are all the connections at the dish secure? Make sure that all connectors are tight and that no water has leaked in. Are drip loops installed in the correct places? Maybe a connector on the LNB has corroded. When a cable is used to carry power, water entry can cause corrosion similar to the build-up that can be seen on a car battery terminal.

Cut the cable back and install a new connector past the point where the corrosion ends. Make sure that coaxial sealant has been used. Clean any suspected connectors with a solvent like carbon tetrachloride, dry them thoroughly and re-seal. Also, if a defective F-connector has a slightly short center conductor, it may perform well until cold weather arrives. The cold could cause this lead to shrink and break contact. This phenomenon is appropriately known as "center connector suck-out." If the center conductor is too long, it may split the shroud on the center pin when inserted into an LNB and cause it to short out against the inner connector wall.

11. If an LNB/feed cover has not been used, check to see if some water might be collecting in one of the waveguides. If a polarizer has been used with the right angle waveguide fitting, water may also have collected and a cure might be to drill a small 1 or 2 millimeter drain hole on its underside. Maybe the gaskets between the various flanges have been omitted to further worsen the problem.

12. Does the system work intermittently? Sometimes a bad LNB functions only when it is cooled down to the point where its gain increases. A test for this malfunction is to use a hair drier to warm it up on a cold day or to use a can of freeze spray on a warm day to bring it back to life.

Electromechanical Subsystem

1. Is the polarity adjustment having any effect on picture quality? If it is a mechanical device, is the probe moving when channels are changed or when the skew adjustment is altered? For a voltage switched device, perhaps inadequate voltage is reaching the PIN diodes. Perhaps they are burned out. Check to see if obstructions caused by moisture collection, birds or even wasps are interfering with signal reception in the throat of the feed.

2. Is the skew adjustment having little or no effect in a mechanical feed? Maybe the probe has been bent or mechanically distorted by freezing and thawing water. Make sure that the waterproofing was ensured by correctly installed gaskets.

3. Is the probe motor bad? Have a spare handheld controller because the 555 timer in the actuator controller or receiver may be at fault, not the motor.

4. Is the motor servohunting? This means that the motor is driving the probe back and forth unsuccessfully looking for the maximum signal. This can cause hum bars visible as wide horizontal regions of disturbances rolling across the screen. Have a 10 Vdc, 1000 microfarad electrolytic capacitor on hand and install it on the polarizer motor at the dish. Also, check that the wire feeding the motor is of sufficient gauge and thus thick enough to avoid this problem. When using two servo motors in a parallel dual-band configuration, a capacitor should be used to provide an extra current when simultaneously switching both probes.

5. Does the dish respond to the actuator? Maybe a voltage transient from a nearby lightning strike has fried the Hall effect or reed switch feedback sensor. If the sensor leads were not shielded, the actuator controller may be miscounting and aiming the dish incorrectly. Possibly water has entered the arm and has rusted the internal assembly so that it does not move.

6. Make sure the motor is pointed up so that the drain holes are pointed down. If there are no drain holes, drill some. Maybe the gasket between the motor cover and housing has been omitted.

7. Is the actuator tube bent or is there wear or scoring on the inner tube? Perhaps the original installer did not use ball joints on the actuator arm supports in order to minimize lateral force on the tube. Maybe he did not set both mechanical and electrical limits to prevent bending.

Mechanical Subsystem

1. Does the dish look as if it is mounted correctly and that all the angles are still set properly? The collar may not be properly tightened, drilled and bolted.

2. Has the concrete shifted or settled under the pole or pad so that the pole is no longer level or plumb?

3. When using fiberglass dishes, perhaps the surface of the dish has delaminated or is blistering resulting in poor reception.

4. Has the owner painted the dish with a roughly applied metallic paint that has affected its reflective surface? Or maybe he has applied a highly reflective paint that is concentrating sunlight enough to overheat the components at the focus.

5. Shake the dish from the outer rim for maximum leverage in order to check for loose nuts and bolts. Maybe the original installer did not use lock washers or lock-tight and the wind has gradually loosened up the assembly.

6. Are any nuts or bolts missing?

7. Is there excessive wear at any attachment points? Scoring on the support pipe caused by such movement may be a telltale sign. If this is the case, set a bolt through the outer to the inner pipe when all the re-adjustments are completed. This will maintain a true north/south tracking axis.

8. Is the dish warped? When sighting along one lip does the other side line up perfectly parallel? Do the two or three string test, if necessary, to check for warping. Maybe the mount support was not designed properly and the reflective surface has twisted.

9. Is the feed system secured and centered? This can be checked with a focal finder or with a tape measure. A loose feed could possibly move in the wind or be jarred out of place as the dish rotates across the arc. Guy wires are recommended.

10. Is the feed opening free of bugs or wasp nests? A quick visual inspection will eliminate this possibility.

11. Does the actuator jack have complete freedom of movement across the entire belt of satellites?

TECHNICAL TROUBLESHOOTING

If the initial visual inspection does not provide a solution to the problem, then more technical approaches can be followed. There are always a number of ways to arrive at the same conclusion. It is important to understand that the intuitive approach is a valid one, not to be diminished in value. Troubleshooting certainly has its share of personal touch.

The tools used in technical troubleshooting are test instruments. A voltmeter, spectrum analyser or portable receiver can greatly simplify and speed up the diagnostic process. Switching components is included in this section as one of the technical troubleshooting methods because it involves more than casual observations of TVRO problems.

Test Instruments

Voltmeters and Multimeters

A voltmeter is by far the simplest, lowest cost and potentially most effective test instrument. Most electrical breakdowns occur in power supplies that provide energy to operate the LNB, actuator and all other signal processing circuitry. A competent technician can isolate power supply and other voltage related problems in minutes with the proper use of a voltmeter (see Figure 11-3).

Voltmeters, like most test instruments, have variable sensitivities. The maximum range for each setting is typically 200 millivolts, 2 volts, 20 volts, 200 volts and 1000 volts dc. Testing should always begin in a range which can manage the maximum voltage expected so the indicator will not be driven off the upper end of its scale.

Most voltmeters are designed to read resistance and ac voltage as well as dc voltage. These capabilities are extremely useful in testing for shorted wires, in checking line voltages or in troubleshooting other areas.

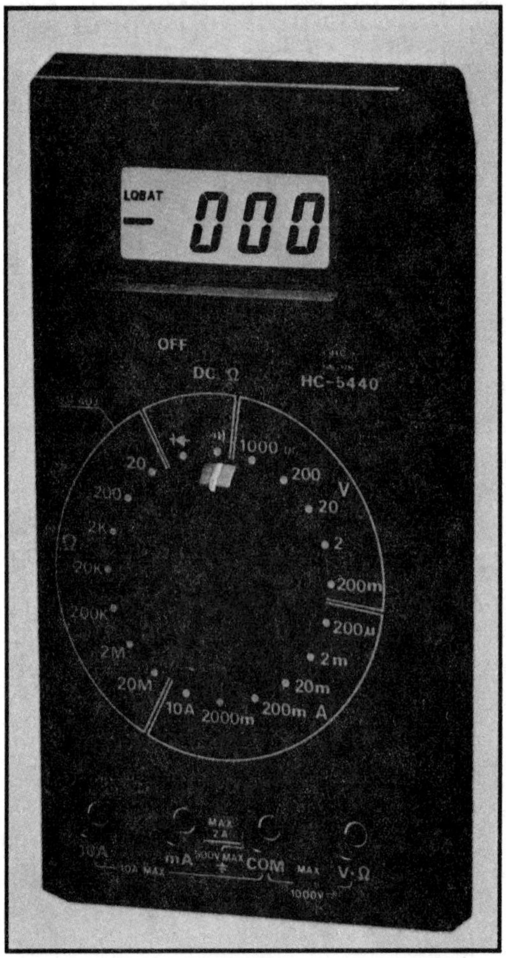

Figure 11-3. Multimeter. *One of the most frequently used pieces of test equipment either in the field or on a test bench is a multimeter. It has the capability of reading dc and ac voltages or currents as well as taking resistance measurements for checking cable continuity.*

Spectrum Analyzers

Spectrum analyzers are powerful diagnostic tools designed so that incoming signals in a range of frequency bands can be observed and measured (see Figure 14-4). This instrument essentially plots a graph of signal power versus frequency. Frequency increases from left to right on a CRT screen while signal amplitude is presented as a series of peaks and valleys. Incoming microwaves located anywhere within the entire band of satellite communication frequencies can be observed at once or fine details of the frequency and amplitude structure of any single carrier can be studied. The spectrum analyser is to an RF designer what an oscilloscope is to an electrical engineer. A spectrum analyser displays the signals plotted versus frequency whereas an oscilloscope displays signals plotted against time.

Portable spectrum analyzers have long been used for cable television applications. Many analyser manufacturers have marketed add-on modules designed to allow coverage of the satellite television IF block frequency ranges. Such analyzers can be used to observe signals at any point in a satellite system. When the signal from an LNB, or preferably an LNB and feed assembly, is fed into this instrument, microwaves of any chosen polarity can be detected and measured. By scanning a feed or a dish across the sky and observing the variation in detected power, the source can also be pinpointed. This capability makes a spectrum analyser invaluable in aiming a dish and in peaking a feed.

It is important to realize that the American LNB output range of 950 to 1450 MHz is more narrow than the LNB output band in African, Middle Eastern and European systems. While various manufacturers around the world create analyzers to handle these different ranges, the upper frequency limit should be specified.

An LNB typically generates a signal having powers ranging from -60 to -80 dBm while the highest level signal at the modulator output is on the order of -40 dBm. One common portable, relatively low cost instrument, the Avcom PSA-35, has a sensitivity spanning the range of zero to -90 dBm and therefore can respond to RF signals produced at any point in a typical TVRO.

Technicians at uplink facilities use on-line analyzers to monitor the power, frequency, stability and composition of signals relayed to communication spacecraft. Powers are generally equalized across all transponders on-board a broadcast satellite. This is seen as a series of equal peaks spread across the entire C- or Ku-band frequency range. A spectrum analyser monitoring the output of an LNB should display the same, equal distribution of transponder power.

The ability of an analyser to create a visual display of signals over a wide range of frequencies makes it a valuable troubleshooting instrument which can pinpoint system or component problems (see Figure 11-5). This tool can also be directly connected into the output of an LNB, the output of the cable leading from the LNB to the satellite receiver, the output of the satellite receiver or any other point in the RF system. Signals can therefore be traced enroute from the dish to the television set in order to identify any faulty components or cable breaks.

Two common symptoms observed on an analyser are tilt and wipe-out. If transponder power appears to decrease across the satellite band on the instrument's display, the feed or LNB may be faulty. If the problem is a bad component, it might be identified by replacing it with a healthy one. Tilt can also be a result of attenuation in coaxial cables that have losses which increase with frequency. An analyser can be used to locate the source of high frequency "roll-off" in distribution systems. Once the source has been isolated a solution might be to use a line amplifier with tilt compensation or to somehow shorten the cable runs. Another common problem, transponder wipe-out, which can also be traced with an analyser, appears as a sharp reduction in power of one or a group of transponders. One source of this problem may be an impedance mismatch that causes signals in a rather narrow frequency band to be reflected back to the dish. Something as simple as a missing terminator on an unused splitter port or a faulty connector could be the cause.

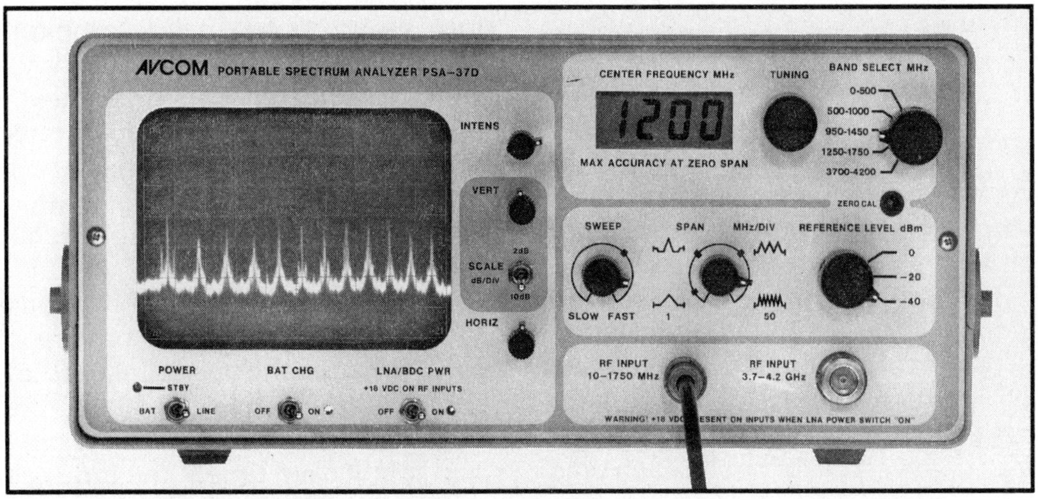

Figure 14-4. Avcom PSA-37D Spectrum Analyzer. *This battery-operated analyzer accepts input signals from 10 to 1750 MHz and from 3.7 to 4.2 GHz in five bands. It has both an LCD digital and CRT output. A built-in supply powers LNAs and LNBs.*

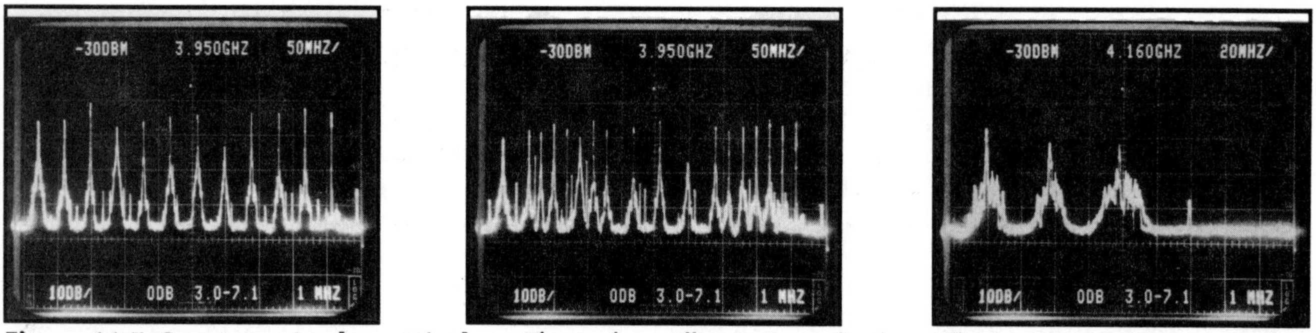

Figure 11-5. Spectrum Analyser Displays. *These three illustrations display different views of transponders received from satellite. The left shows a view of 12 transponders and a control beacon on the American satellite Galaxy I. The middle displays all 24 cross-polarized transponders. The right photo shows three transponders and a control beacon. Its amplitude is approximately one third that of the satellite broadcast signals. (Courtesy of Tektronix)*

The potential uses of an analyser are limited only by the creativity of the technician (see Table 11-2). Some unusual tasks can be created. For example, this instrument can even be used to test a hand-held remote which emits RF signals. As buttons are pushed on the remote, spikes should appear on the analyzer's screen. Probably one of the most important uses of an analyser is to locate faint Ku-band satellites on narrow beamwidth dishes during alignment. An analyser can also allow visualization of weaker signals that would not normally be seen with a satellite receiver and monitor. Thus an installer can detect those satellites that may not yet be transmitting television signals by observing their telemetry beacons.

Portable Receivers

Aligning a dish onto the satellite arc is most easily accomplished with a portable receiver or other sensitive signal detection instruments. While most installers simply connect the customer's receiver or an extra unit from the shop, using a self-contained portable satellite receiver avoids the difficulty of hooking up three or four components as well as lugging a TV monitor or receiver out to the site each time a test is conducted.

Most portable receivers today are battery operated and are equipped with either a color or a black and white monitor all packaged in a relatively lightweight container (see Figure 11-6). The only extra equipment required to receive a satellite broadcast is a dish and LNB/feed. This portable tool eliminates the risk of forgetting to take some necessary component from the shop or of dropping a customer's prized set in the mud during testing.

Testing with a portable receiver proceeds as follows. The LNB output is first connected to the portable receiver input. If an adequate picture is seen on the small TV monitor, the LNB is fine and any problem would lie downstream towards the customer's TV set. Next, the output of the coaxial cable can be connected to the portable test unit. In this fashion, by working from the dish towards the television, the faulty component can ultimately be identified.

Testing with a portable receiver can also follow along the lines of a process called "half-split" troubleshooting. Instead of checking each component sequentially, the diagnosis begins somewhere in the middle of the system. This might save time if the input components seemed to be working fine but nothing was happening at the output, the television set.

TABLE 11-2. SPECTRUM ANALYZER ANALYSIS REPORT

Customer Name______________________________Date____________
Address______________________________Telephone______________
____
City______________________________State____________ZIP________

System Description

Dish: Manufacturer____________Diameter____________Model____________
Polarization/Feed: Manufacturer____________________Model____________
LNB: Manufacturer____________Max. Noise Temperature________Gain________
Coaxial Cable: Length from LNB to Receiver______________Type____________
Receiver(Main): Manufacturer______________________Model____________
Serial Number______________Location______________
Receiver(Slave): Manufacturer______________________Model____________
Serial Number______________Location______________

System Performance Statistical Data

Satellite-Targeted______________________________Transponder#____________
Output Levels: LNB________dBm @ 3.7 GHz ________dBm @ 3.9 GHz ______dBm @ 4.2 GHz
Coax_______dBm @____MHz ________dBm @____MHz _______dBm @____MHz
Terrestrial Interference: Frequency ______MHz Level ________dBm Bandwidth _______MHz

Notes

____

____

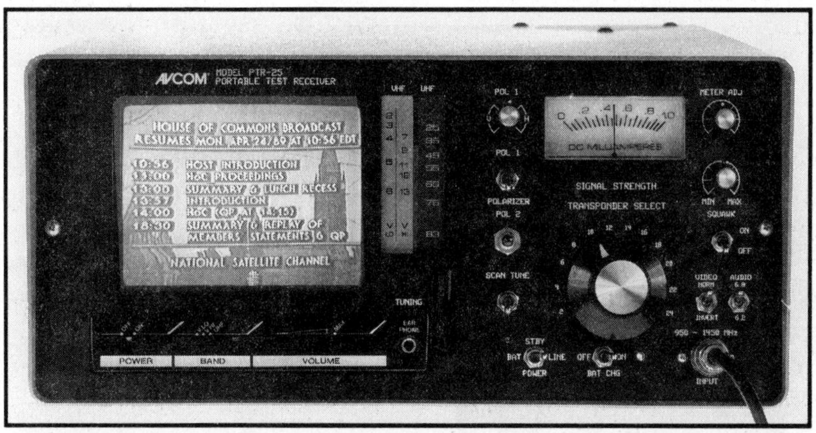

Figure 11-6. THe Avcom PTR-25 Portable Test Receiver. *Acvom's COM-2R and COM-3R satellite receiver circuits as well as polarizer controls are included in this battery-operated unit. Outputs include connections to drive TV monitors, video recorders and audio amplifiers, an audible tone and a signal strength meter. It accepts an input frequency of 950 to 1450 MHz at a –70 to –25 dBm level. (Courtesy of Avcom)*

Switching Components

One simple method of troubleshooting a satellite system is to switch out components, one by one. Thus, if the LNB is suspected of being faulty, substitute a healthy one and see what happens. Remember, in this and all other similar situations when any component is switched, always be sure that the power is turned off and the receiver is unplugged to avoid causing further damage. Some brands of receivers continue to send voltages to the system even when the power switch is off.

This diagnostic method is direct but can be time-consuming. Also, if switching components is attempted during a new installation and more than one component happens to be bad, it may prove to be a frustrating endeavor. Fortunately, there are other creative troubleshooting options, such as using a portable fluorescent light to test LNBs (see Figure 14-7).

Interchanging or shotgunning components is used mainly to diagnose problems in the RF system. It is easy to check if the actuator or polarizer is not functioning properly by visual inspection and then to find the problem by voltage checks and other troubleshooting methods.

Power Supplies

Most electronic malfunctions can be traced to satellite receiver and actuator power supplies. Unless adequate measures have been taken to suppress voltage spikes, it is common to have a fuse blow or a voltage regulator fail. Some brands of equipment have external, screw-in and internal fuses so both must be checked. If nothing happens when a receiver or actuator is connected, the unit should be unplugged and the fuses examined. To be absolutely certain that fuses are blown, if this is not obvious by a visual inspection, check their continuity with an ohmmeter. When replacing fuses it makes good sense to install spike suppressers so this problem does not reoccur.

If fuses are repeatedly blown, an inexpensive current limiter can be constructed with a 40 or 50 watt light bulb and a 1-ohm resistor connector in series with the power line into the receiver. The ac voltage is measured across the resistor. If there is a direct short in the receiver, the bulb will glow brightly as it limits current to the receiver and protects the fuses. Make sure that this test setup is well insulated or severe shocks could occur.

Integrated circuit (IC) voltage regulators are often not adequately protected and can malfunction. Excess voltages can damage the receiver, actuator or LNB. In unusual cases, the dish could even be harmed by being driven into the ground by a runaway actuator. However if a regulator does fail, identification of the faulty part is easy with the voltage tracing methods outlined below. Replacement is just as simple.

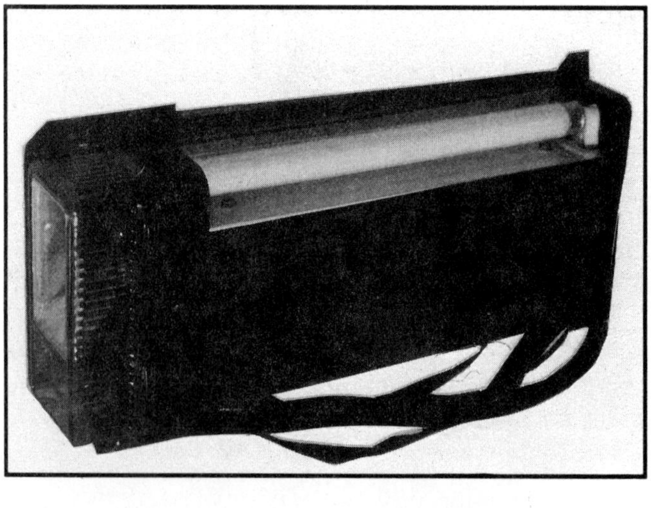

Figure 11-7. A Portable Fluorescent Light. *A portable fluorescent light can be used as a simple tool to diagnose the condition of an LNB. It is, in effect, a portable signal simulator. In addition to generating light, it emits microwaves covering a broad frequency band. When waved in front of a healthy LNB, the noise level on the TV screen should change and the signal level meter should indicate full scale reading. If this does not occur, the LNB may not be receiving adequate power or it may be defective. Realize that this action will overpower and certainly blank out any signals received from satellite.*

The 7800 and LM340 series of positive regulators are capable of supplying up to 1 amp of positive current and are often mounted directly onto the receiver chassis. The 7900 and LM320 series are the complementary negative components but must be isolated from the chassis and therefore mounted by use of a mica insulator and insulated screw. The output voltage is identified by the last two digits of the 7800 and 7900 regulators. For example, a 7818 regulator provides a positive 18 Vdc. Two other adjustable regulators, the LM317T and LM723, are also often used in satellite receivers.

Voltage Checking

Tracing voltages with a voltmeter is an effective method to rapidly identify a faulty component. When examining an RF problem, all voltage checking should begin at the receiver. It is the brains of a satellite system and it (1) powers the LNB, polarizer and modulator, (2) selects channels by sending a correct voltage to an internal VTO and (3) communicates polarity, channel selection and audio format information to the user.

The satellite receiver relays from 15 to 24 Vdc to the LNB and typically 13 to 17 Vdc to a voltage switched and 5 to 6 Vdc to a mechanical polarizer. If this power does not reach its destination for one reason or another, the system will not function.

Voltage checking should begin at the receiver output. The voltmeter is set on an appropriate range, from 20 to 200 Vdc. An output of 36 Vdc is common on the actuator circuit so the 200 Vdc scale would be used here. If the range is set too low, the indicator will be pushed off scale. In general, the voltmeter red lead is connected to power and the black lead to ground. Voltage is measured between the central conductor and ground at one port of a two-way, power passing splitter. The input coax is connected to the second port. It is important not to leave the splitter in the line because 3.5 dB will be lost by its insertion. An open port is also an invitation to ingress interference and causes signal losses because of reflection at this discontinuity. If no voltage or a low voltage is measured at the receiver IF input, turn the power off, unplug the receiver and disconnect the cable connected to the LNB. If a second reading obtained after powering back on indicates the correct voltage, there is a short in the system upstream towards the dish. If the voltmeter still reads zero or near zero, 99% of the time an internal or external fuse or a voltage regulator is blown in the receiver.

If the disconnected receiver was delivering full power, but when connected to the cable the voltage was too low, there is probably a short or a bad component further upstream towards the dish. Unplug the receiver and disconnect the coax at the LNB input. Then re-plug the receiver and turn it back on. If full voltage is measured across the disconnected cable, the fault lies in the LNB. If zero voltage is measured, then there is certainly a shorted connector or coax. Perhaps the cable was crushed somewhere along its length.

Always carry extra fuses and voltage regulators. Inexpensive, standard 7812, 7815, 7912 or 7915 regulators for the receiver or 7805 regulators for the polarizer control circuit are available at electronic supply stores. Also, have an extra 555 timer chip used to send timing pulses to many brands of mechanical polarizers. These can easily be soldered into a receiver in the field, saving many hours of travel time. 555 timers are 8-pin chips which look like a small centipede. They control the position of the polarizer probe.

Whenever leads are disconnected or connected it is important that the satellite receiver be unplugged in order to avoid inadvertently destroying a good component during the troubleshooting procedure. Just because the receiver is turned off does not mean that power is not being sent to the various components.

It is also wise to have an extra receiver, actuator and LNB, as well as appropriate connectors on hand so a bad one can be temporarily replaced. Most users appreciate the loaned equipment during the time their faulty ones are being repaired.

Similar techniques can be used to ensure that the proper voltages reach both the polarizer and the actuator motor. The path can be traced from the controller or, if voltage is present at the peripheral equipment, in reverse from the dish to either the receiver or actuator control box. When checking voltages on the actuator motor two people are required, one to activate the motor via the control box and one to take the reading at the motor.

The voltmeter can also be used to check both continuity and shorting of cables. If a coax or other wire is suspected of having a break, set the volt/ohm meter on the ohms x 10 scale. Then use alligator clips to join the central conductor and ground together at one end. Make sure that the other end is disconnected from power and attach the meter at this end. If the meter reads zero ohms, then the circuit is continuous; otherwise the cable is broken. If the alligator clips are not attached and a resistance reading is taken on the disconnected cable, a zero resistance indicates a short between the central conductor and the ground shield.

When a connector is shorted, first check if there is any mechanical cause. Water can be soaked into a cable at a leaky connector just as wax is drawn up a wick on a candle. This "wicking" can pull water into a cable as far as 30 cm or more. If this is so, the sign is usually a discoloration of the center conductor. If the cable contains aluminum, it can be severely corroded. Cut back beyond the discoloration and attach a new connector.

Grounding Problems and Hum Bars

Sometimes problems cannot be traced to a single faulty component but are related to how all the components and wiring fit together. Diagnosing such difficulties can be more time consuming but also more rewarding than simply replacing a bad LNB or finding a shorted cable.

Ground loops are a good example of this type of system problem. A ground loop results from having two grounding points within a TVRO that offer slightly different resistances to ground. In most installations all the electrical components are grounded to a common point through the wall outlet. However, if the cable run between the dish and the indoors equipment is long enough, the resistance between the ground potential at the earth near the dish and the coax's ground resistance can be substantially different. When a ground loop occurs, it may be necessary to "float" the indoors components from the wall outlet and use only a single common ground at the dish.

Ground loops can create a series of problems including "hum bars" on the screen, difficulties with polarity control when a pin diode controller has been installed, actuator positioning errors or a low-level audio hum. Hum bars appear as messy horizontal regions of varying width that may be stationary or may move up or down on the TV screen. In effect, the difference in ground potential causes the picture to lose some of the necessary synchronizing ability.

Ground loops can be broken into two categories: those that appear long after the installation is completed and are usually caused by bad connections; and those that begin during installation that can, on occasion, be tricky to cure. However, one of two methods can usually clear up the difficulty. The first cure is to systematically eliminate any bad connections that might exist in the TVRO wiring or, perhaps, in the home's ground connection wiring. The second is to tie all of the interconnected satellite equipment together via a heavy gauge wire with a common ground at the dish by "floating" the indoor equipment. Or the dish and LNB can be tied directly into the electrical service ground at the point where it enters the home with a heavy gauge (e.g. #6 gauge) or a flat welding cable.

Inadequate system voltages can have the same effect and cause hum bars by not providing enough power to generate the necessary signal to adequately power the receiver. This could happen if under-sized wire is used on a long run to the LNB so that the cumulative cable resistance causes unacceptably large voltage drops. Note that the same symptom may be seen if the feed probe servo hunts. This is also caused by inadequate voltages in undersized cables spanning long distances.

UNDERSTANDING COMPONENT FAILURES

One approach to troubleshooting is to examine the operation of all components in order to understand how each may fail. This background knowledge is a useful aid in following the methods described earlier.

Satellite Receiver

The function and operation of a receiver must be clearly understood by a competent troubleshooter. It is crucial to be versed in the alignment procedures for each unit installed to avoid many potential problems. Technicians should also be familiar with all front panel controls such as skew adjustment, if any, subcarrier selection and video scanning.

A visual inspection of all feedback lights, fuses and connections and measurement of voltages will often reveal a simple problem. If this is not the case, more technical approaches can be used to determine any fault. One simple alternative is to swap an identical working unit and see if the trouble disappears.

Some of the problems more commonly encountered are listed below and in Table 11-3.

1. A high signal strength meter reading but no picture. Assuming that the TV is tuned to the correct channel and all other components are functioning properly, this fault can be caused by a bad modulator. Note that many receivers use VCR modulators and replacements that are available in electronic part stores. Be sure that a crystal controlled, "linear" modulator is purchased. Non-linear brands are intended for use in computers.

2. A zero signal strength indication with no picture. This may be caused by a shorted IF/voltage cable, lack of powering voltages or a problem somewhere else in the system such as poor dish alignment.

3. The receiver shuts off when it is hot. A thermal overload is often caused by a voltage regulator overheating. A can of cold spray or a hair dryer may help isolate the defective component. The problem may be solved by simply clearing enough room to allow air to freely flow around the receiver vents. Occasionally a defective component may have to be repaired in the shop.

4. A herringbone pattern in the picture. This can be caused by a bad modulator or unwanted feedback from the IF line into the modulator. A strong, nearby FM station or transmitter may also cause the same problem if signals leak into the system through poorly shielded cables or components.

5. No audio. This may be a problem in either the receiver or modulator. Check all connections and make sure the right audio subcarrier is being selected. If a faint sound can be heard when listening with a pair of headphones at the receiver's audio output port, the modulator is probably bad.

6. A buzz in the audio. This can be caused by a faulty modulator, incorrect audio tuning or an unduly high video level in the modulator. The appearance of saturated colors on the screen is also associated with an excessive video level.

7. Thick black bars across the TV screen. This is often a symptom of a faulty power supply caused by an irregular voltage or a ground loop. A 4700 microfarad, 25 volt capacitor applied across the dc output and ground may quickly solve this problem, but the underlying cause should also be found and corrected.

Aside from swapping out a voltage regulator or visually inspecting internal components, repair work on a satellite receiver of any complexity should be done in the shop or in a portable van equipped as a shop. There is nothing worse than scattering parts from a receiver all over a customer's home and then not having the components necessary to fix it when the problem is discovered.

TABLE 11-3. COMMON RECEIVER PROBLEMS

- Blown fuses
- Poor or opened cable connections on rear panel
- Failure in power supply to LNB
- Power supply output to LNB high (25-30 Vdc) or low (10-16 Vdc)
- Failure in power supply to polarizer
- Power supply output to polarizer high (5-10 Vdc) or low (0-3 Vdc)
- Polarizer wires incorrectly connected
- Power supply to actuator motor absent or low, less than 15 Vdc
- Actuator wires incorrectly connected
- Front panel function buttons inoperative
- Memory back-up battery dead
- Switches on rear panel incorrectly set
- All or part of remote control section not functioning
- LNB input connected to incorrect output port
- TV output connected to incorrect port
- IF cable loop not installed when necessary
- Defective connection from an external power supply to receiver
- Defective external power supply
- Defective circuit breaker
- Not all receiver sections functioning
- Receiver does not accept programming information
- Receiver loses memory
- Polarizer incompatible with LNB or feed
- Pulse to polarizer missing

Cable and Connectors

Coaxial cable and connectors can easily be overlooked as potential sources of trouble (see Table 11-4). But incorrect installation procedures and improper selection of cables can often cause substantial deterioration in overall performance.

A poorly grounded cable or a connector that leaks moisture can introduce unwanted noise or can severely attenuate signals. A cable that is bent too sharply can result in signal losses because its impedance changes at the bend and causes signals to be partially reflected. Improperly installed direct burial cables and leaky connectors can corrode in surprisingly short times.

When installing cables and connectors, the casing must be securely crimped onto the shield wires with the proper tool, a cable crimper. All locations where water may enter must be well sealed and protected. Coaxial sealant can be used to prevent water ingress at all outside, exposed locations.

Troubleshooting cables and connectors can be as simple as a visual inspection. If necessary, a continuity test will make sure that signals can pass without serious or complete attenuation. If wire or cable having too fine a gauge has been used for long runs, insufficient voltages reaching the LNB or polarizer may cause faulty operation.

LNB

A faulty LNB can cause problems ranging from excessively sparklie pictures to complete whiteout. Some common symptoms and their causes are listed below and in Table 14-4.

1. Lack of voltage to the LNB. This can result from a blown fuse or other fault in the receiver, a corroded connector or a shorted or broken coax. A simple method to check that the full 15 to 24 Vdc is reaching the LNB is to disconnect the input line and measure voltage between ground and its central conductor. Be careful not to short the leads together.

2. An open circuit at the LNB output. One or more of the LNB stages may be blown. There will be no signal strength reading at the receiver and the TV picture will be blank white.

3. Excessively sparklie pictures. One or more LNB stages may be faulty, resulting in noise temperatures in excess of rated values. Switching in a good LNB or using a signal simulator is a quick troubleshooting procedure if this is suspected.

4. Presence of hum bars. Hum bars, horizontal lines of varying thickness moving up or down across the screen, can be caused by insufficient voltage at the LNB. Most LNBs require a minimum of at least 15 Vdc and 150 milliamps of current to function correctly.

5. A blank TV screen. This may be caused by a faulty LNB. To determine if the LNB is powered and operating, connect and disconnect the LNB cable while observing the television screen. If the noise level on the picture does not change, this amplifier is probably at fault or is not receiving power. Another check for a faulty LNB is to wave a fluorescent light in front of the feed. If the noise level does not change, the LNB is probably faulty or is not receiving power.

6. Picture deteriorates in hot or cold weather. This problem can be isolated by using a can of freeze spray or a hair dryer to change the external case temperature of the LNB.

TABLE 11-4. COMMON CABLE & CONNECTOR PROBLEMS

- Cable not the correct type
- Cable run to dish too long
- Cable has been damaged underground or above ground by rodents
- Cable cut with a shovel or other utensil
- Cable impregnated and shorted out with water
- Connectors inadequately installed
- Incorrect type of connectors used
- Poorly waterproofed connectors corroded
- Cable crushed or pinched
- Nails or staples driven through cable
- Cable has been burned by lying across a heater exhaust vent or attic lamp housing
- Cable crushed or cut by movement of the antenna
- Center conductor of coax shorted by a strand of its shield
- Coax center cable in connector was ringed during stripping and has since broken off
- A splice pulled apart or not making adequate contact
- Sharp kink in coaxial cable
- One or more wires in cable run are opened
- One or more wires shorted together or to the shield

LNBs are finely tuned pieces of equipment. A faulty unit should be returned to the manufacturer for repair. Although they are carefully encased to eliminate water ingress, if a leak does develop, the moisture can destroy sensitive circuits by oxidizing copper tracks on circuit boards. The early signs of water damage can be an overly noisy signal and the resulting sparklies.

A faulty LNB can show up as a blown fuse in the receiver. However, if fuses keep on blowing even after the LNB is disconnected, the cable or connectors are probably at fault.

TABLE 11-5. COMMON LNB PROBLEMS

- Channels not on their assigned number on receiver
- Channels 1,2 and 3 missing
- Channels 1 and 24 missing
- Channels 23 and 24 missing
- TV screen all white, no sparklies
- TV screen all black, no sparklies
- TV screen has stationary horizontal bars
- Excessive sparklies
- Black horizontal bars floating slowly up or down TV screen
- Smeared picture
- TV screen snowy with a faint picture in background
- Corrosion at input connector
- F-connector cross-threaded onto LNB, no contact made
- Picture disappears as temperature changes during day
- Picture improves during evening hours

Feed

Feeds are precisely machined waveguides designed to efficiently capture microwaves. Any obstruction in their throats can detune the waveguide and seriously impair performance. Some commonly encountered symptoms of problems are listed below and in Table 11-6.

1. Sparklie pictures. This may be caused by wasps, water, ice or other obstacles lodged in the waveguide throat. If water is not removed it could freeze and crack the housing. When using the 90° waveguide elbow with the polarizer, a cure is to drill a very small hole, less than 2 mm, in the lower corner of the right angle guide to allow water drainage.

   Some poor quality units have a base that is not perfectly round or scalar rings that are not exactly at right angles to the body of the feed. This feed may then not "see" the focal point as if the dish were distorted. The symptom of these problems is sparklies on just a few channels.

   Such performance impairments may also be caused by a bent or mechanically distorted probe. But never readjust the probe position even if it appears to be bent. These are finely tuned devices.

2. Signals of only one polarity are received. This probably indicates a bad PIN diode, voltage not reaching the PIN diode or a bad servo motor. Burnout in mechanical polarizers can be caused by improperly installing servo motors without allowing for full 270° freedom of motion at the correct position. This can also be caused by using a wire gauge that is too small for the cable length used or by not using shielded wires.

3. Oscillation or servo hunting of the mechanical probe. A random oscillation in a servo motor polarizer can appear in the form of black bars or white lines on a TV screen. In severe forms, it may show up as jumping between two adjacent channels of opposite polarity as the probe hunts intermittently around its axis. This differs from channel drifting because the same two channels are always seen. An easy method to prove the problem is caused by servo hunting is to disconnect the polarizer connections. If the problem disappears, its cause is evident.

TABLE 11-6. COMMON FEED PROBLEMS

- Servo motor and feed not aligned correctly with polar axis
- Servo motor dead
- No voltage or excessive voltage at motor
- Motor stuck in position
- Probe stuck in housing
- Insect nest in boresight causing signal attenuation and sticking of probe
- Incorrect connection of +5 Vdc, pulse and ground wires
- Feed type not matched to receiver
- Inadequate voltages causing servo hunting
- Buzz from failing motor
- Scalar type feed not tightened into a position parallel to reflector
- LNB waveguide incorrectly installed
- Water in waveguide causing corrosion
- Waveguide loose on feed housing
- Scalar rings on backwards
- Adjustable scalar rings incorrectly set

This most often occurs when cable runs exceed 25 meters and an under-sized wire or parallel servo motors are installed. The solution is to install a 1000 microfarad, 10 Vdc electrolytic capacitor between the red +5 volt wire and black ground on a servo motor. The capacitor should be connected so the plus on the capacitor is connected to the +5 volt and the minus side to ground. To prevent oscillation, the suggested minimum wire sizes outlined in Chapter 6 should be used.

Actuators

A well-designed and well-manufactured dish actuator is usually quite reliable when correctly installed. However, there can be large quality variations between different brands, and proper equipment selection will save troubleshooting and repair time.

The actuator consists of a mechanical assembly and its electronic control circuitry. The mechanical component serves to hold the dish in place and to provide movement across the polar arc while the controller keeps track of the dish position. Some common symptoms and underlying causes for mechanical failures are listed below and in Table 11-7.

1. Inaccurate dish pointing. If there is too much play in the movement of either the inner or outer tubes (or gears or a chain drive in the case of a horizon-to-horizon assembly) then some wearing or loosening of parts has occurred. If the dish has more than 1 cm of play in either direction at its rim, the mount needs attention. The bolts supporting the pivot points or jack attachment should have double nuts, lock washers and lock-tight, and be tightly set.

2. Wear or scoring on the jack tube. This may result when the internal O-rings are damaged by lateral stress being applied when the arm is not installed using ball joints. Waterproofing should be adequate so that water does not leak into the casing and motor assembly.

3. The jack tube is binding or even bending as satellites are tracked. This may occur if both ends of the jack arm are not properly aligned and mounted on self-adjusting ball joints. A single ball joint on one end is not enough. The jack should be able to track through its entire range of possible motion without binding on either the dish or mount. So if the electrical east/west limits fail, damage to the arm will not occur. The motor may also have failed possibly due to water damage. It makes sense to have a spare motor or brushes for placement around the armature.

4. Water entering the motor housing or the tube. Water can collect inside the arm or housing if they have not been properly sealed by gaskets or a rubber boot. Even so, condensation can cause water to collect, freeze, seize the motion and even damage the housing. Drain holes must be in their correct position to prevent this chain of events from occurring. A neoprene rubber accordion sleeve with a hole at either end to allow condensed water to escape is an excellent method to prevent water ingress.

5. A grid of dots appears on the TV screen. Some control boxes cause interference on channels 2 or 3. The internal 3 to 6 MHz clock crystal that runs the microprocessor makes a grid of evenly spaced dots appear on the screen. This effect sometimes occurs when the actuator is set on top of the TV set and beats directly into the tuner. Or if 300-ohm flat, twin lead cable is used to feed over-the-air broadcasts into the TV set, it may also pick up the signal from the clock crystal. These dots can also appear on satellite TV channels, but can be easily eliminated by using a 10 dB line amplifier on the receiver modulator output to overpower the interference.

The most common types of feedback circuits used for controlling actuator position are Hall effect sensors, reed relay switches, 10-turn potentiometers and photo optical read-outs. Most are coupled to a central microprocessor. Each has its own set of potential problems and solutions.

Feedback potentiometers receive a constant voltage and divide it into a smaller portion depending upon the dish position. This returned voltage is calibrated and fed onto a read-out on the face of the controller.

If the potentiometer, also known as a pot, breaks or becomes disengaged from the drive shaft in the motor, a dish can be driven too far and the jack tube can be bent. A preset electronic and mechanical limit is a crucial feature in such control systems.

A method to troubleshoot this type of actuator is to use a spare 5 K-ohm potentiometer. It can be temporarily installed in place of the permanent one by simply disconnecting it and reconnecting the two or three wires using alligator clips. If read-out numbers change on the control box when the temporary pot is adjusted, then the problem is in the motor; if not, the problem is in the control box.

To troubleshoot a unit with a reed switch or Hall effect transistor sensor, attach an alligator clip between the pulse and ground connector at the control box after disconnecting the pulse input line. As the drive arm is moved in either east or west direction, make and break the connection on the pulse and ground wires to simulate drive pulses. If the controller works properly and does not display an error, then the problem is in the sensor or the cable connection between the motor and control. Troubleshoot for cable continuity.

One method of troubleshooting a Hall effect sensor at either the controller or motor is by reading voltage between the pulse and ground points. A drop of about 0.75 Vdc should be measured when the east or west button is pressed. If not, then the Hall effect transistor is bad. If the sensor itself is not faulty, check that the magnets on the motor have not moved or fallen out of place. A spare Hall sensor and reed switch are useful additions to a troubleshooter's spare parts kit. Simulated pulses can also be produced by hooking the polarizer pulse and ground outputs from a receiver or a hand-held polarizer controller to the sensor and ground inputs on the positioner. This will cause the actuator motor to move and will indi-

TABLE 11-7. COMMON ACTUATOR PROBLEMS

- Rust or ice binding inner tube
- Tube slightly bent
- Ball joint sticking
- Bolts attaching actuator to dish loose
- Actuator-to-mount clamp installed incorrectly
- Actuator clamp/collar set too tight, causing tube to score
- Bolt broken inside tube
- Actuator tube driven past end and loose from sleeve
- Actuator does not extend to full length of travel or completely contract
- Sensor count mechanism defective
- Motor case full of water or corroded from water intrusion
- Magnet wheel wobble
- Mechanical stops broken
- Brushes in motor defective
- Motor case sections coming apart because of loose bolts
- Sloppy tube-sleeve bolt or clutch
- Motor running but tube does not extend or contract

cate a bad sensor. Pulses will continue until this connection is broken. Be careful not to drive the dish into the ground.

It is important to always use shielded sensor cable on Hall effect or reed sensors. These devices will give spurious readings if they pick up outside noise. Even the inductive action of a starting or stopping motor can be detected as a "spike" in the sensor. This added to the control count would cause a cumulative error in measuring dish position. After the problem has been corrected, the actuator control must be re-synchronized because each satellite will probably be missed by the same amount.

Make sure that the magnets surrounding these two types of sensors have not been dislodged. If they have, remember that magnets have a positive and negative end and should be reinserted facing in the correct direction.

The ground lug on the controller power cord must never be cut to try to fit the three prong plug into a two prong outlet. The microprocessor requires a reliable ground to accurately set its counting circuitry. Always use some type of surge protection on the power line into an actuator control box to protect the microprocessor. Note that if a problem does develop in the internal circuitry of an actuator controller, the best bet is to return it to the manufacturer for repair.

Dish and Mount

Servicing a dish and mount is a relatively easy matter of conducting an educated visual inspection (see Table 11-8). Reflectors should be checked for warping and surface imperfections. Mounts must be examined for stability, alignment and secure connections. The feed/LNB support structure can be checked for centering and stability. Methods of performing all these checks have been outlined earlier in Chapter 8.

Television Set

A television is a complex piece of electronic equipment. However, troubleshooting a TV is easy when over-the-air channels of adequate signal strength are present. If the set works well with conventional reception, then about the only problem encountered is when the television is not tuned onto the modulated channel.

TABLE 14-7. COMMON DISH/ MOUNT PROBLEMS

- Buttonhook bent or too short
- Tripod or quad feed supports bent
- Quad supports not flush against scalar
- Dish warped
- Panel(s) missing
- Pivot points sticking or rusted
- Elevation or declination angles not correctly set
- Spacers missing or bolts not tightened
- Missing hardware on antenna or mount
- Incorrect replacement parts installed on dish or mount
- Mount cap not correctly bolted to ground pipe or cracked
- Diameter of ground pipe incorrect causing a rocking of dish
- PVC, EMT (electrical metallic tubing) or plastic coated gas pipe used instead of steel in ground pipe
- Cracked or broken parts
- Mount not moving to maximum east or west positions
- Turnbuckles installed in wrong holes
- Elevation bar bolt holes tapered out and not correctly tightened
- Saddle for elevation bar spread apart so that dish rocks or shifts

There are major differences in quality between different brands of televisions sets. A well-equipped troubleshooter/installer is wise to carry a small, portable, excellent quality television or monitor to demonstrate if there ever is a question about the performance of a satellite system. After all, most judgments are made on the basis of picture quality. Also, if a monitor is used, a faulty modulator internal to a receiver can easily be diagnosed since it would be bypassed.

SPARE PARTS KIT

If a site visit is necessary, it is essential to have the proper tools on hand. Driving 100 miles to discover that a 50 cent voltage regulator is missing can be expensive.

All the light tools of the trade should be taken on a service call. This usually excludes concrete mixers and other large items. In addition, a complete, healthy spare system, possibly including a small test dish, should be part and parcel of the troubleshooting kit. This kit is composed of:

- Feed with polarizer
- Extra servo or dc motor for mechanical polarizer
- Extra reed switches or Hall sensors
- Inexpensive receiver
- Ample supply of applicable fuses
- Appropriate voltage regulators
- Spare motor or a complete actuator
- 5 K-ohm potentiometer
- Fluorescent light
- 1000 microfarad electrolytic capacitor
- Scotch locks
- Electrical tape, solder
- Splitters
- Assorted connectors
- 10 dB line amplifier
- Coaxial sealant and shrink tubing

ADJACENT SATELLITE INTERFERENCE and DISH SIZING

As satellite power levels have risen and as reception and compression technology have improved, ever smaller dishes can receive excellent quality video, audio and data signals. However, there is a lower limit to the size of reflectors that can be used since beamwidth increases as reflector diameter decreases. Beyond a certain beamwidth, more than one satellite may be detected by a dish, especially since many spacecraft are now spaced as closely as 2° apart in the Clarke belt.

If, for example, the half power or 3 dB beamwidth of a dish is 4°, then at half of this value or 2° on either side of the main axis, signals are received at 3 dB below those detected along the boresight. If a satellite happens to be located just 2° to either side of the target spacecraft and is transmitting on the same channel and polarity, two garbled pictures would appear on a TV screen.

A dish may not behave as its theoretical beamwidth would suggest. One with an accurate surface and a well-matched feed assembly is closer to optimal performance and has a smaller beamwidth than one of lesser quality. Real-world beamwidths are partially determined by dish efficiency and almost always larger than would be expected.

At what level is a picture judged unwatchable when receiving interference from an adjacent satellite? This depends to some extent upon whether the intending signal is an analog or digital one and, to a lesser extent, upon whether the interfering signal is digital or analogue. When viewing an analog broadcast, subjective tests have shown that an interfering signal was first noticeable when its level was 11 dB below the target signal, was annoying when received at 5 dB and quite unacceptable at 4 dB.

While similar considerations as outlined in the previous section on TI apply to the effect of interfering signals from adjacent satellites on digital broadcasts, satellite interference is wide band, not narrow band as in the case of most terrestrial signals. Past a certain threshold level all programs would be lost. In the worst case, the center frequency of the interfering signal would be the same as that of the target broadcast and its bandwidth would exceed that of the desired signal.

CHAPTER 12

LARGE DISH SYSTEMS

In the earlier days of satellite broadcasting, dishes often 15 feet or more in diameter were used by cable TV companies and other commercial ventures. With advances in technology, dish size has decreased dramatically. However, there are still situations where very large dishes, perhaps up to 20 or 30 feet in diameter, are required.

The NEED for LARGE DISHES

There are two main reasons for using large diameter reflectors. First, if the installation location is far removed from satellite boresight, the EIRP is so low that an especially large reflector is needed to capture enough signal to adequately drive the receiver. For example, those in Central America wishing to directly view American broadcast satellites often require dishes as large as 20 feet in diameter. Also, if a TVRO designed to receive broadcasts from an Intelsat satellite transmitting a global beam were installed on the east coast of North or South America, a very large antenna would also be required. In these cases, a 16 foot (5 meter) or larger reflector is generally required for high-quality, sparklie-free reception.

Second, large dishes must often be installed in commercial installations where signal levels must be substantially above receiver threshold to ensure perfectly clean reception. Cable TV companies or SMATV headends, for example, require video and audio signals of sufficient strength so that reception remains excellent even after the signals are distributed to thousands of subscribers.

INSTALLING LARGE DISH SYSTEMS

Satellite TV earth stations with extra large dishes are similar in many respects to more conventional designs. There are, however, some major points of difference in their installation.

Mechanical Assembly

A 20 or 30 foot reflector is a very large structure weighing much more than a typical home system. Total dish and mount weights in excess of 5000 pounds are not uncommon. Consequently, the supporting structure cannot simply be a pole planted on a small pad or set a few feet into the ground. Excavation is often required before pouring the four or five cubic yards of concrete necessary as a foundation for a double A-frame support (see Figures 12-1, 12-2 and 12-3).

Figure 12-1. Comtech 7.3 Meter Antenna. *This 7.3 meter dish is mounted on an adjustable tripod assembly which is itself supported by a large recessed concrete pad. (Courtesy of Comtech Antenna Corporation)*

Figure 12.-2 Three Pier Support. *This 20-foot ADM dish with horizon-to-horizon mount is supported by a three pier foundation. In general, when installing tripod supported mounts the two rear legs should be set on an east-west line. Fine adjustments to establish the final position can then be made by*

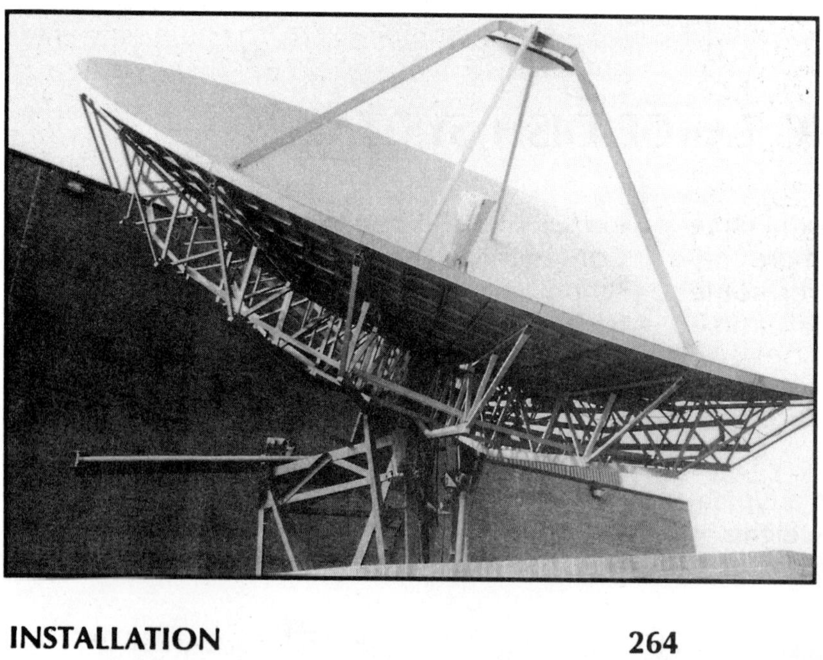

Figure 12-3. Uplink/Downlink. *This 10 meter (33 foot) Cassegrain dish has a carefully designed rib structure to support the weight of its reflector. This uplink monitors its signal by also receiving the matching downlinked satellite signals. (Courtesy of Brent Gale)*

Figure 12-5. Maneuvering a Large Dish. *Five men and one observer were needed to lift this 4-piece, 16-foot dish into its final position. Using a crane proved to be too difficult on top of this five story building in Jerusalem. (Courtesy of Brent Gale)*

Figure 12-4. Use of Crane. *A special crane and scaffolding was necessary to lift this 25 foot Hero dish onto its mount. This Sri Lankan installation is supported by a combination of pole and pad planted in over 5 cubic yards of concrete. (Courtesy of Hero Communications)*

The safety of the installation crews is an important consideration. Much of the work is performed far above the ground and very heavy components are involved. The danger of injury is always present. The effects of a bright, hot sun should also not be ignored. Protective headgear and shoes are necessary. Technicians must consider the intense heat and potential for sunburn when large dishes are installed in regions near the equator. Working at the focus of a 30 foot reflector can be more than just uncomfortable if the dish happens to be pointed at or near the sun. Polarizer caps have been known to melt when sunlight is focused by dishes as small as six feet in diameter. When a fixed dish is being installed, work inside the structure should be planned only for those times of day when the sun is far off the main axis. Or, if the surface can be rotated, it should be positioned away from the sun whenever crews are working within the reflector.

Heavy equipment is often needed to install large dishes (see Figures 12-4 and 12-5). If one is pieced together on the ground, a crane or special hoist is generally needed for lifting it onto the supporting structure. However, there are cases where this is not necessary. For example, the Paraclipse 16-footer can be assembled piece by piece on the mount (see Figure 12-6).

Shipping a very large dish is not simply a matter of having UPS or another freight company deliver two or three small boxes. Specially designed and often heavy crates must be used. Some manufacturers of 16 foot or larger spun aluminum brands spend a sizable portion of their sales revenue on simply crating the dish!

Electronic Components

The electronic components used in large dish installations are generally quite similar to those used in home systems. But there are some important differences.

When some international satellites are being targeted, either a specially designed or a modified feed must be used to properly detect circularly polarized signals. As described in Chapter 5, a teflon dielectric slab inserted into the feed throat permits a conventional unit to receive signals having this different polarization format. Although the focal length-to-diameter ratio of large dishes also ranges between approximately 0.3 and 0.4, reflector diameter is much larger than normally encountered, so the feed assembly can be at a substantial distance from the reflector. Either a tripod or a buttonhook supported by guy wires must be used to prevent excessive movement in winds or when the dish is tracked across the arc. Today, few large dishes incorporate buttonhooks.

Various video formats, NTSC, PAL, SECAM, B-MAC and a few others, are transmitted in various places so satellite receivers must be properly equipped to decode these signals. In addition center frequencies vary and half transponder formats, often broadcast via circularly polarization signals, are frequently encountered.

All electronic components should be tested before being installed. Once a feed and LNB are in place, replacement often involves a fair amount of effort such as using a cherry-picker to reach the focus. Tracking large dishes with narrow beamwidths in locations where satellite EIRP is low can be difficult enough without having to worry about replacing a loose connector or a faulty component. Furthermore, in many locations around the globe this job is made even more difficult because only a handful of satellites have transponders that are active 24 hours a day.

When attempting an installation in a remote area, having spare parts on hand can be crucial. It is generally easier to bring a spare component than to convince customs officials to admit a replacement part at a later date. All components should be tested and burned-in for at least 48 hours prior to shipment. Remember that an export license is often required to move many types of microwave components between countries because some parts, for example the VideoCipher technology, may potentially be used for military purposes. Similarly, all special tools or test instruments that may be needed should be taken on the trip.

Figure 12-6. Arthur C. Clarke's Dish. *A 16-foot Paraclipse dish was installed by one of the authors at Arthur C. Clarke's residence in Sri Lanka. The small tower support was firmly bolted into the structural beams under the floor of this patio. (Courtesy of Paraclipse Antenna and Brent Gale)*

CHAPTER 13

DESIGNING SATELLITE TV SYSTEMS

The performance quality of a satellite reception system depends upon how well the uplink-to-TVRO link has been designed. Many factors in this process such as the power level uplinked to the satellite or the type of modulation technique or transponder power used are out of the control of all but a select few. However, the design of the receive site equipment can have a great effect on the quality of the audio, video or data signal received.

A well-designed system must perform well not only when first installed on a sunny day but also on a rainy day six years in the future. The choice of components is a crucial factor. A poorly built dish may work well for the first year or two but literally fall apart later. Or an LNB assembled from cheap electronic components may become noisy.

The objective here is to present methods to determine how to size a receive dish or choose an LNB or satellite receiver so a high quality signal can be detected anywhere within the footprint of a satellite.

OVERALL DESIGN CONSIDERATIONS

The method to size satellite receive dishes and choose LNB noise temperature/noise figure is based on the relatively straightforward analysis that follows. This method can be applied whether the underlying signal is analog or digital. There is no difference. However, the actual value of the carrier-to-noise ratio, the C/N ratio, required to achieve the same result differs between analog and digital systems. Digital systems require lower input C/N ratios to achieve high quality pictures and sound.

The Link Equation

The main goal in satellite downlink analysis is to design a reception system that produces a signal-to-noise ratio (S/N) to yield an acceptable quality television picture. The most important factor is that the signal entering a satellite receiver should have a carrier- to-noise ratio of at least at or above threshold.

Standards for acceptable S/N ratios vary from 45 dB which produces an excellent picture to 30 dB for just a passable one. This ratio depends upon a number of factors, some of which including satellite EIRP, path losses and the design of the modulator in the satellite receiver are beyond the control of a downlink system designer. However, the designer can select the type and size of dish, feed, LNB and coaxial cable, all of which directly impact TVRO performance.

The link equations are used to calculate C/N ratio at the input to a satellite receiver. The value of C/N ratio depends upon both satellite link parameters, namely the EIRP, as well as upon the choice of dish, LNB, cable, connectors and satellite receiver. For example, a digital IRD imposes different requirement on input C/N ratio than does an analog receiver.

The performance of the dish/LNB system depends upon a number of uncontrollable factors that include variations in satellite output power, satellite stationkeeping accuracy, atmospheric attenuation of the signal, aiming and deformation error of the receive dish caused by wind loading or even signal scattering due to obstructions caused by wind-blown branches or airplane crossings. However, the effective isotropic radiated power (EIRP) at any particular location is a good measure of parameters that reflect satellite operation.

It is important to realize that the EIRP calculated from satellite design and location may not correspond exactly to what is actually received, especially because the output power of a transponder slowly decreases as it ages. Calculated EIRP is based upon an assumption that the downlink antenna is perfect. In addition, published EIRPs do not account for the fact that the downlink antenna has side lobes and nulls between side lobes. Therefore, areas far off axis may receive more power than expected, especially if they happen to be on the target of a side lobe peak (see Figure 13-1). This phenomenon is called antenna spillover. For example, a satellite transmitting a spot beam to the southern United States may have its first side lobe relaying power far off boresight to northern South America. Such effects are usually detected only by on-the-spot tests.

The downlink C/N ratio is determined using the "link" equation. The components of this equation include the EIRP, the signal power aimed towards any geographic location as found from a footprint map, gain of the receive dish and system noise temperature. The analysis accounts for the weakening of earthbound microwaves as they spread out in space, the receive dish gain as it captures and concentrates these signals, and the effect of noise, in large part determined by receive dish and LNB characteristics. All these factors are included in the link equation:

Carrier-to-noise power ratio = Signal power leaving the satellite, EIRP

less

free space path and atmospheric losses

plus

Receive system G/T

less

Dish pointing errors

In a more condensed form, in which the above words, excluding pointing losses, are expressed in algebra, the link equation expressed in decibels becomes:

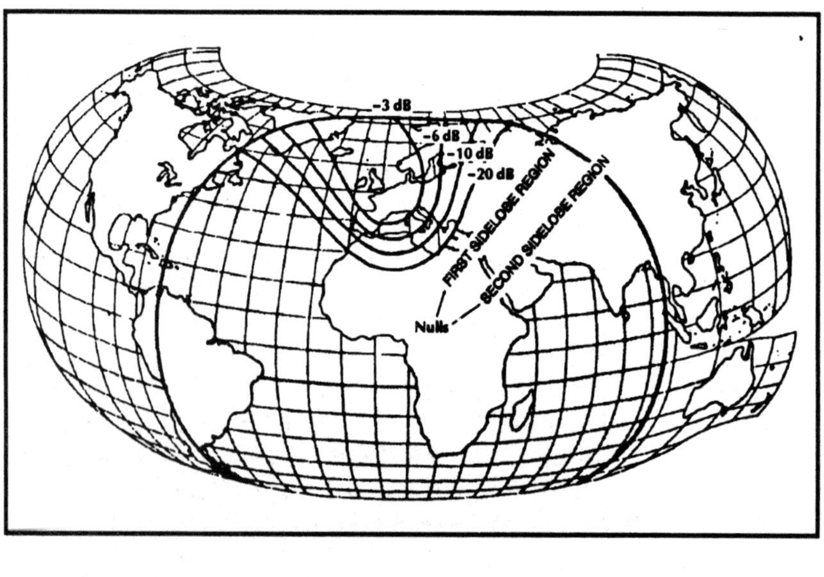

Figure 13-1. Off-Axis Reception of Satellite Broadcasts. *Signals can sometimes be received via side lobes as indicated here. As the boresight EIRP increases, side lobe levels also increase in power. This phenomenon, known as dish spillover, allows signals that are high enough in power to be detected at an off-boresight location with use of relatively large receive dishes.*

$$C/N = EIRP - PL + G/T_{sys} - 10 \log B + 228.6$$

In this equation, T_{sys} equals the noise temperature of the dish, low noise amplifier and all other system components. The factor B is the demodulator bandwidth. As either noise temperature or bandwidth is increased, more noise enters the system, so 10 log B is subtracted to show its effect on the C/N ratio. Bandwidth is expressed in Hertz in this equation. The constant, 228.6, is known as Boltzman's constant.

Thus, if EIRP, path loss and bandwidth are known, the G/T_{sys} required to obtain a desired C/N ratio can be calculated by addition and subtraction. G/T_{sys} is shorthand for:

$$G - 10 \log T_{sys}$$

where gain is expressed in decibels and noise temperature in degrees Kelvin. The G/T_{sys} is the "figure of merit" of an dish/feed/LNB combination. The "bottom line" G/T_{sys} defines the combination of minimum dish diameter and LNB noise temperature required so that input C/N ratio is just at receiver threshold. Most analog satellite receivers have C/N thresholds of 8 dB or slightly lower. Digital receivers have C/N thresholds of 6 dB and lower.

EIRP

The EIRP of a particular satellite can be easily determined for any geographic location from a published footprint map. While calculated footprint maps are available for most communication satellites, on-site measurements of EIRPs are available for only a limited number of spacecraft. (The *World Satellite Yearly,* published each year, lists footprints of nearly all the world's geostationary satellites.

EIRPs can differ somewhat from theoretical expectations. In addition, footprint maps show only the power levels radiated from a downlink antenna's main lobe. Power levels of signals from downlink antenna side lobes, antenna spillover, are rarely published because even small movements in a satellite's orbital position can radically alter the aiming of a narrow side lobe.

Free Space Path Loss

Free space path loss, the second term in the link equation above, expresses how the power of the downlinked signal is attenuated in traveling from a satellite to a receive dish. There are two effects at work. First, microwaves spread out and are weakened on their earthbound journey. It is evident that this must occur since the energy radiated from the small surface area of a downlink antenna ends up blanketing a portion or all of a continental land mass. This free space path loss is simply a function of the distance between a satellite and the receive site as the signal frequency (see Figure 13-2). Formulas for calculating path loss and slant range, the distance from any site to any satellite, are outlined in Appendix B.

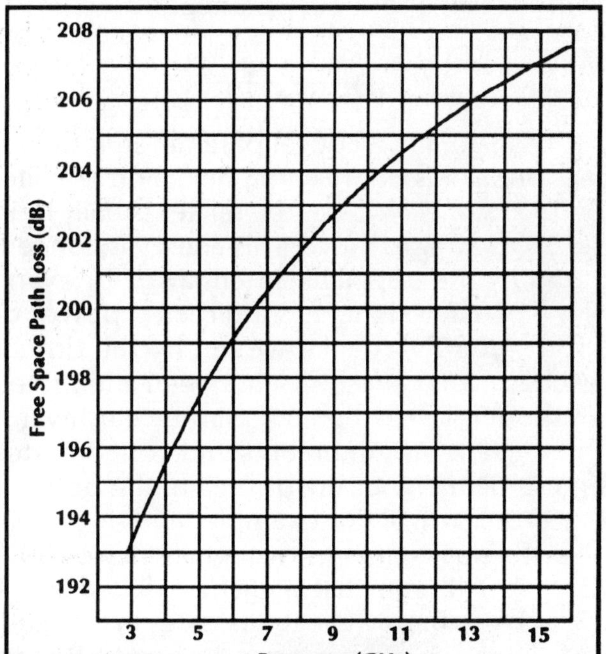

Figure 13-2. Free Space Path Loss. *This graph shows the free space path loss for a site at the equator, elevation of 90°. As frequency increases so does pass loss. On a clear day it is 195.6 dB at 4 GHz, while it increases to 205.6 dB at 12.75 GHz. Signal losses increase as the path traversed through the atmosphere increases. Consequently, losses are greatest for nearly horizontal paths at low elevation angles. Additional free space path losses at 4 GHz incurred at 45° and 0° elevations are 0.44 dB and 1.33 dB, respectively, because the dish "looks" at a thicker layer of atmosphere.*

The second, much less predictable, cause of path loss is absorption of signal by the earth's atmosphere. At Ku-band frequencies by far the greatest reduction in power is caused by water vapor and rain; this attenuation is approximately 9 to 10 times greater at Ku-band than at C-band for a given amount of water in the downlink path.

Typical Values for Path Loss

The path loss at 4 GHz on a clear day for a satellite directly overhead as viewed from the equator is −195.6dB. This is equivalent to a reduction in signal power by a factor of over 2.75 x 10^{-20} (1 in 275 billion billion)! This number very forcefully reminds us that communication spacecraft are over 22,300 miles away in outer space. Detecting a C-band satellite signal is similar to attempting to observe a 20 watt light bulb from such an enormous distance!

Dish Gain

Dish gain is a critical parameter in the reception signals from a communication satellite. TWTA outputs on C-band satellites usually lie in the 10 to 20 watt range. This energy is concentrated by the downlink antenna so, for example, a 36 dBw signal has a directed power of about 4,000 watts. However, by the time it reaches the receive dish it has been attenuated by about 195.6 dB on the journey earthward. The receive dish intercepts and concentrates signals. The higher the dish gain, the better a TVRO performs. For example, an 8-foot dish operating at 4 GHz with an efficiency of 60% has a 38.0 dBi gain that might be reduced by 0.1 dB to 37.9 dBi by waveguide losses in the feed. Below a minimum gain, even the best LNB or receiver cannot compensate enough to adequately capture the signal essential to reconstruct a satellite broadcast. The C/N ratio would simply be too low.

System Noise Temperature

System noise temperature is a measure of how much noise is added to a signal as it is processed by all components of a TVRO. These sources of noise include that detected by the dish and introduced by the LNB, waveguide losses and the downstream electronic circuitry.

Unlike terrestrial microwave links, satellite systems receive extremely low signal levels, not far above the power of both galactic noise and microwave radiation from the warm earth. Galactic noise, which is all that an orbiting antenna pointed into deep space would detect, typically contributes about 3K to a TVRO.

Noise from the warm earth at an average temperature of 290K is detected via dish side lobes. It depends upon both the dish's side lobe patterns and its angle of elevation. As a dish is inclined more towards the horizon it faces closer to the ground and consequently detects more ground noise. Since the elevation to satellites at the center of the geosynchronous arc is higher than that of those at the easterly and westerly portions, dish noise is lower when the higher elevation satellites are targeted.

The noise performance of a dish can be minimized in a number of ways. A higher quality dish of a given size has lower level side lobes and is thus susceptible to less noise. A larger diameter dish also has a "tighter" side lobe pattern and a more narrow main lobe so it generally contributes lower amounts of noise. Similarly, a deeper dish having a lower f/D, all else being constant, detects lower ground noise power.

The noise introduced by the LNB is simply expressed by its noise temperature or figure. Therefore, for example, a 20K LNB adds 20K of noise to a TVRO.

Noise is also contributed to a receive system as a result of losses in the waveguide between the feed and the LNB as well as VSWR mismatch or reflection losses. This may amount to a few tenths of a decibel.

The noise contributed by the cables and satellite receiver is low by comparison with that added within the LNB because the LNB amplifies both the noise and signal detected by the dish/feed system. Its output signal is thus substantially higher in power than the noise added by downstream components, namely cables, connectors and the satellite receiver. So the balance of system components has only a small effect, at most about 10K to 15K, on system noise temperature. G/T_{sys}, the figure of merit for an earth station, can therefore be closely approximated by replacing T_{sys} with the noise contribution of just the dish and LNB. The LNB output signal is the key determinant of overall performance quality of a TVRO.

Nevertheless, noise generated in cable runs can be a major factor in some situations. For example, when signal attenuation in long cables increases to the point that the signal power is not substantially above the system background noise, performance can suffer. This may occur in the connection between the LNB and receiver or in an extensive SMATV distribution system. If, for example, line amplifiers are not used in correct locations, a dish/LNB having an acceptable G/T_{sys} may still result in poor quality pictures.

SIZING a DISH

In order to clarify how the link equation can be used to determine required dish size and LNB noise temperature, some calculations are presented here. These assume that the dish is perfectly aimed, not always the case. Slight misalignments or wind loading can cause losses. A safe assumption is that 0.5 dB will be lost to pointing errors. This decrease in C/N ratio should also be factored into the performance margin.

The first analysis shows how to calculate what C/N ratio will be delivered to the input of a satellite receiver for a dish/LNB with given characteristics. The second calculation shows how the figure of merit, G/T_{sys} and then dish diameter can be obtained from an assumed C/N.

Determining C/N

Assume that the free space path loss on a clear day is -195.6 dB. Substituting this value into the link equation yields:

$$C/N = EIRP + -195.6 + G - 10 \log T_{sys} - 10 \log B + 228.6$$

$$= EIRP + G - 10 \log T_{sys} - 10 \log B + 33.0$$

Assume that the satellite receiver bandwidth is 27 MHz, a choice that should allow reception of a video signal with good fidelity. Substituting B equal to 27,000,000 Hertz into the link equation:

$$C/N = EIRP + G - 10 \log T_{sys} - 41.31$$

The next step is to include the dish and LNB noise temperatures in this equation. Assume that a 25K LNB is used with a 10 foot dish having a noise temperature of 50K at an elevation angle of 20° and a gain of 40.65 dB. These values are taken from results presented in Table 5-2. The noise temperature at this low elevation angle is used because it presents a near worst case value for dish noise. A reflector would point at such a low elevation in high northern or southern latitudes or near the ends of the satellite arc even at a site on the equator. In order to present the worst case result, assume that an extra 10K is added to the system noise temperature by the feed, coaxial cable and other system components.

Then plugging the above values for noise temperatures and gain into the link equation:

$$C/N = EIRP + 40.65 - 10\log(25 + 50 + 10) - 41.31$$

$$C/N = EIRP + 40.65 - 10\log(85) - 41.31$$

$$= EIRP - 19.95$$

Therefore, if the EIRP is 30 dB, the TVRO will deliver a C/N ratio of 10.0 dB to the receiver. This is about 2 dB above the threshold of most analog receivers. However, if the EIRP happened to be 28 dB on the edge of a downlink footprint, the C/N would be 8.0 dB which is just at threshold with no margin for pointing inaccuracies.

Determining G/T_{sys}

Begin with the same assumptions made above for path loss and receiver bandwidth. The link equation can be expressed as:

$$C/N = EIRP + G - 10\log T_{sys} - 41.30$$

$$C/N = EIRP + G/T_{sys} - 41.3$$

Rearranging the algebraic terms yields:

$$G/T_{sys} = C/N - EIRP + 41.3$$

The G/T can be determined from this equation if both the EIRP and required C/N are known. For example, assume that a particular location has an EIRP of 32 dB and that a C/N of 10 dB is acceptable. Then

$$G/T_{sys} = 10 - 32 + 41.3 = 19.3 \text{ dB}$$

Assuming a worst case system noise temperature, like above, of 85K leads to:

$$G - 10\log(85) = 19.3$$

or

$$G = 38.6 \text{ dB}$$

Referring again to Table 5-2, such a gain can be obtained using a 9.5 foot dish having a 50% efficiency or a 8.5-footer having a 60% efficiency.

Recommended Dish Sizes

The link equation has been used to construct Table 13-1 which lists recommended minimum reflector diameters necessary to attain a given C/N. The assumptions here are that the free space path loss on a clear day is −195.6 dB, dish efficiency is 60%, system noise temperature is 80K and satellite receiver bandwidth is 27 MHz. If the noise temperature is lower, if bandwidth is reduced or if efficiency is increased, the minimum acceptable reflector diameter will decrease. These recommendations can therefore be regarded as the worst case scenario.

Six values for the target C/N ratio were chosen. The lowest, C/N = 4 dB, is a typical minimum value for a digital receiver threshold; 8 dB is a typical value for analog receiver threshold. This allows no margin for signal fading due to pointing inaccuracies, rain, wind or other factors. The highest value of C/N ratio listed allows for a 12 dB margin on analog broadcasts.

The effect of increasing the necessary performance margin or the EIRP is quite clear from this table. A very powerful satellite would allow the use of small, easy-to-install dishes. However, in weak footprint areas, a relatively large, high performance reflector is necessary to obtain adequate reception.

How Small Can A Dish Be?

Although the link equation is a powerful tool in choosing dish size and LNB noise temperature given the EIRP of a communication satellite, one factor not considered in this type of analysis is reflector beamwidth. Even if the EIRP were hundreds of watts and the link equation suggested that an extremely small parabolic dish would perform well, dish beamwidth limits minimum size that can be successfully installed. For example, a 3-foot reflector would have a 5.2^{o} calculated beamwidth, much too wide for reception of C-band satellites spaced two degrees apart. Such a small receive dish would detect relatively strong signals from five satellites that were spaced two degrees apart, two targeted one and two on either side!

TABLE 13-1. RECOMMENDED DISH SIZE versus SATELLITE EIRP C-BAND (4.00 GHz)

EIRP (dBw)	Diameter (feet)					
	C/N=4	C/N=6	C/N=8	C/N=10	C/N=15	C/N=20
30	5.25	6.63	8.33	10.53	18.70	33.23
31	4.69	5.90	7.45	9.38	16.66	29.62
32	4.20	5.28	6.63	8.36	14.06	26.40
33	3.74	4.69	5.90	7.45	13.25	23.52
34	3.31	4.20	5.28	6.63	11.81	20.96
35	2.95	3.74	4.69	5.90	10.50	18.70
36	2.66	3.31	4.20	5.28	9.35	16.66
37	2.36	2.95	3.74	4.69	8.36	14.86
38	2.10	2.62	3.31	4.20	7.45	13.22
39	1.87	2.36	2.95	3.74	6.63	11.81
40	1.67	2.10	2.62	3.31	5.90	10.50
41	1.48	1.87	2.36	2.95	5.28	9.38
42	1.31	1.57	2.10	2.66	4.69	8.36
43	1.18	1.48	1.87	2.36	4.20	7.51
44	1.05	1.31	1.67	2.10	3.71	6.63
45	0.95	1.18	1.48	1.87	3.31	5.90

EIRP (dBw)	Diameter (meters)					
	C/N=4	C/N=6	C/N=8	C/N=10	C/N=15	C/N=20
30	1.60	2.02	2.54	3.21	5.70	10.13
31	1.43	1.80	2.27	2.86	5.08	9.03
32	1.28	1.61	2.02	2.55	4.53	8.05
33	1.14	1.43	1.80	2.27	4.04	7.17
34	1.01	1.28	1.61	2.02	3.60	6.39
35	0.90	1.14	1.43	1.80	3.20	5.70
36	0.81	1.01	1.28	1.61	2.85	5.08
37	0.72	0.90	1.14	1.43	2.55	4.53
38	0.64	0.80	1.01	1.28	2.27	4.03
39	0.57	0.72	0.90	1.14	2.02	3.60
40	0.51	0.64	0.80	1.01	1.80	3.20
41	0.45	0.57	0.72	0.90	1.61	2.86
42	0.40	0.51	0.64	0.81	1.43	2.55
43	0.36	0.45	0.57	0.72	1.28	2.29
44	0.32	0.40	0.51	0.64	1.13	2.02
45	0.29	0.36	0.45	0.57	1.01	1.80

Assumptions:

Path Loss = −195.6 dB
Receiver Bandwidth = 27 MHz
Dish Efficiency = 60%
System Noise Temperature = 80K

Why is a Performance Safety Margin Needed?

The concept of a safety margin is simple. If receiver threshold is 8 dB and a satellite system is designed to deliver a C/N of 8 dB in calm weather what happens when strong winds blow? The resulting deflection of the dish off target can result in a loss of signal and picture deterioration. Or during a torrential rainstorm as much as 0.5 dB can be lost at higher frequencies as rain absorbs the microwaves. Then a flurry of sparklies would appear on the screen. The same result can occur as a satellite ages and its transponders weaken and EIRP drops or as a dish or mount sags and falls slightly off target.

A safety margin should be at least 1 dB for home systems and a minimum of 3 dB for commercial systems. The bottom line is clear. While a well designed system may cost a little more, user satisfaction more than warrants the extra expense.

BEAMWIDTH and SATELLITE SPACING

There is no mistaking the fact that smaller dishes are more appealing to all except the rare individual who thinks bigger always means better. As satellite power levels have risen and as earth station technology has improved, ever smaller dishes have become capable of producing excellent quality video. However, the spacing between satellites puts a lower limit on reflector size because beamwidth increases as diameter decreases. Beyond a certain beamwidth, more than one satellite can be detected by a dish.

In the United States, satellite spacing was not an issue for C-band systems until 1980. RCA Americom had been operating two satellites, Satcom I and II, Western Union had three vehicles, Westar I, II and III, and AT&T/GTE operated Comstar I, II and III. There was more than ample orbital space and spacing between satellites was wide. In 1981, the American Federal Communications Commission (FCC) surprised the industry by requesting comments on the possibility of authorizing spacing of satellites along the North and South American portion of the geosynchronous arc at tight 2° intervals. Until then, the accepted policy was 4° and 2° degree spacing for C and Ku-band satellites, respectively.

In April, 1983 the American FCC approved 2° orbital spacing for satellites operating in both frequency bands. This decision, implemented gradually by reducing spacing to 3° at first, has allowed a potential doubling of the orbital arc capacity.

The Importance of Beamwidth

Dish quality and beamwidth are critical factors in satellite spacing. For example, if the half power (3 dB) beamwidth of a reflector is 2°, then at half of this value or 1° on either side of the main axis, signals are received at 3 dB lower than those detected along the boresight. If a satellite happens to be located just 1° off the target and is transmitting on the same channel and polarity, two garbled pictures and the resulting interference would appear on the television screen.

However, the situation would be quite different if the system detected signals from 1° off boresight at powers reduced by 10 dB, or by 15 dB. At what level is a picture judged un-watchable when interference from an adjacent satellite is just barely detectable? Subjective tests have shown that an interfering signal is first noticeable when its level is 11 dB below the desired signal. The interference is annoying when received at -5 dB and quite unacceptable at -4 dB.

Thus a 10 foot dish with a 0.52 degree 3 dB beamwidth would detect interfering signals reduced by only 3 dB at 0.26° off its main axis. Levels will even be higher at, for example, 0.20°. However, a good quality system would have unwanted signals reduced by substantially more than 3 dB at 1.0° off-axis.

Most prime focus parabolic dishes have sufficiently low side lobes so that off-axis signals are not appreciably detected. However, smaller diameter Cassegrain fed dishes can have side lobes only 10 dB lower in power than the main lobe. Such reflectors are more susceptible to interference from adjacent satellites as well as from terrestrial sources.

The Role of Dish Quality

A dish with an accurate reflective surface and a well-matched feed assembly would function more closely to optimal performance and have a smaller beamwidth than one of lesser quality. This is because measured beamwidths are also determined by dish efficiency. Also, because smaller dishes require substantially less material for construction than comparable larger reflectors, leftover manufacturing dollars can be spent in achieving better surface accuracy. The other side of the coin is that better accuracy can be achieved with more manageable, smaller reflectors. The net result is that a well made 6-footer may have less than half the beamwidth of a poorly constructed 12 foot dish, contrary to theoretical expectations.

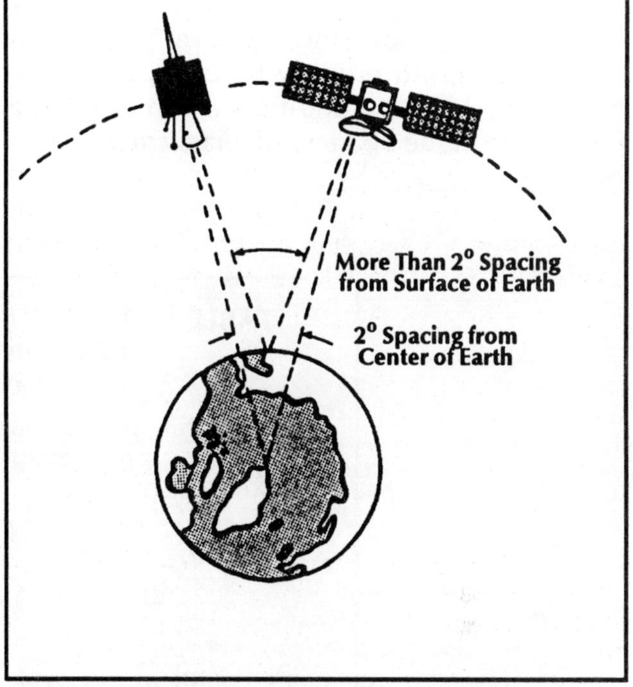

Table 13-3. Variation of Spacing with Site Latitude. *Satellites spaced 2° apart are really separated by a slightly larger angle. While not drawn to scale, this polar view of the earth illustrates the point. The published separation between satellites is measured from the center of the earth but a dish actually lies closer to the arc and therefore subtends a smaller angle. As the site becomes closer to either pole, the angle decreases towards the published angle.*

Is One Degree Spacing Really One Degree?

The published separation between satellites is measured by their difference in longitude. From the perspective of an observer at the center of the earth, the angular separation between two satellites directly overhead would be the same as their longitude difference. But a dish located on the earth's surface pointing straight upwards would see these satellites separated by more than 1° or 2°. This spacing decreases as satellites closer to the horizon are targeted but still remains in excess of the nominal value (see Figure 13-3).

When a satellite dish is located at other points on the surface of the earth, the perceived separation between satellites will also have values in excess of or equal to the satellite spacing. For example, from Honolulu, Hawaii satellites at 131° and 132° longitude will appear to be not 1.0° but 1.14° apart. It is important to realize that these angles are simply related to the geometry of the situation. The maximum spacings occur when the reflector is aimed due south, while the minimum ones occur when it points towards either horizon.

Thus if satellites are spaced at 1° intervals it does not necessarily mean that the power levels at the one degree point on a beamwidth diagram tell the whole story. Performance results can be determined only by comparing the exact angular separation between satellites. For example, the beamwidth diagram may indicate that power levels are down by 18 and 20 dB at 1° and 1.2°. Table 13-2 shows that if a TVRO located at a 40° latitude is targeting satellites due south, the 18 dB figure should be used, but if they are near the horizon, signals from the off-target spacecraft would be detected at a 20 dB lower power level.

TABLE 13-2. EFFECTIVE SATELLITE SPACING versus LATITUDE (at 1° Satellite Spacing)

Latitude	Maximum Angular Angular Separation	Minimum Separation
81.00	1.000	1.000
64.83	1.053	1.000
47.60	1.101	1.000
40.67	1.120	1.000
39.73	1.121	1.000
33.75	1.138	1.000
25.75	1.154	1.000
00.00	1.178	1.000

Maximum and minimum angular separation occur when oriented at the horizon and due south, respectively.

APPENDIX A. THE DECIBEL NOTATION

Decibels (dB) are used to express the relative values of two signals. In order to compress large differences in numbers to a more manageable range the definition includes use of the logarithm:

Decibel difference = 10 log (signal A/signal B)

For example, if signal A is 1000 watts and signal B is 10 watts, then signal A is 20 dB stronger than signal B because:

Decibel difference in power = 10 log (1000/10)
= 10 x 2
= 20 dB

Therefore, if an amplifier received a signal of 10 watts and increased its strength by a factor of 100 to 1000 watts, it would have a gain of 20 dB. Similarly, if a 10 watt signal was increased by a factor of 1,000,000 to 10 million watts, the gain would be 60 dB.

Decibels are also expressed relative to a reference value such as watts, milliwatts or millivolts. The abbreviations dBw, dBm and dBmv mean the relative increase in power relative to one watt, to one milliwatt and to one millivolt, respectively. For example, 60 dBw means a power of 1 million watts.

The definition of decibels relative to one millivolt (or ampere) differs from that used to express power differences. A signal is expressed in dBmv by

20 log (signal in millivolts / 1 millivolt).

Therefore, for example, 20 dBmv is equal to a signal of 10 millivolts. The difference in the definition arises because power levels are proportional to the square of the voltage (or current).

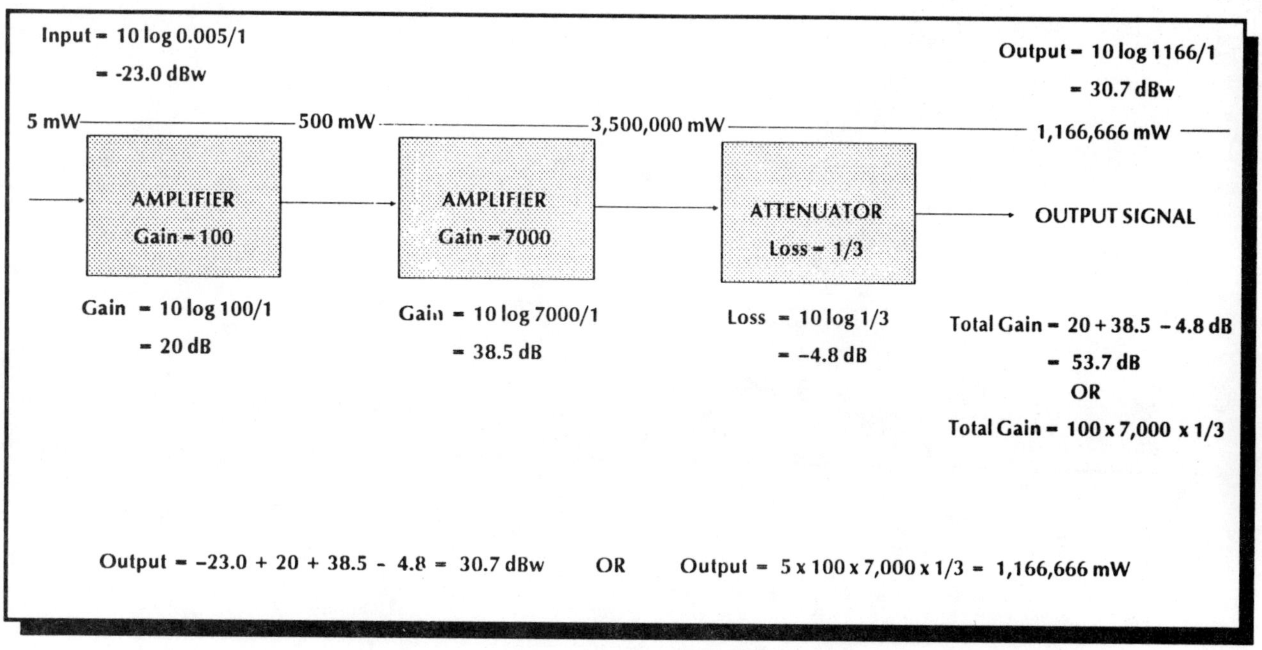

APPENDIX B. SATELLITE TV EQUATIONS

The performance of a satellite TV system can be theoretically calculated by using some general equations. Some of these are presented below.

Link Equations

The link equations are used to calculate the ratio of carrier-to-noise (C/N) power reaching the input of a satellite receiver. The link equation is:

$$C/N = EIRP - Path\ Loss + G/T_{sys} - 10 \log B + 228.6$$

EIRP is the effective isotropic radiated power directed by a downlink antenna to a location below. It is expressed in dBw, decibels relative to one watt.

Path loss measures how much signal is lost on the journey from the communication satellite to the receiving antenna. Losses are mainly due to the "spreading out" of the signal on its long journey. The amount of spreading out is determined by the distance the signal travels from a satellite to a receive dish. Path loss in decibels is given by:

$$Path\ Loss = 20 \log 4\pi S f$$

where S is the slant range and f is the signal frequency in Hertz. Slant range, the distance to the satellite is given by:

$$S = [R^2 + (R+h)^2 - 2R(R+h)\cos\Phi \cos\Delta]^{1/2}$$

where R equals 6,367 km, the radius of the earth, h equals 35,803 km, the distance of a satellite above the center of the earth, Φ is the site latitude and Δ is the absolute difference between the site and satellite longitude. Substituting R and h in this equation gives:

$$S = 1818.2\,[58.32 - 53.69 \cos\Phi \cos\Delta]^{1/2}$$

Substituting this value of S into the above equation we find path loss equals:

$$10 \log[1 - 0.295 \cos\Phi \cos\Delta] + 20 \log f + 185.05$$

where f is expressed in GHz. At 4 GHz the path loss equals 195.6 dB at an earth station located on the equator directly below a satellite. This equation also shows that signals from a satellite 10 degrees in longitude away from an earth station at 40 degrees latitude would suffer a 196.0 dB free space path loss.

Atmospheric absorption causes additional path losses. Absorption increases with slant range because the satellite signal must pass through a greater thickness of atmosphere. An absorption loss of 0.3 to 0.5 dB is assumed in most calculations for reception on average clear days.

G/T_{sys}, the ratio of dish gain to system noise temperature, is the figure of merit of an dish/feed/LNB system. In decibels:

$$G - 10 \log T_{sys}$$

System noise temperature primarily depends upon both the dish and LNB noise temperatures. However, components further downstream towards the receiver also contribute small amounts of noise. It is given by:

$$T_{sys} = T_{ant/feed} + T_{LNB}/G_{feed} + \frac{T_{rec/coax}}{G_{LNB} + G_{feed}}$$

where G refers to gain. The gain of a typical feed is in the vicinity of 0.99 while LNB gain is usually about 50 dB, a factor of 100,000. This equation clearly shows why the noise contributed to T_{sys} by the satellite receiver and coaxial cable are negligible. The LNB amplifies both the signal and noise so much that any later noise contributions are of minor importance.

The second of last term in the link equations adjusts for the system bandwidth. The last is a constant, called Boltzman's constant.

Dish Gain

Dish gain relative to an "isotropic" antenna, one that radiates equally in all directions, is given by

$$G = E(\pi D/\lambda)^2$$

where E is dish efficiency, D is diameter and λ is wavelength of the incoming radiation. The wavelength in centimeters can be calculated by dividing 30 by the frequency expressed in GHz. The wavelength of 4 GHz microwaves is 7.48 cm; at 12.5 GHz it is 2.4 cm.

For example, a 2 meter reflector with an efficiency of 55% operating at 4 GHz has a gain of:

$$G = 0.55 \times (3.14 \times 200\text{ cm} / 7.48\text{ cm})^2 = 3878$$

Expressed in decibels this gain or signal concentration of a factor of 3878 relative to an isotropic antenna is given by:

$$G = 10 \log 3878 = 35.9\text{ dBi}$$

Loss of Gain with Surface Irregularities

The decrease of gain relative to a perfect dish having no surface irregularities can be calculated in decibels from:

$$\text{Loss of Gain} = 0.762\,[\text{RMS} \bullet f]^2$$

where RMS is the root mean square deviation from a perfect geometrical shape and f is the frequency of the incoming signal. The RMS deviation is a measure of the "tightness" of the surface or its average tolerance.

For example, a C-band dish operating at 4 GHz having a RMS tolerance of 0.15 cm has a decrease in gain relative to a perfect dish given by:

$$\begin{aligned}\text{Loss of Gain} &= -0.762\,[0.15 \times 4]^2 \\ &= -0.27\text{ dB}\end{aligned}$$

Dish Beamwidth

An approximate but very useful formula for 3 dB dish beamwidth is:

$$\text{Beamwidth} = 70\,\lambda / D$$

where λ is the wavelength of the microwave radiation and D is reflector diameter. For example, a 2 meter dish has a 3 dB beamwidth given by:

$$\text{Beamwidth} = 70 \times 7.48 / 200 = 2.62^\circ$$

Similarly, a 1 meter dish would have a calculated beamwidth of:

$$\text{Beamwidth} = 70 \times 7.48 / 100 = 5.24^\circ$$

Noise Temperature and Figure

The noise any system generates is proportional to its ambient temperature and the bandwidth of the signal it processes. The larger either of these two quantities, the greater the contributed noise:

$$\text{Noise} = kTB$$

where k is Boltzman's constant, T is the ambient temperature and B the system bandwidth.

A quantity called noise factor is defined by the ratio of the noise at the output on an electronic component to the noise at its input. This quantity measures, in essence, the amount of noise internally generated in any device. In a perfect device whose electronic circuits added no extra noise to a signal, the noise factor would be one.

$$\begin{aligned}\text{Noise Factor} &= \frac{(\text{Ideal Noise} + \text{Internal Noise})}{\text{Ideal Noise}} \\ &= (kBT_{Ideal} + kBT_{Eq}) / kBT_{Ideal} \\ &= (T_{Ideal} + T_{Eq}) / T_{Ideal} \\ &= 1 + T_{Eq} / T_{Ideal} \\ &= 1 + T_{Eq} / 290\end{aligned}$$

T_{Eq}, termed the equivalent noise temperature. The reference noise temperature, T_{Ideal}, is usually taken to be 290K, equal to an average room temperature of about 17°C.

Noise figure is the decibel equivalent of noise factor as is given by

$$\text{Noise Figure} = 10 \log (\text{Noise Factor})$$

For example, if the noise figure is 1.2 dB, the equivalent noise temperature is

$$1.2 = 10 \log (1 + T_{Eq} / 290)$$

turning this equation inside out

$$1 + T_{Eq} / 290 = 10^{0.12} = 1.32$$

$$T_{Eq} / 290 = 0.32$$
$$\therefore T_{Eq} = 93\text{K}$$

The Effect of Bandwidth on System Noise Power

The noise power in any communication system is given by:

$$\text{System noise power} = kT_{sys}B$$

where T_{Sys} is the system noise temperature in degrees Kelvin mainly determined by antenna and LNB noise, k is Boltzman's constant equal to 1.381×10^{-23} and B is the communication bandwidth. The change in noise power between two systems can be computed as:

$$\text{Change in noise power} = kT_1B_1/kT_2B_2$$

$$= T_1B_1/T_2B_2$$

Thus, if the noise temperature remains constant the change in noise power is simply the ratio of bandwidths. If the bandwidth were cut from 36 to 18 MHz as would be the case in half transponder formats, the noise power would be reduced by 50 percent or 3 dB. The resulting doubling of the signal-to-noise ratio sometimes makes the difference between a watchable picture and a sparklie, faint ghost of a picture. But reducing the bandwidth will also result in a "softening" of the video as well as smearing and streaking of pictures having fast changes.

Declination Angle

The declination angle for a polar mount can be found from the tables and figures in the early chapters. It can also be calculated from:

$$\text{Declination} = \text{Tan}^{-1} \frac{3964 \sin L}{22300 + 3964(1 - \cos L)}$$

where L is the site latitude. The two numbers in this equation are the radius of the earth and the distance from the surface of the earth to the arc of satellites. For example, at 40 degrees latitude

$$\text{Declination} = \text{Tan}^{-1} \frac{3964 \sin 40}{22300 + 3964(1 - \cos 40)}$$
$$= \text{Tan}^{-1} 0.11$$
$$= 6.26^{\circ}$$

Azimuth and Elevation Angles

Dish pointing angles can be calculated in degrees from true north from the following equations:

$$\text{Azimuth Angle} = \cos^{-1}[-\tan\Phi/\tan Y]$$

$$\text{Elevation Angle} = \tan^{-1}[(\cos Y - 0.15116)/\sin Y]$$

$$Y = \cos^{-1}[\cos\Phi\cos\Delta]$$

where Δ is the absolute value of the difference between satellite and TVRO site longitudes and Φ is the site latitude.

Voltage Standing Wave Ratio

The voltage standing wave ratio, VSWR, is a measure of the amount of input signal reflected back and lost. A perfect device would have no reflective losses and a VSWR of 1:1. Table B-1 shows how reflected signal power and transmission losses vary with VSWR.

TABLE B-1. VSWR

VSWR	Percentage of Reflected Signal	Transmission Loss(dB)
1.0:1	0	0
1.1:1	0.2	0.01
1.2:1	0.9	0.03
1.3:1	1.6	0.07
1.5:1	4.0	0.18

Reflector Parabolic Geometry

The basic equation for a parabolic reflector is given by:

$$y = x^2/4f$$

where f is the focal distance. Another useful formula gives the focal distance f in terms of the dish diameter and depth:

$$f = \text{diameter}^2/16 \times \text{depth}$$

Rain and Increase in System Noise Temperature

The noise temperature of a TVRO increases when it rains because the water droplets are warm and thus radiate energy which is detected by a dish. The increase in system noise temperature is given by:

$$T_{atm} = (1 - 10^{-R/10})\ 10^{-W/10}\ T_{rain}$$

where R is the rain attenuation expressed in decibels, W is the waveguide loss between the feed and LNB and T_{rain} is the temperature of the rain in degrees Kelvin.

If R is zero, namely if there is no rain and thus no rain attenuation, T_{atm} becomes zero. Thus system noise temperature does not increase over its clear sky value as would be expected. Also, if there are no waveguide losses so that W is zero, $10^{-W/10}$ is one and losses depend only upon the amount of rain attenuation.

For example, assume a typical value of 0.5 dB for waveguide losses, a rain fade of 2 dB and a rain temperature of 17°C or 290K. Then the increase in system noise temperature becomes:

$$T_{atm} = (1 - 10^{-2/10})\ 10^{-0.5/10} \times 290K$$

$$= (1 - 10^{-0.2})\ 10^{-0.05} \times 290K$$

$$= 95.4K$$

Wind Loading

Winds can have dramatic effects on dishes and their supporting structures. The following tables, kindly calculated and provided by Dick Zlotky indicate the strength and direction of the forces expected in a real-world environment.

Design wind loads on parabolic reflectors were calculated from wind tunnel coefficient data complied by the Jet Propulsion Laboratory at the California Institute of Technology as presented in report number JPL 78-16. Wind tunnels tests were performed on several models of parabolic reflectors, both solid and porous, at azimuth angles ranging from 0°, wind directly along the dish axis, to 180°, as well as at elevation angles from zero to 90° at each azimuth angle. Axial, normal, side force, moment and torsional coefficients were determined at each position of the reflector.

All coefficients compiled in JPL 78-16 act at the apex of the reflector as shown in Figure B-1. These are all positive in sign. Since all the force coefficients are perpendicular and parallel to the longitudinal axis of the reflector, it is necessary to resolve these forces and moments in planes perpendicular and parallel to the ground. These must be then transferred to the centerline of the pole in order to use them for the pole and foundation structural design. The resolution of these forces and moments is shown in Figure B-2. It is assumed that the center-line of the pole is at a distance of 15.2 cm from the apex of the reflector. The forces and moments presented in Tables B-2, B-3 and B-4 are centered about point "A" in Figure B-2. This point is located at the intersection of the pole center-line and a horizontal plane through the apex of the reflector at a height H above ground level.

A dish having an f/D of 0.313 was used in the tests. Reflectors having surface porosities of 0% (solid) and 25% were tested. The wind loads shown in the tables below are calculated from these two different surfaces.

JPL 78-16 states that the amount of reduction of wind loads offered by a porous panel such as a perforated reflector is primarily affected by its t/d ratio, where t is the thickness of the panel and d is the diameter of its holes. As t/d approaches 3, the efficiency nears zero and the drag characteristics approach that of a solid reflector. The t/d for models used in the JPL wind tunnel tests approached 100% efficiency in reducing drag. In other words, the t/d values were low and therefore representative of the performance of today's perforated and mesh dishes.

Figure B-1. Forces and Moments are Positive as Shown per JPL 78-16)

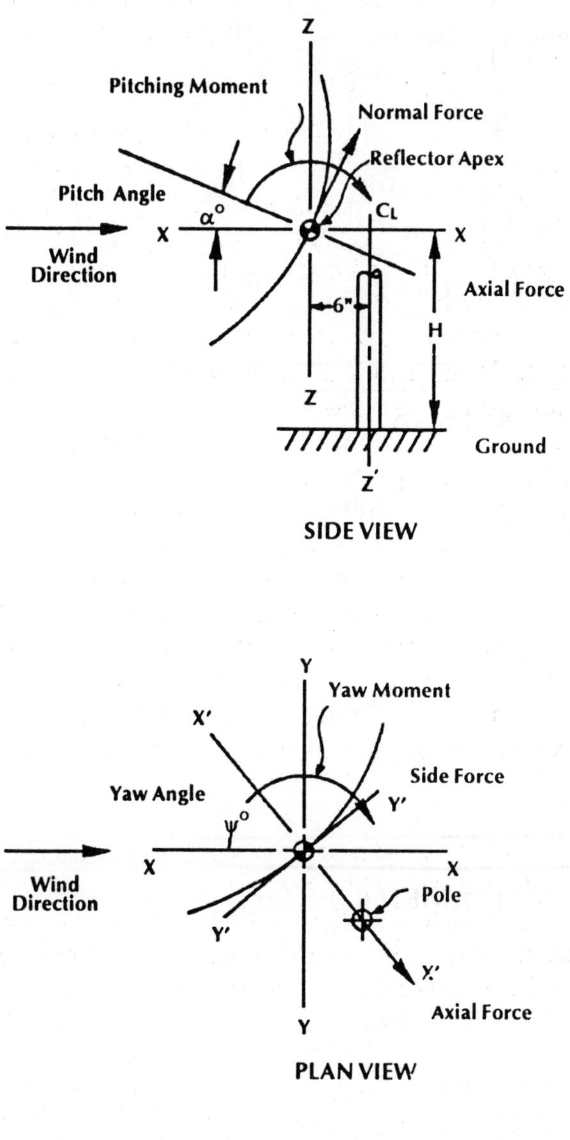

Additional Calculations

The forces and moments given in Tables B-2 through B-4 are for 160.9 km/hour winds exerted at a height H above ground level. Forces and moments for other wind velocities are easily calculated by multiplying the 160.9 km/hour velocity by the square of the new velocity divided by 10,000, namely:

$$(v/100)^2$$

For example, to find $F_{X'}$ exerted by a 112.6 km/hour wind on a 1.8 meter solid antenna, multiply the 499 kg found in Table B-2 by $(112.6/100)^2$. The result is a force of 245 kg at the lower velocity.

To find the forces at the point where a pole attaches to the foundation, point B in Figure B-2, it is necessary to transfer the forces and moments from point A to point B. All components of the forces have the same values at both points A and B.

Figure B-2. Forces and Moments Resolved at Center of Pole

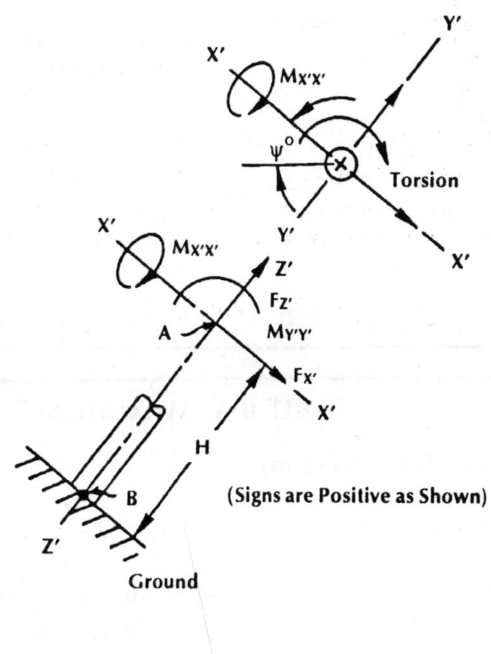

(Signs are Positive as Shown)

For example, from Table B-2 the following values are found for a 1.8 meter solid dish at a height, H, of 121.9 cm above ground level:

$$F_{X'} = 499 \text{ kg}$$
$$F_{Y'} = -16 \text{ kg}$$
$$F_{Z'} = -180 \text{ k}$$
$$\text{Torsion} = 02432 \text{ kg-cm}$$
$$M_{Y'Y'} = 2257 \text{ kg-cm}$$
$$M_{X'X'} = -807 \text{ kg-cm}$$

The loads at ground level (point B) are found by:

$$M_{Y'Y'} = F_{X'} \times H + M_{Y'Y'} = 499 \times 121.9 + 2257$$
$$= 63085 \text{ kg-cm}$$
$$M_{X'X'} = F_{Y'} \times H + M_{X'X'} = -16 \times 121.9 - 807$$
$$= -2757$$

The resultant overturn moment at point B is then given by the resultant of these two moments

$$= [(63085)^2 + (-2757)^2]^{1/2}$$
$$= 63145 \text{ kg-cm}$$

TABLE B-2. SOLID DISH

Forces & Moments	Dish Diameter (meters)					
	1.0	**1.2**	**1.8**	**2.4**	**3.0**	**3.7**
$F_{X'}$ (kg)	154	222	499	886	1384	2107
$F_{Y'}$ (kg)	-5	-7	-16	-28	-44	-66
$F_{Z'}$ (kg)	-55	-80	-180	-320	-500	-761
Torsion (kg-cm)	-450	-756	-2432	-5589	-10759	-19946
$M_{Y'Y'}$ Moment (kg-cm)	9	263	2257	6952	15478	31766
$M_{X'X'}$ Moment (kg-cm)	-138	-238	-807	-1898	-3721	-6977

Assumptions: Wind Velocity = 160.0 k/hr
Azimuth = 60°
Elevation = 20°

TABLE B-3. 25% POROUS ANTENNA

Forces & Moments	Dish Diameter (meters)					
	1.0	**1.2**	**1.8**	**2.4**	**3.0**	**3.7**
$F_{X'}$ (kg)	74	110	248	439	686	1044
$F_{Y'}$ (kg)	15	21	48	84	132	201
$F_{Z'}$ (kg)	-24	-34	-78	-137	-215	-327
Torsion (kg-cm)	616	1006	3026	6728	12647	23024]
$M_{Y'Y'}$ Moment (kg-cm)	207	467	2160	5779	12135	23927
$M_{X'X'}$ Moment (kg-cm)	147	255	866	2036	3991	7469

Assumptions: Wind Velocity = 160. 0 km/hr
Azimuth = 60°
Elevation = 20°

TABLE B-4. MAXIMUM TORSION DISH DIAMETER (meters)

Maximum Torsion (kg-m)	**1.0**	**1.2**	**1.8**	**2.4**	**3.0**	**3.7**
Solid Dish	1647	2735	8776	16646	38533	71289
25% Porous Dish	1467	2450	7754	17716	33895	62629

Assumptions: Wind Velocity = 160.0 km/hr
Azimuth = 120°
Elevation = 20°

APPENDIX C. CHANNEL WAVELENGTHS and BANDWIDTHS

Channel	Video Frequency (MHz)	Wavelength (Inches)
	SUB CHANNELS	
01	15.75	749.7
02	21.75	542.9
03	27.75	425.5
04	33.75	349.9
05	39.75	297.1
	VHF-LO CHANNELS	
2	55.25	213.7
3	61.25	192.8
4	67.25	175.6
5	77.25	152.9
6	83.25	141.8
FM	88.0-108.0	134.2-109.3
	VHF-HI CHANNELS	
7	121.25	97.4
8	127.25	92.8
9	133.25	88.6
10	139.25	84.8
11	145.25	81.3
12	151.25	78.1
13	157.25	75.1
	SUPERBAND CHANNELS	
J	217.25	54.5
K	223.25	52.9
L	229.25	51.5
M	235.25	50.2
N	241.25	49.0
O	247.25	47.8
P	253.25	46.6
Q	259.25	45.6
R	265.25	44.5
S	271.25	53.5
T	277.25	42.6
U	283.25	41.7
V	289.25	40.8
W	295.25	40.0
	HYPERBAND CHANNELS	
AA	301.25	39.2
BB	307.25	38.4
CC	313.25	37.7
DD	319.25	37.0
EE	325.25	36.3
FF	331.25	35.7
GG	337.25	35.0
HH	343.25	34.4
II	349.25	33.8
KK	361.25	32.7
LL	367.25	32.2

Channel	Video Frequency (MHz)	Wavelength (Inches)
MM	373.25	31.6
NN	379.25	31.1
OO	385.25	30.7
PP	391.25	30.2
QQ	397.25	29.7
RR	403.25	29.3
SS	409.25	28.9
TT	415.25	28.4
UU	421.25	28.0
VV	427.25	27.6
WW	433.25	27.3
	UHF CHANNELS	
14	471.25	25.1
15	477.25	24.7
16	483.25	24.4
17	489.25	24.1
18	495.25	23.8
19	501.25	23.6
20	507.25	23.3
21	513.25	23.0
22	519.25	22.7
23	525.25	22.5
24	531.25	22.2
25	537.25	22.0
26	543.25	21.7
27	549.25	21.5
28	555.25	21.3
29	561.25	21.0
30	567.25	20.8
31	573.25	20.6
32	579.25	20.4
33	585.25	20.2
34	591.25	20.0
35	597.25	19.8
36	603.25	19.6
37	609.25	19.4
38	615.25	19.2
39	621.25	19.0
40	627.25	18.8
41	633.25	18.7
42	639.25	18.5
43	645.25	18.3
44	651.25	18.1
45	657.25	18.0
46	663.25	17.8
47	669.25	17.6
48	675.25	17.5
49	681.25	17.3
50	687.25	17.2
51	693.25	17.0
52	699.25	16.9
53	705.25	16.7
54	711.25	16.6

APPENDIX D. GLOSSARY

A/B Switch

A switch that selects one of two inputs (A or B) for routing to a common output while providing adequate isolating between the two signals.

Absolute Zero

The coldest possible temperature at which all molecular motion ceases. It is expressed in degrees Kelvin as measured from absolute zero. Zero degrees Kelvin equals minus 273.16°C or minus 459.69°F.

Adjacent Channel

An adjacent channel is immediately next to another channel in frequency. For example, PAL channels 5 and 6 as well as 8 and 9 are adjacent.

Agile Receiver

A satellite receiver which can be tuned to any desired channel.

Alignment

The process of fine tuning a dish or an electronic circuit to maximize its sensitivity and signal receiving capability.

AM

An abbreviation for amplitude modulation

Ambient Temperature

The existing dry bulb temperature.

Amplifier

A device used to increase the power of a signal.

Analog

A system in which signals vary continuously in contrast to a digital system in which signals vary in discrete steps.

Analog-to-Digital Converter

A circuit that converts analogue signals to an equivalent digital form. The varying analogue signal is sampled at a series of points in time. The voltage at each of these points is then represented by a series of numbers, the digital value of the sample. The higher this sampling frequency, the finer are the gradations and the more accurately is the signal represented.

Antenna

A device that collects and focuses electromagnetic energy, i.e., contributes an energy gain. Satellite dishes, broadband antenna and cut-to-channel antennas are some types of antennas encountered. In the case of satellite dishes, gain is proportional to the surface area of the microwave reflector.

Antenna Efficiency

The percentage of incoming satellite sign il actually captured by an antenna system.

Aperture

The collection area of a parabolic dish.

Aperture Blockage

An obstruction such as the feed assembly which causes a blockage of the incoming signal.

Aspect Ratio

The ratio of television screen width to height. The standard aspect ratio is 4 to 3.

Attenuation

The decrease in signal power that occurs in a device or when a signal travels to reach a destination point (path loss).

Attenuator

A passive device which reduces the power of a signal. Attenuators are rated according to the amount of signal attenuation.

Audio Subcarrier

The carrier wave that transmits audio information within a video broadcast signal. Satellite transmissions can relay more than a single audio subcarrier in the frequency range between 5 and 8.5 MHz.

Automatic Fine Tuning

A circuit that automatically maintains the correct tuner oscillator frequency and compensates for drift and for moderate amounts of inaccurate tuning. Similar to AFC.

Automatic Frequency Control (AFC)

A circuit that locks onto a chosen frequency and will not drift away from that frequency.

Automatic Gain Control (AGC)

A circuit that locks the gain onto a fixed value and thus compensates for varying input signal levels keeping the output constant.

Azimuth-Elevation (Az-El) Mount

A dish mount that tracks satellites by moving in two directions: the azimuth in the horizontal plane and elevation up from the horizon.

Azimuth

A compass bearing expressed in degrees of rotation clockwise from true north. It is one of the two coordinates, azimuth and elevation, used to align a satellite dish.

Back Match

The matching of the resistive values of the input and output of electronic devices to reduce signal reflection and ghosting. Also known as impedance matching.

Band

A range of frequencies.

Band Separator

A device that splits a group of specified frequencies into two or more bands. Common types include UHF/VHF, Hi/Lo-band and FM separators. This device is essentially a set of filters.

Bandpass Filter

A circuit or device that allows only a specified range of frequencies to pass from input to output.

Bandwidth

The frequency range allocated to any communication circuit.

Baseband

The raw audio and video signals prior to modulation and broadcasting. Most satellite headend equipment utilizes baseband inputs. More exactly, the composite unclamped, non-de-emphasized and unfiltered receiver output. This signal contains the complete set of FM modulated audio and data subcarriers.

Beamwidth

A measure used to describe the width of vision of a dish. Beamwidth is measured as degrees between the 3 dB half power points.

Bit Error Rate – BER

The number of errors in a data stream usually expressed a ratio to the total number of bits in which an error occurs. For example, 1 in 10^7 or 10^{-7}

Bits per Second – BPS

The number of bits transmitted each second

Block Downconversion

The process of lowering the entire band of frequencies in one step to some intermediate range to be processed inside a satellite receiver. Multiple block downconversion receivers are capable of independently selecting channels because each can process the entire block of signals.

Boresight

The direction along the principle axis of either a transmitting or a receiving dish.

Broadband

A device that processes a signal(s) spanning a relatively broad range of input frequencies.

Buttonhook Feed

A rod shaped like a question mark supporting the feed and LNA. A buttonhook feed for use with commercial grade dishes is often a hollow waveguide that directs signals from a feed to an LNA behind the dish.

CATV

An abbreviation for Community Antenna Television - another name for cable TV.

C-Band

The 3.625 to 4.2 GHz band of frequencies at which some broadcast satellites operate.

Carrier

A pure-frequency signal that is modulated to carry information. In the process of modulation it is spread out over a wider band. The carrier frequency is the center frequency on any television channel.

Carrier-to-Noise Ratio – C/N

The ratio of the received carrier power to the noise power in a given bandwidth, expressed in decibels. The C/N is an indicator of how well an receive system will perform in a particular location, and is calculated from satellite power levels, dish gain and the system noise temperature.

Cassegrain Feed System

A dish feed design that includes a primary reflector, the dish, and a secondary reflector which redirects microwaves via a waveguide to a low noise amplifier.

Channel

A segment of bandwidth used for one complete communication link.

Characteristic Impedance

The impedance in ohms of a device in the path of a communication signal such as a cable, a connector or the input of an amplifier.

Chrominance

The hue and saturation of a color. The chrominance signal is modulated onto a 4.43 MHz carrier in the PAL television system and a 3.58 MHz carrier in the NTSC television system.

Circular Polarity

Electromagnetic waves whose electric field uniformly rotates along the signal path. Broadcasts used by Intelsat and other international satellites use circular, not horizontally or vertically polarized waves as are common in North American and European transmissions.

Clamp Circuit

A circuit that removes the dispersion waveform from the downlink signal.

Clamped Outputs

Satellite receiver outputs that have the energy dispersal waveform removed. Unclamped outputs are often required as input to a decoder.

Clarke Belt

The circular orbital belt at 35 786 kilometers above the equator, named after the writer Arthur C. Clarke, in which satellites travel at the same speed as the earth's rotation. Also called the geostationary orbit.

Color Bars

A test pattern of specifically colored vertical bars used as a reference to test the performance of a color television.

Coaxial Cable

A cable for transmitting high frequency electrical signals with low loss. It is composed of an internal conducting wire surrounded by an insulating dielectric which is further protected by a metal shield. The impedance of coax is a product of the radius of the central conductor, the radius of the shield and the dielectric constant of the insulation. In most satellite and SMATV systems, coax impedance is 75 ohms.

Color Sync Burst

A "burst" of 8 to 11 cycles in the 4.43361875 MHz (PAL) or 3.579545 MHz(NTSC) color subcarrier frequency. This waveform is located on the back porch of each horizontal blanking pulse during color transmissions. It serves to synchronize the color subcarrier's oscillator with that of the transmitter in order to recreate the raw color signals.

Composite Baseband Signal

The complete audio and video signal without a carrier wave. Satellite signals have audio baseband information ranging in frequency from zero to 3400 Hertz. NTSC video baseband is from zero to 4.2 MHz. PAL video baseband ranges from 0 to 5.5 MHz.

Composite Video Signal

The complete video signal consisting of the chrominance and luminance information as well as all sync and blanking pulses.

Compressor

A unit that accepts uncompressed video, audio and data and then digitizes and compresses these signals

Compression System

A collection of compressors, multiplexers and modulators that generate one multiplex signal

Conditional Access – CA

The encryption system

Cross Modulation

A form of interference caused by the modulation of one carrier affecting that of another signal. It can be caused by overloading an amplifier as well as by signal imbalances at the headend.

Cross Polarization

Term to describe signals of the opposite polarity to another being transmitted and received. Cross-polarization discrimination refers to the ability of a feed to detect one polarity and reject the signals having the opposite sense of polarity.

Crosstalk

Interference between adjacent channels often caused by cross modulation. Leakage can occur between two wires, PCB tracks or parallel cables.

dc Power Block

A device which stops the flow of dc power but permits passage of higher frequency ac signals.

Decibel (dB)

The logarithmic ratio of power levels used to indicate gains or losses of signals. Decibels relative to one watt, milliwatt and millivolt are abbreviated as dBw, dBm and dBmV, respectively.

Declination Offset Angle

The adjustment angle of a polar mount between the polar axis and the plane of a satellite dish used to aim at the geosynchronous arc. Declination increases from zero with latitude away from the equator.

Decoder

A circuit that restores a signal to its original form after it has been scrambled.

De-emphasis

A reduction of the higher frequency portions of an FM signal used to neutralize the effects of pre-emphasis. When combined with the correct level of pre-emphasis, it reduces overall noise levels and therefore increases the output S/N ratio.

Demodulator

A device which extracts the baseband signal from the transmitted carrier wave.

Detent Tuning

Tuning into a satellite channel by selecting a preset resistance.

Digital

Describes a system or device in which information is transferred by electrical "on-off," "high-low," or "1/0" pulses instead of continuously varying signals or states as in an analog message.

Digital-to-Analog Converter

A circuit that converts digital signals into their equivalent analog form.

Direct Broadcast Satellite (DBS)

A term commonly used to describe Ku-band broadcasts via satellite directly to individual end-users. The DBS band ranges from 11.7 to 12.75 GHz.

Direct-to-Home – DTH

Dish Illumination

Describes how a feed "sees" the surface of a dish as well as the surrounding terrain.

Distribution System

A communication system consisting of coax but occasionally of line-of-sight microwave links that carries signals from the headend to end-users.

Downconverter

A circuit that lowers the high frequency signal to a lower, intermediate range. There are three distinct types of downconversion used in satellite receivers: single downconversion; dual downconversion; and block downconversion.

Downlink Antenna

The antenna on-board a satellite which relays signals back to earth.

Drifting

An instability in a preset voltage, frequency or other electronic circuit parameter.

DTH

Direct-to-home satellite broadcasts.

Dual-Band Feed

A feed which can simultaneously receive two different bands, typically the C and Ku-bands.

Earth Station

A complete satellite receiving or transmitting station including the dish, electronics and all associated equipment necessary to receive or transmit satellite signals. Also known as a ground station.

Effective Isotropic Radiated Power (EIRP)

A measure of the signal strength that a satellite transmits towards the earth below. The EIRP is highest at the center of the beam and decreases at angles away from the boresight.

Elevation Angle

The vertical angle measured from the horizon up to a target satellite.

Energy Dispersal

The modulation of an uplink carrier with a triangular waveform. This technique disperses the carrier energy over a wider bandwidth than otherwise would be the case in order to limit the maximum energy compared to that transmitted by an unclamped carrier. This triangular waveform is removed by a clamp circuit in a satellite receiver.

F-connector

A standard RF connector used to link coax cables with electronic devices.

f/D Ratio

The ratio of an dish's focal length to diameter. It describes dish "depth."

Feed

A device that collects microwave signals reflected from the surface of a dish. It is mounted at the focus of all prime focus parabolic dishes.

Field

One half of a complete TV picture or frame, composed of 325 scanning lines. In the PAL broadcast system there are 50 fields per second.

Filter

A device used to reject all but a specified range of frequencies. A bandpass filter allows only those signals within a given band to be communicated. A rejection filter, the mirror image of a bandpass filter, eliminates those signals within a specified band but passes all other frequencies.

FM

An abbreviation for frequency modulation

Focal Length

The distance from the reflective surface of a parabola to the point at which incoming satellite signals are focused, the focal point.

Footprint

The geographic area towards which a satellite downlink antenna directs its signal. The measure of strength of this footprint is the EIRP.

Forward Error Correction – FEC

FEC is a technique for improving the accuracy of data transmission. Excess bits are included in the out-going data stream so that error correction algorithms can be applied upon reception. On satellite links this is in the form of Reed-Solomon and convoluted Viterbi coding implemented at modulator/demodulator level.

Frame

One complete TV picture, composed of two fields and a total of 525 and 625 scanning lines in NTSC and PAL systems, respectively.

Frequency

The number of vibrations per second of an electrical or electromagnetic signal expressed in cycles per second or Hertz.

Gain

The amount of amplification of input to output power often expressed as a multiplicative factor or in decibels.

Gain-to-Noise Temperature Ratio (G/T)

The figure of merit of a dish and LNA. The higher the G/T, the better the reception capabilities of an earth station.

Geostationary Orbit

See Clarke Belt.

Ghosting

A term used to describe the appearance of multiple TV images that is usually caused by reception of a signal via two different paths.

GigaHertz (GHz)

1000 MHz or one billion cycles per second.

Global Beam

A footprint pattern used by communication satellites targeting nearly 40% of the earth's surface below. Many Intelsat satellites use global beams.

Ground Noise

Unwanted microwave signals generated from the warm ground and detected by a dish.

Hall Effect Sensor

A semiconductor device in which an output voltage is generated in response to the intensity of a magnetic field applied to a wire. In an actuator, the varying magnetic field is produced by the rotation of a permanent magnet past a thin wire. The pulses generated serve to count the number of rotations of the motor.

Hard-line

A low-loss coaxial cable that has a continuous hard metal shield instead of a conductive braid around the outer perimeter. This type of cable was used in the pioneer days of satellite television.

Headend

The portion of an SMATV or MMDS system where all desired signals are received and processed for subsequent distribution.

Heliax

A thick low-loss cable used at high frequencies; also known as hard-line.

Hertz

An abbreviation for the frequency measurement of one cycle per second. Named after Heinrich Hertz, the German scientist who first described the properties of radio waves.

High Definition Television (HDTV)

An innovative television format having approximately twice the number of scan lines in order to improve picture resolution and viewing quality.

High Power Amplifier (HPA)

An amplifier used to amplify the uplink signal.

Horizontal Blanking Pulse

The pulse that occurs between each horizontal scan line in an analogue television signal and extinguishes the beam illumination during the retrace period.

Horizontal Sync Pulse

A 4.7 microsecond (in the PAL system) rectangular pulse riding on top of each horizontal blanking pulse. It synchronizes the horizontal scanning at the television set with that of the television camera.

Hum Bars

A form of interference seen as horizontal bars or black regions passing across the field of a television screen.

Impedance

The resistance to alternating current flow in an electrical circuit.

Impulse Pay-Per-View

Impulse pay-per-view (IPPV) is a feature of a decoder that allows an authorized subscriber to purchase a one-time scrambled program at will. IPPV shows are selected by a button on the decoder or its remote control unit.

Inclinometer

An instrument used to measure the angle of elevation to a satellite from the surface of the earth.

Interference

An undesired signal intercepted by a TVRO that causes video and/or audio distortion.

Insertion Loss

The amount of signal energy lost when a device is inserted into a communication line. Also known as "feed through" loss.

Intermediate Frequency (IF)

A middle range frequency generated after downconversion in any electronic circuitry including a satellite receiver. The majority of all signal amplification, processing and filtering in a receiver occur in the IF range.

INTELSAT

The International Telecommunication Satellite Consortium, a body of countries working towards a common goal of improved worldwide satellite communications.

Isolator

A device that allows signals to pass unobstructed in one direction but which attenuates their strength in the reverse direction.

Isolation Loss

The amount of signal energy lost between two ports of a device. An example is the loss between the feed through port and the tap/drop of a top-off device.

Kelvin Degrees (K)

The temperature above absolute zero, the temperature at which all molecular motion stops, graduated in units the same size as degrees Celsius (°C). Absolute zero equals -273 °C or -459 °F.

Kilohertz (kHz)

One thousand cycles per second.

Ku-Band

The microwave frequency band between approximately 11 and 13 GHz used in satellite broadcasting.

Latitude

The measurement of a position on the surface of the earth north or south of the equator measured in degrees of angle.

Line Amplifier

An amplifier in a transmission line that boosts the strength of a signal.

Line Splitter

An active or passive device that divides a signal into two or more signals containing all the original information. A passive splitter feeds an attenuated version of the input signal to the output ports. An active splitter amplifies the input signal to overcome the splitter loss.

Local Oscillator

A device used to supply a stable single frequency to an upconverter or a downconverter. The local oscillator signal is mixed with the carrier wave to change its frequency.

Longitude

The distance in degrees east or west of the prime meridian, located at zero degrees.

Low Noise Amplifier (LNA)

A device that receives and amplifies the weak satellite signal reflected by a dish via a feed. C-band LNAs typically have their noise characteristics quoted as noise temperatures rated in degrees Kelvin. Ku-band LNA noise characteristics are usually expressed as a noise figure in decibels.

Low Noise Block Downconverter (LNB)

A low noise microwave amplifier and converter which downconverts a block or range of frequencies at once to an intermediate frequency range, typically 950 to 1450 MHz, 950 to 1750 MHz or 950 to 2050 MHz.

Low Noise Converter (LNC)

An LNA and a conventional downconverter housed in one weatherproof box. This device converts one channel at a time. Channel selection is controlled by the satellite receiver. The typical IF for LNCs is 70 MHz.

Magnetic Variation

The difference between true north and the north indication of a compass.

Master Antenna TV (MATV)

Broadcast receiving stations that use one or more high-quality centrally located UHF and/or VHF antennas which relay their signals to many televisions in a local apartment/condo or group- housing complex.

Match

The condition that exists when 100% of available power is transmitted from one device to another without any losses due to reflections.

Matching Transformer

A device used to match impedance between devices. A matching transformer is used, for example, when connecting a 75 ohm- coax to a television 300 ohm input terminal.

MegaHertz (MHz)

One millions cycles per second.

Microwave

The frequency range from approximately 1 to 30 GHz and above.

Mixer

A device used to combine signals together.

Modulator

A device that modulates a signal, for example an analog signal or an MPEG-2 transport stream onto a radio frequency carrier

Modulation

A process in which a message is added or encoded onto a carrier wave. Among other methods, this can be accomplished by frequency or amplitude modulation, known as AM or FM, respectively.

Monochrome

A black and white television picture.

Mount

The structure that supports a receive dish. Polar and az-el mounts are the most common variety.

MPEG-2

Motion Pictures Expert Group-Two

Multiplexing

The simultaneous transmission of two or more signals over a single communication channel. The interleaving of the luminance and chrominance signals is one form of multiplexing, known as frequency multiplexing. MultiChoice transmissions use time division multiplexing (TDM) whereby data streams are divided in time into interspersed data packets.

Multiplexer – MUX

A device that takes the outputs from a number of encoders and multiplexes them together to form one data stream

N-Connector

A low-loss coaxial cable connector used at the elevated microwave frequencies.

NTSC

The National Television Standards Committee which created the standard for North American TV broadcasts.

NTSC Color Bar Pattern

The standard test pattern of six adjacent color bars including the three primary colors plus their three complementary shades.

Noise

An unwanted signal which interferes with reception of the desired information. Noise is often expressed in degrees Kelvin or in decibels.

Noise Figure

The ratio of the actual noise power generated at the input of an amplifier to that which would be generated in an ideal resistor. The lower the noise figure, the better the performance.

Noise Temperature

A measure of the amount of thermal noise present in a system or a device. The lower the noise temperature, the better the performance.

Offset Feed

A feed which is offset from the center of a reflector for use in satellite receiving systems. This configuration does not block the dish aperture.

Orthomode Coupler

A waveguide, generally a three-port device, that allows simultaneous reception of vertically and horizontally polarized signals. The input port is typically a circular waveguide. The two output ports are rectangular waveguides.

Quadrature Phase Shift Keying – QPSK

A modulation technique used on satellite transmissions that uses phase shifts of a carrier wave to relay 4 symbols per cycle

PAL

Phase Alternate Line. The European/African color TV format which evolved from the American NTSC standard. PAL-I version used in South Africa.

Pad

A concrete base upon which a supporting pole and dish can be mounted.

Path Loss

The attenuation that a signal undergoes in traveling over a path between two points. Path loss varies inversely as the square of the distance traveled.

Parabola

The geometric shape that has the property of reflecting all signals parallel to its axis to one point, the focal point.

Pay-Per-View

Pay-per-view is a method of purchasing programming on a per-program basis.

Pay-per-View (PPV) Event

A PPV event is similar to a "normal" event, except that the subscriber is charged for viewing. In this respect, it is similar to a product. However, there can be many PPV events in a short time and registering them all as products is not practical

Pay-TV Transactions

Access control transactions such as enable, hold, etc...

Phase

A measure of the relative position of a signal relative to a reference expressed in degrees.

Phase Distortion

A distortion of the phase component of a signal. This occurs when the phase shift of an amplifier is not proportional to frequency over the design bandwidth.

Polar Mount

A dish mount that permits all satellites in the geosynchronous arc to be scanned with movement of only one axis.

Polarization

A characteristic of the electromagnetic wave. Four senses of polarization, determined by the direction of the electric field, are used in satellite transmissions: horizontal; vertical; right-hand circular; and left-hand circular.

Preamplifier

The first amplification stage. In a terrestrial receive system, it is the amplifier mounted adjacent to an antenna to increase a weak signal prior to its processing at the headend.

Pre-emphasis

Increases in the higher frequency components of an FM signal before transmission. Used in conjunction with the proper amount of de- emphasis at the receiver, it results in combating the higher noise detected in FM transmissions.

Prime Focus Dish

A parabolic dish having the feed/LNA assembly at the focal point directly in the front of the dish.

Radio Frequency (RF)

The approximately 10 kHz to 100 GHz electromagnetic band of frequencies used for man-made communication.

Reed Switch

A mechanical switch which uses two thin slivers of metal in a glass tube to make and break electrical contact and thus to count pulses which are sent to the dish actuator controller. The position of the slivers of metal is governed by a magnetic field applied by a bar or other type of magnet.

SAW (Surface Acoustic Wave) Filter

A solid state filter that yields a sharp transition between regions of transmitted and attenuated frequencies.

Satellite Receiver

The indoors electronic component of an earth station which downconverts, processes and prepares satellite signals for viewing or listening.

Scanning

The organized process of moving the electron beam in a television picture tube so an entire scene is drawn as a sequential series of horizontal lines connected by horizontal and vertical retraces.

Scrambling

A method of altering the identity of a video or audio signal so it cannot be received intelligibly in order to prevent its reception by persons not having authorized decoders.

Screening

A metal, concrete or natural material that screens out unwanted TI from entering a dish or a metal shield that prevents the ingress of unwanted RF signals in an electronic circuit.

Servo Hunting

An oscillatory searching of the feedhorn probe when use of inadequate gauge control cables results in insufficient voltage at the feedhorn.

Side Lobe

A parameter used to describe the ability of a dish to detect off-axis signals. The larger the side lobes, the more noise and interference a dish can detect.

Single Channel Per Carrier (SCPC)

A satellite transmission system that employs a separate carrier for each channel, as opposed to frequency division multiplexing that combines many channels on a single carrier.

Signal Dropout

The loss of signal that occurs when the signal becomes too weak to be usable.

Signal-to-Noise Ratio - S/N

The ratio of signal power to noise power in a specified bandwidth, usually expressed in decibels.

Skew

A term used to describe the adjustment necessary to fine tune the feed polarity detector when scanning between satellites.

Slant Range

The distance that a signal travels from a satellite to a TVRO.

Snow

Video noise or sparklies caused by an insufficient signal- to-noise input ratio to a television set or monitor.

Sparklies

Small black and/or white dashes in a television picture indicating an insufficient signal-to-noise ratio. Also known as "snow."

Spherical Dish

A dish system using a section of a spherical reflector to focus one or more satellite signals to one or a series of focal areas.

Splitter

A device that takes a signal and splits it into two or more identical but lower power signals.

Subcarrier

A signal that is transmitted within the bandwidth of a stronger signal. In satellite transmissions a 6.8 MHz audio subcarrier is often used to modulate the C-band carrier.

Sun Outage

The loss of reception that occurs when the sun is positioned directly behind a target satellite. When this occurs, solar noise drowns out the satellite signal and reception is lost.

Surface Acoustic Wave - SAW

A sound or acoustic wave traveling on the surface of the optically polished surface of a piezoelectric material. This wave travels at the speed of sound but can pass frequencies as high as several gigahertz. See SAW Filter.

TVRO

A television receive-only earth station designed only to receive but not to transmit satellite communications.

Tap

A device that channels a specific amount of energy out of the main distribution system to a secondary outlet.

Television Receive-Only (TVRO)

A satellite system that can only receive but not transmit signals.

Terrestrial Interference (TI)

Interference of earth-based microwave communications with reception of satellite broadcasts.

Tilt

The uneven attenuation of a broadband signal as it travels through a coaxial cable. In general, attenuation increases as signal frequency increases.

Thermal Noise

Random, undesired electrical signals caused by molecular motion, known more familiarly as noise.

Trace

The movement of the electron beam from left to right on a television screen.

Threshold

A minimal signal to noise input required to allow a satellite receiver to deliver an acceptable picture.

Transponder

One circuit on a satellite that receives, modulates, amplifiers and re-transmits an uplinked signal

Trap

An electronic device that attenuates a selected band of frequencies in a signal. Also known as a notch filter.

UHF

Ultrahigh frequencies ranging from 300 to 3,000 MHz. North American TV channels 14 through 83. African and European TV channels 21 to 69.

Upconverter

A device that increases the frequency of a transmitted signal.

Uplink

The earth station electronics and antenna which transmit information to a communication satellite.

VSWR (Voltage Standing Wave Ratio)

The ratio between the minimum and maximum voltage on a transmission line. An ideal VSWR is 1.0. Ghosting can result as the VSWR increases. It is also a measure of the percentage of reflected power to the total power impinging upon a device.

Vertical Blanking Pulse

A pulse used during the vertical retrace period at the end of each scanning field to extinguish illumination from the electron beam.

Vertical Sync Pulse

A series of pulses which occur during the vertical blanking interval to synchronize the scanning process at the television with that created at the studio. See also Serrated Vertical Pulse.

VHF

Very high frequencies. The lower frequency range for terrestrial television broadcasts.

Video Signal

That portion of the transmitted television signal containing the picture information.

Voltage Tuned Oscillator (VTO)

An electronic circuit whose output oscillator frequency is adjusted by voltage. Used in downconverters and satellite receivers to select from among transponders.

Video Monitor

A television that accepts unmodulated baseband signals to reproduce a broadcast.

Wave Length

The length in one complete vibration of an electromagnetic signal. Wavelength decreases as signal frequency increases.

Wind Loading

The force exerted by the wind on the surface of a dish which can cause misalignment or, in the worst case, damage to the system.

APPENDIX E. MANUFACTURERS of SATELLITE TV EQUIPMENT

A.D.L.
2216 Agate Court, Unit B
Simi Valley, CA 93065
Telephone: 805-526-5249
FAX: 805-584-0634
Feeds

AJAK INDUSTRIES, INC.
121 South Robinson
Florence, CO 81226
Telephone: 719-784-6301
FAX: 719-784-6763
Horizon-to-Horizon Antenna Mounts

ALL SYSTEMS
782 East 93rd Street
Brooklyn, NY 11236
Telephone: 800-836-4590
FAX: 718-498-0182
Dish and roof mount systems

AMSTRAD
Brentwood House
169 Kings Road
Brentwood, Essex, U.K.
Telephone: 0277-228888
Satellite Receivers

ASTROTEL COMMUNICATIONS
17906 Crusader Avenue
Cerritos, CA 90703
Telephone: 310-403-7036
FAX: 310-403-7040
C/Ku-band feeds, LNBs and servomotors

AVCOM OF VIRGINIA
500 Southlake Blvd.
Richmond, VA 23236
Telephone: 804-794-2500
FAX: 804-794-8284
Portable spectrum analyzers, test equipment

BAIRD SATELLITE SYSTEMS
3160 Logan Ave., Highway 63 North, Box 2366
Waterloo, IA 50704
Telephone: 319-233-3561
FAX: 319-235-7653
Non-penetrating and penetrating roof bases and wall, in-ground and mounting brackets

BELDEN WIRE & CABLE
P.O. Box 1980
Richmond, IN 47375
Telephone: 800-235-3364
Electronic wire & cable, including high temperature, fiber optics, flat cable, coax and multi-conductor cables

BESTMADE RV SATELLITES
2356 South Sarah Street
Fresno, CA 93706
Telephone: 209-266-1043
FAX: 209-266-1045
RV satellite reception systems

BIRDVIEW SATELLITE SERVICE
151 East Lexington Avenue
Independence, MO 64050
Telephone: 815-252-2030
FAX: 816-252-2034
Service and Repair of Equipment

BLONDER-TONGUE LABORATORIES
One Jake Brown Rd.
Old Bridge, NJ 08857
Telephone: 908-679-4000
FAX: 908-679-4353
Dishes, receivers, LNBs, downconverters, SMATV

CALIFORNIA AMPLIFIER
460 Calle San Pablo
Camarillo, CA 93010
Telephone: 805-987-9000
FAX: 805-987-5359
Full-line of TVRO and MMDS products

CHANNEL VISION TECHNOLOGY
23322 Madero Road, Suite J
Mission Viego, CA 92691
Telephone: 714-840-0288
FAX: 714-770-2085
1Multi-room components

CHAPARRAL COMMUNICATIONS
2450 North First St.
San Jose, CA 95131-1195
Telephone: 408-435-1530
FAX: 408-435-1429
Feeds and receivers

COMMSCOPE, INC.
3642 Highway 70 East
Claremont, NC 28610
Telephone: 704-459-5000
FAX: 704-459-5099
Cox cable and wire

DX COMMUNICATIONS, INC.
10 Skyline Drive
Hawthorne, NY 10532-2120
Telephone: 914-347-4040
FAX: 914-347-3953
IRDs, LNBs, modulators

ECHOSPHERE CORP.
90 Inverness Circle East
Englewood, CO 80112
Telephone: 303-799-8222
Satellite reception components

EUROSAT
Unit 27A, Popin Building
Southway, Wembley HA9 0HB, U.K.
Telephone: 081-903-1188
IRDs and DBS components

E-Z TRENCH MANUFACTURING
1665 Highway 701 South
Loris, SC 29569
Telephone: 803-756-6444
FAX: 803-756-6442
Walk-behind trenchers

GARDINER COMMUNICATIONS
3605 Security Street
Garland, TX 75042
Telephone: (214) 348-4747
FAX: (214) 341-1933
LNBs, feeds, VSAT, LNCs

GENERAL INSTRUMENT CORP.
6262 Lusk Blvd.
San Diego, CA 92121
Telephone: 619-455-1500
FAX: 619-597-3954
Satellite and encryption systems

GOURMET ENTERTAINING
3915 Carnavon Way
Los Angeles, CA 90027
Telephone: 213-666-2728
FAX: 213-666-2442
Alignment tools

HOUSTON TRACKER SYSTEMS.
90 Inverness Circle East
Englewood, CO 80115
Telephone: 303-790-4445
FAX: 303-790-4447
Satellite Receivers

INT'L ELECTRONIC WIRE & CABLE
89-1/2 O'Leary Drive
Bensenville, IL 60106
Telephone: 800-323-0210
FAX: 708-860-0305 *Coaxial Cables*

ITT NOKIA (SALORA)
Bridegmead Close, Westmead
Swindon, Wiltshire SN5 7YG, U.K.
Telephone: 0793-644223

J.G.S. ENGINEERING, INC.
9380 Watson Industrial Drive, Suite 1
St. Louis, MO 63126
Telephone: (314) 968-8788
FAX: (314) 968-3030
Linear actuators and horizon-to-horizon mounts

KAUL-TRONICS INC. (KTI)
1140 Sextonville Road
Richland Center, WI 53581
Telephone: (608) 647-8902
FAX: (608) 647-7394
Satellite dishes

LEITCH VIDEO INTERNATIONAL, INC.
220 Duncan Mill Road, Suite 301
North York, ON, Canada, M3B 3J5
Telephone: 416-445-9640
800-387-0233 (Canada)
800-231-9673 (USA)
FAX: 416-445-0595
Digital and video encryption products

LENSON HEATH
Unit 75, Woodburn Green
Wooburn Green Industrial Park
Bucks, U.K.
Telephone: 06285-25887

LUNAR INDUSTRIES INC.
9353 Activity Rd. Suite I
San Diego, CA 92126
Telephone: (619) 549-9555
LNBs and related products

MICROWAVE FILTER COMPANY
6743 Kinne Street
East Syracuse, NY 13057
Telephone: (800) 448-1666
(315) 437-3953
FAX: (315) 463-1467
Microwave, IF, block filters for TI

MULTIPLEX TECH./CHANNEL PLUS
3200 East Birch
Brea, CA 92621
Telephone: 800-999-5225
714-996-4100
FAX: 714-996-4900
Multi-room distribution/control equipment

NII – NORSAT INTERNATIONAL
302-12886 78th Avenue
Surrey, BC, Canada V3W 8E7
Telephone: (604) 597-6200
FAX: (604) 597-6214
Satellite products and turnkey systems

ODOM ANTENNAS
2502 DeWitt Henry Drive
Beebe, AR 72012
Telephone: 501-882-6495
FAX: 501-882-6198
Satellite dishes

ONTOP SYSTEMS, INC.
1243 West 7th Avenue
Eugene, OR 97402
Telephone: 503-485-7020
FAX: 503-343-7064
Business management software

ORBITRON ANTENNAS
351 South Peterson Street
Spring Green, WI 53588
Telephone: (608) 588-2923
FAX: (608) 588-2257
Dishes from 1 to 7:3 meters

PANAMAX
150 Mitchell Blvd.
San Rafael, CA 94903
Telephone: 415-499-3900
FAX: 415-472-5540
Surge protectors

PANAREX ELECTRONICS
11672 Tuxford Street
Sun Valley, CA 91352
Telephone: (818) 768-5161
FAX: (818) 768-5191
Ku-band system package

PACE
Victoria Road, Saltaire
Shipley BD 18 3LF, U.K.
Telephone: 0274-532000

PANASONIC
1 Panasonic Way - Panazip 3E-7
Secaucus, NJ 07094
Telephone: (201) 392-4108
FAX: (201) 392-6821
Integrated receiver decoders

PARACLIPSE, INC.
2271 29th Avenue East
Columbus, NE 68601
Telephone: 402-563-3625
FAX: 402-564-2109
Hydroformed and ribbed antennas

PAULLIN INDUSTRIES, INC.
1446 State Route 60
Ashland, OH 44805
Telephone: 800-821-7384
FAX: 419-281-3890
Mounts, LNA covers, actuator motors

PERFECT 10 ANTENNA
2301 North Redmond Road
Jacksonville, AR 72076
Telephone: 501-982-2354
FAX: 501-985-9768
Aluminum dishes

PRECISION R.V. SATELLITES
14520 East Ashlan Avenue
Sanger, CA 93657
Telephone: 800-272-1780
RV Systems

PRO BRAND INTERNATIONAL, INC.
1900 West Oak Circle
Marietta, GA 30062
Telephone: (404) 423-7072
FAX: (404) 423-7075
Actuators, mounts, LNBs

PRODUCTOS METSA, S.A. de C.V.
Chihuahua 131-1, Col. Republica
Saltillo, Coahuila, Mexico
Telephone: 52-841-63311
Satellite Dishes

PROMAX
Francesc Moragas, 71-75
Apartado Correos, 118
08907 L'Hospitalet de Llobregat
Barcelona, Spain
Telephone: (343) 337 90 08
FAX: (343) 338 11 26
Test Equipment

RCA/THOMSON CONSUMER ELECTRONICS
10330 North Meridan Street
Indianapolis, IN 46290-1024
Telephone: 317-587-4289
FAX: 317-587-6712
Electronic components

R.L. DRAKE COMPANY
9111 Springboro Pike, P.O.Box 112
Miamisburg, OH 45342
Telephone: (513) 866-2421
FAX: (513) 866-0806
Satellite receivers, modulators, LNBs

RF-LINK TECHNOLOGY, INC.
411 Ampola Avenue
Torrance, CA 90501
Telephone: 310-787-2328
FAX: 310-787-2323
Video/audio remote sender

SUPERIOR ANTENNA MFG.-SAMI
1009 North First Street
Judsonia, AR 72081
Telephone: 501-729-3103
FAX: 501-729-3671
Satellite dishes

TEE-COMM ELECTRONICS
775 Main Street East
Milton, ON L9T 3Z3, Canada
Telephone: 416-878-8181
FAX: 905-878-4747
Satellite Receivers

THOMPSON SAGINAW BALL SCREW
628 North Hamilton Street
Saginaw, MI 48608
Telephone: (517) 776-5111
FAX: 517-776-3632
Actuators

TIMES FIBER COMMUNICATIONS
358 Hall Avenue
Wallingford, CT 66492
Telephone: 800-243-6904
Cable, fiber optics,off-premises
Addressable Converters

TOSHIBA AMERICA INC.
1010 Johnson Drive
Buffalo Grove, IL 60089
Telephone: 708-541-9400
FAX: 708-541-9401
Satellite receivers

TRAVEL-SAT
74-877 Joni Drive, #11
Palm Desert, CA 92260
Telephone: 619-568-0666
FAX: 619-346-6955
RV Satellite Reception Systems

UNDERCOVER SATELLITE SYSTEMS
83-616 Avenue 45, Suite 6
Indio, CA 92201
Telephone: 619-775-9962
Patio Satellite dishes

UNIDEN AMERICA CORPORATION
4700 Amon Carter Blvd.
Ft. Worth, TX 76155
Telephone: 817-858-3300
FAX: 817-858-3657
Satellite receivers

UNIVERSAL ANTENNA MFG.
702 East 2nd Street, Box 338
Ward, AR 72176
Telephone: 501-843-6517
FAX: 501-843-9755
Satellite Reception Antennas

UNIVERSAL ELECTRONICS
4555 Groves Road, Suite 12
Colombus, OH 43232
Telephone: 614-866-4605
800-241-8171
FAX: 614-866-1201
SCPC receivers, LNBs, coax sealant

VENTURE MANUFACTURING CO.
3636 Dayton Park Drive
Dayton, OH 45414
Telephone: 513-233-6832
FAX: 513-233-8485
Actuators and components

VIDEO LINK
12950 Bradley Avenue
Sylmar, CA 91342
Telephonr: 818-362-0353
Remote Controllers

VON WEISE GEAR COMPANY
100 St. Clair Industrial Drive
St. Clair, MO 63077
Telephone: 314-629-1010
FAX: 314-629-3339 *Actuators & Gearmotors*

WINEGARD COMPANY
3000 Kirkwood
Burlington, IA 52601
Telephone: (319) 754-0600
FAX: (319) 754-0787
TVRO satellite dishes and mounts

WINNERS SATELLITE ELECTRONICS
4F, No.2, Lane 263
Chung Yang Road
Taipei 11527, Taiwan R.O.C.
Telephone: 886-2-782-0711
FAX: 886-2-782-8490
Receivers, positioners

XANTECH
12950 Bradley Avenue
Sylmar, CA 91342
Telephone: 818-362-0353
FAX: 818-362-9506
Multi-room/multisource infrared repeaters

ZENITH ELECTRONICS CORP.
1000 Milvaukee Avenue
Glenview, IL 60025
Telephone: 708-391-7000
FAX: 708-391-8889
Satellite receivers

APPENDIX F. TRADE and REFERENCE PUBLICATIONS

ARGUS PUBLISHING COMPANY INC.
6300 S. Syracuse Way, Suite 650
Englewood, CO 80111
303-220-0600
FAX: 303-773-9716
Communications trade magazines

BAYLIN PUBLICATIONS
1905 Mariposa
Boulder, CO 80302, USA
Telephone: 303-449-4551
FAX: 303-939-8720

BAYLIN PUBLICATIONS
24 River Gardens
Purley, Reading RG8 8BX, U.K.
Telephone/FAX: 44 (0) 734-414468
Mobile Phone: 44 (0) 836-582785
Books, videos and software

CABLE & SATELLITE TV GUIDE
23, Jenai Road, Section 3, #5F-1
Taipei, Taiwan, R.O.C.
Telephone: 886-2-778-5818
FAX: 886-2-778-5815
Satellite TV programming guide for the Far East

COMMTEK COMMUNICATIONS CORP
8330 Boone Blvd. #600
Vienna, VA 22180
Telephone: 703-827-0511
Satellite Orbit magazine – TV Guide

DBS DIGEST
140 West 29th Street, Suite 163
Pueblo, CO 81008
Telephone: 719-545-1986
FAX: 719-545-1881
Weekly news and trends in DTH industry

DESIGN PUBLISHERS INC.
800 Siesta Way
Sonoma, CA 95476
Telephone: 707-939-9306
FAX: 707-939-9235
Publishes the International Satellite Directory, What's on Satellite and the Satellite Training Series video tapes

FORTUNA COMMUNICATIONS CORP.
P.O. Box 308
Fortuna, CA 95540
Telephone: (707) 725-6951
FAX: 707-725-4311
Satellite TV Week – Satellite TV Guide
VRO Dealer Magazines

MAGAZIN, NB FORTAG
Kjelsasvn. 51
0488 Oslo 4
Norway
Publishes satellite TV guide

McCORMAC'S HAKCWATCH NEWS
MC2 Publications, 22 Viewmont
Waterford, Ireland
Telephone: 353-51-73640
Publishes European Scrambling Systems, the "black book" and runs BBS

TRIPLE-D PUBLISHING
P.O. Box 2384
Shelby, NC 28151, USA
TEL: 800-234-0023
704-482-9673
On-Sat – Satellite TV guide
Satellite Retailer

ONSAT, TRIPLE-D PUBLISHING – CANADA
9780 Bramalea Road, North, Suite 406
Brampton, Ontario L6S 2P1, Canada
TEL: 416-458-9429
Satellite TV guide for North America

PRIVATE CABLE MAGAZINE, NSP
1909 Avenue G
Rosenburg, TX 77471, USA
Telephone: 800-6212-5990
FAX: 713-342-7016

SATELLITE BUSINESS NEWS
1050 17th Streert, N.W., Suite 400
Washington, DC 20036
Telephone: 202-785-0505
FAX: 202-785-9291

SATELLITE ENTERTAINMENT GUIDE
Vogel and Son Publishing Company
1109 TD Tower, Edmonton Centre
10205 - 101 Street
Edmonton, Alberta T5J 2Z1, Canada
TEL: 403-425-1169
800-665-5543
FAX: 403-425-8392
Satellite TV guide

SIGNAL MAGAZINE
FERNWOOD PUBLISHING
12 Automatic Road, Unit 101
Brampton, Ontario L6S 5N3, Canada
Telephone: 416-759-6639
Canadian home satellite TV trade publication

SUPERGUIDE, TRIPLE-D PUBLISHING
P.O. Box 2384
Shelby, NC 28151
TEL: 800-234-0139
Satellite TV guide for North America via satellite

THE TRANSPONDER, – TERRA PUBLISHING
P.O. Box 460
Salamanca, NY 14779-0460
TEL: 800-992-3488

21st CENTURY PUBLISHING
21st Century Publishing, 533 Kings Road
London, SW10 0BR, UK
TEL: 071-351-3612
Publishes Cable and Satellite Europe
Satellite Trader

WHAT SATELLITE, WV PUBLICATIONS
57/59 Rochester Place
London NW1 9JU, UK
TEL: 071-485-0011
What Satellite Magazines – U.K. & Africa

U-VIEW
4 South Parade, Bawtry
Doncaster, South Yorkshire DN10 6JH
U.K.
Telephone: 44-302-719997
FAX: 44-302-719995
Technical books

WORLD SATELLITE TV AND SCRAMBLING METHODS
The Technician's Handbook – 3rd Revised Edition

by Frank Baylin, Richard Maddox, John McCormac

This thorough text is a must-buy for technicians, satellite professionals and do-it-yourselfers. The design, operation and repair of satellite antennas, feeds, LNBs & receivers/modulators are examined in detail. An in-depth study of scrambling methods and broadcast formats is the backdrop to a discussion of all current American and European satellite TV technologies including the VideoCipherII, Oak Orion, FilmNet, Sky Channel, EuroCypher, D2 MAC, BSB and Teleclub Payview III. Circuit and block diagrams of all components are presented and clearly explained throughout the handbook. This information is a prelude to the chapters on troubleshooting and setting up a test bench. This expert guidance on testing, servicing and tuning is complimented by a wealth of detailed illustrations.

PARTIAL CONTENTS: The Satellite System. Early Designs. **Antennas. Feeds** Polarization Formats. Polarity Selection & Control. Troubleshooting. **Low Noise Amplifiers.** Waveguides. Troubleshooting. Downconversion Methods. **Cables and Connectors.** Coax. Insulated Wire. Custom Cables. Sealing. **Positioning Systems**. Basic Motor Controllers. Feedback Circuits. End Point Limiters. Troubleshooting. **Power Supplies.** Actuator Controllers. IC Regulators. Troubleshooting. **IF Circuits.** Amplifiers. Bandpass Filters. Limiter Circuits. Troubleshooting. **Video Processing.** Demodulator Circuits. Video Processing. **Audio Processing.** Audio Subcarrier Specifications. Typical Audio Circuitry. Other Audio Detection Circuits. Stereo Broadcast Methods. Audio Companding. Dolby Noise Reduction. Troubleshooting & Aligning Audio Circuits. **RF Modulators.** Broadcast Formats. RF Modulator Circuits. Crystal Modulators. RF Interference. Troubleshooting. **Miscellaneous Receiver Circuits.** Indicator & LED Circuits. LED Read-out Displays. Remote Controls. **Complete Circuit Descriptions** European & American Receiver Circuits and Comparisons. **Television Operation. Broadcast Formats** NTSC, PAL, SECAM & MAC. Digital Audio. The Digit 2000 TV Receiver. Teletext. **Basic Scrambling Methods.** Video Techniques. Audio Techniques. Digital & Analogue Methods. DES Algorithm. Smart Cards. **Pioneer Scrambling Systems.** Telease/SAVE. Zenith SSAVI. **Case Studies.** RITC Discrete. OAK Orion. IRDETO. Sound-in-Sync. Lorentz PCM2. Filmnet. Telease SAVE. Payview III, VideoCrypt. VideoCipher II and II Plus. MAC Variants **Decoder Connections.** Loopthough and Chaining. **Future Developments.** European & Worldwide Standards. Security & Cost. Legalities. **Setting Up a Test Bench.** Equipment. Synthesized Tuned TV. **Troubleshooting.** Field Checking Microwave Components. Substitution. Troubleshooting Microprocessors. **Specialized Components.** Diodes. Transistors. ICs. Hybrid Components. SAWs Filters. **Appendices.** Active Component Guide. Equations. Glossary. Channel Allocations. Manufacturers.

356 pages / 8.5 x 11" / over 200 photos, diagrams, wiring schematics / 16 tables / appendices, index / 0-917893-15-8

SATELLITE & CABLE TV
Scrambling & Descrambling - 2nd edition

by Brent Gale and Frank Baylin

An important beginners aid in understanding encryption technology for satellite and cable television broadcasts. Fundamental issues and concepts are carefully explained in a comprehensive view of how televisions operate, a topic critical to understanding modern scrambling systems. Various methods used to relay stereo over satellite links and cable lines, including the over-the-air MTS stereo standard, are studied. The principles and techniques of encryption systems are examined in detail. A clear explanation of the innovative B-MAC broadcast system in relation to world-wide television standards is also provided. A wide selection of scrambling systems and equipment is explored, with particular attention to the more sophisticated cable television addressable systems and to the Oak Orion and M/A COM VideoCipher encoding/decoding devices.

PARTIAL CONTENTS: Introduction. Underlying Concepts. Principles of Television. Scrambling Techniques. Cable TV Scrambling Systems. Satellite TV Scrambling Systems. APPENDICES: Worldwide Broadcast Formats & Television Channel Structures. The Decibel Notation. Glossary. Reference Materials. Manufacturers/Distributors.

280 pages / 7 x 9-1/4" / 120 photos and illustrations / appendices, glossary / 0-917893-07-7

:ongratulations to the authors of ... this much needed and excellent publication! ... the information contained in this 1ew book is accurate, concise and timely ... the authors have created a work that artfully guides the reader through ıll facets of this industry. *Steve Johnson, American Wireless Systems*

WIRELESS CABLE and SMATV

(Replaces "Satellite, Off-Air & SMATV")

ɔy Frank Baylin and Steve Berkoff

\ comprehensive study of the new broadcast method, Wireless Cable, and the :losely related field of satellite master antenna TV (SMATV) systems. This thorɔugh manual clearly presents the concepts behind private cable systems as well ıs technical details of construction and operation. Private cable systems are nstalled in apartment complexes, hotels and motels, condominiums, hospitals, nobile home parks, and auditoriums as well as in many other multi-user ɜnvironments.

The book explores the background and history of this rapidly evolving field and ɔutlines the steps required to legally purchase and resell satellite entertainment 'or profit. Three chapters are devoted to the details of the site survey, planning ınd design phases of a private cable system. Off-air and satellite headends and ıll components from antennas to processing and mixing electronics are studied n detail. The chapter on distribution systems explores the components required o supply a high quality signal to every television set. Numerous examples are ɔrovided as illustrations of each stage of design. Following a treatment of the ɔasics of bidding projects, the construction and installation tasks are detailed ınd the critical choice between in-house or subcontracted labor is discussed. ᵓroven methods of locating and managing competent subcontractors are ɜxplained. Complex design issues such as inserting locally originated signals, wo-way services and satellite audio reception are also studied. The chapter on ystems operations presents methods to manage one or more systems as well as ı logical approach to troubleshooting.

SECTION I. BASICS and SMATV SYSTEMS. Overview and History of Private Cable TV. A Brief History of Satellite Television. The Development of Community Antenna TV Systems. SMATV versus Franchised Cable TV Systems. The Development of SMATV and Wireless Cable TV. The Advantages of SMATV. **Economics, Sales Contracts and Regulations**. Economic Options for Installers, Operators and Owners. Sales Contracts. Regulations. **Programming.** Available Satellite Programming. The Mechanics of Purchasing Programming. Current Trends and Issues in the U.S. **The Site Survey and Planning**. The Site Survey Process. Background Technical Information. The Satellite Programming Survey. The TVRO Site Survey. Avoiding Terrestrial Interference. The Off-Air Site SUrvey. Off-Air Interference and Reception Problems. Headend Survey. Distribution System Survey. **Headend System Design.** The Basics of Headend Design. Off-Air Antennas. Off-Air Electronic Components. Solving Off-Air Signal Level Problems. Satellite TV Reception Components. Designing the Satellite Headend. A Strategy for Combating TI. Examples of Headend Designs. **Distribution System Design..** The Basic Components of Distribution Systems. Cable TV Design Issues Relating to SMATV Systems. Television Input SIgnal Requirements. Television Input Signal Requirements. Midband and Higher Frequency Distribution Systems. Sample System Designs and Calculations. 18 GHz Microwave Line-of-Sight Transmission Systems. Laser Links. **Project Bidding**. Bid Preparation. The Bid Document. Negotiations. **Construction and Installation**. The Permit Process. Subcontracting the Construction. In-House Construction. Local Origination and Additional SMATV Services. Stereo Audio Services. **System Operations.** The Customer Relations Office. Computer Systems. The Technical Support Group. The Overall Management Structure. Marketing The Programming. **Troubleshooting and Test Instruments.** Troubleshooting. A Catalog of System Problems. The Field Strength Meter. The Spectrum Analyzer. Point-by-Point Diagnostics. **SECTION II. WIRELESS CABLE /MMDS SYSTEMS. Wireless Cable System Overview.** Technical Overview. History. Frequency Allocations, Channel Availability and Regulations. **Wireless Cable Components**. The Transmitter Facility. Signal Processing and Transmission. Operations Facilities. Receive Sites. Single and Multi-Unit Receive Site Configurations. Viewer Quality Assessment and S/N Ratio. **Wireless Signal Coverage.** Wireless Coverage and Limitations. Coverage Predictions. Extended Reception Techniques. **Receive-Site Installation.** Installation Planning and Site Survey. Mast, Antenna and Downconverter Installation. Grounding and Weatherproofing. Cable and Converter Installation. Customer Education. Service Procedures and Safety. Safety. **System Planning and Operation.** The Business Plan. System Operations Planning. **A Legal Perspective on Wireless Cable in the United States.** An Historical Perspective. The Application Process for an MDS Operation. Recent Trends. The WIRELESS DEAL in the U.S. The Wireless Deal. The Wireless Cable Association International, Inc.**SECTION III. PRIVATE CABLE SECURITY SYSTEMS. Private Cable Security**. Wireless Cable Security Systems. SMATV Security Systems. **Appendix A. Additional Example of Distribution Systems. Appendix B. Equations and Technical Details.** Satellite and Cable TV Equations. Wireless Cable System Parameters.**Appendix C. Programmers. Appendix D. Satellite TV Guides and Publications. Appendix E. Off-Air and Cable TV Channels. Appendix F. Glossary. Appendix G. Manufacturers.** Manufacturers of SMATV Components. Manufacturers of Wireless Components.

386 pages / 8.5 x 11 / photos, illustrations, diagrams, tables / appendices, glossary / 0-917893-17-4

One of the best books on the subject that I've read to date...
John McCormac – well-known author, columnist and hacker

THE "HOW-TO" OF SATELLITE COMMUNICATIONS

by Dr. Joseph Pelton

This excellent book by a seasoned veteran of the industry is a brilliant excursion through the world of satellite communications. For any serious user of satellite services, it gives even a novice a knowledge of how to design, evaluate and purchase any type of data, audio or video services. It covers topics including:

- How geosynchronous satellites work
- The design, construction or purchase of satellite capacity
- The fundamentals of earth station operation
- Global and ISDN standards
- One-way and two-way spread spectrum VSATs and interactive technology
- How and why to buy satellite networks and services
- Voice and data networks
- Video services and digital television
- The digital communication revolution
- LEO (low earth orbit) and MEO (medium earth orbit) satellites
- The future of satellite communications

250 pages / photos / 5.5 x 8.5" / illus / index / 0-936361-24-7

HIDDEN SIGNALS ON SATELLITE TV
Revised 3rd Edition

by Thomas P. Harrington

The third, updated edition is devoted entirely to the hidden, non-video services relayed by communication satellites, a topic of interest to satellite buffs worldwide. Most satellites and their transponders are known by their video transmissions and only enthusiasts are aware that satellites also transmit large amounts of non-video information including

- audio world news services
- news teletypewriter channels
- high speed, multiplex data systems
- long distance telephone circuits
- teletext and teletype services
- stock market and commodity news reports
- telephone and business data channels
- audio subcarrier and single channel per carrier (SCPC) audio services

While many of these services, such as SCPC, can be received with simple, readily-obtainable equipment, others require microprocessors to detect intelligent signals. Most of the well-known audio signals as well as the hi-tech services are outlined and explained in this book.

PARTIAL CONTENTS: 1. Understanding the Satellite Signals. 2. Audio Subcarriers on Satellite TV (SBC/FM). 3. Satellite Telephone Systems (SSB/FDM). 4. Single Channel Per Carrier (SCPC). 5. Satellite Networks. 6. Teletex Basics. 7. Other VBI Services. 8. Misc. Satellite Services. 9. Ku-band and Hidden Signals. Glossary. Appendix: Position charts for world satellites, C- and Ku-band.

234 pages / 8.5 x 11" / over 200 photos, illustrations, tables / appendix, glossary / 0-91661-04-0

Learn All About Small-Dish DBS Technology

A comprehensive exploration into the rapidly evolving Ku-band industry – SATELLITE RETAILER MAGAZINE

Ku-BAND SATELLITE TV – Theory, Installation & Repair
Completely Revised 4th Edition

y Frank Baylin, Brent Gale and John McCormac

clear presentation and explanation of all aspects of worldwide Ku-band tellite television. This fourth edition has been substantially expanded with lditional sections on small dishes, European DBS satellite TV, flat plate itennas, actuators, LNB and satellite receiver electronic design, worldwide rambling technologies, link analysis, fixed antenna installation, interfacing ceivers and decoders and aligning a polar mount without a compass.

ie book first explores the background of Ku-band reception equipment with a etailed survey of frequency allocations for broadcast satellites around the obe. The equipment used to detect signals from orbiting spacecraft is then amined in depth. This includes the design of Ku-band antennas. The chapter i selecting TVRO components includes a study of scrambling and encryption ethods. An explanation of proper installation of Ku-band systems as well as trofitting Ku-band components onto C-band TVROs is presented in considerable detail together with a step-by-step examination of multiple-receiver systems id distribution networks. Existing North American, Soviet, European, Japanese id Australian Ku-band broadcast systems are outlined and, finally, a consistent id comprehensive method of troubleshooting and repairing TVROs and a etailed description of different types of test equipment complete the text. The ppendices are packed with a variety of useful information including a list of anufacturers around the globe, footprints and satellite TV equations.

CONTENTS: Basics of Satellite Communication. The Satellite Circuit. Microwaves. Communication Fundamentals. Microwaves & Satellite Communication. Uplink Operation. Worldwide Frequency Allocations. Satellite Design & Operation. Satellite Launching & Maintenance. **Component Theory and Operation.** Antennas. Antenna Mounts. Actuators. Feedhorns. Low Noise Block Downconverters. Downconversion Methods. LNB Design. Coaxial Cable and Connectors. Satellite Receivers. Modulators. Television Receivers & Monitors. Monophonic & Stereo Audio Reception. **Designing and Testing Ku-Band Systems.** Overall Design Considerations. Beamwidth & Satellite Spacing. Interpreting Television Test Signals. **Selecting Equipment.** Evaluating Equipment. Multistandard Televisions. Ku-Band Terrestrial Interference. Scrambling Technology. **Installing Ku- Band Systems.** Site Survey. Antenna Support Structures. Cable Runs & Trenching. Assembling the Antenna, Feedhorn & LNB. Installing Actuators. Electrical Connections. Decoder Interfacing. Aligning the Antenna onto the Arc. Fine Tuning the Receiver. Programming the Receiver/Actuator. Waterproofing. Connecting Stereo Processors & Accessories. Tools Required. Cold Weather Installations. Customer Relations. Documentation. **Retrofitting C-Band Systems for Ku-Band Reception.** Performance Requirements for Ku-Band Components. Installation Procedures. **Multiple Receiver Satellite TV and Distribution Systems.** Basic Components. Headends & Distribution Networks. Headend and Distribution System Design. **Worldwide Ku-Band Satellite Television.** The Americas. Europe. Australia, New Zealand and New Guinea. Japan. Soviet Union. Intelsat Ku-Band Satellites. **Troubleshooting and Repair.** Systematic Approach. Visual Inspection. Technical Troubleshooting. Understanding Component Failures. Spare Parts Kit. Symptom-Cause Chart. **APPENDICES:** The Decibel Notation. Satellite TV Equations. Footprints and Aiming Charts. Glossary. Lists of Manufacturers. References.

426 pages / 8.5 x 11" / Over 400 photos, illustrations, tables, charts / appendices, glossary / 0-917893-14-X

Nuestro Libro de Mayor Venta en Inglés y en Español

Si Ud. está interesado en sistemas de satélites domésticos, ya sea para la venta o para su uso personal, le recomiendo que compre y lea este libro. Los autores han producido una obra que es fácil de leer para los aficionados sin prescindir la información técnica que necesitan los instaladores e ingenieros. – JLC VIDEO SPECIALIST NEWSLETTER

TELEVISION DOMESTICA VIA SATELITE
Manual de Instalación y de Localización de Averías

Una fuente de información muy valiosa tanto para los dueños de sistemas de televisión por satélite domésticos así como para los instaladores profesionales. Con más de 45.000 ejemplares publicados, este completo libro de referencia se ha transformado en el estándar de la industria. Examina en forma detallada los principios de la comunicación por satélite y la operación de las estaciones terrestres, explica cómo seleccionar y evaluar componentes de TV por satélite, expone los procedimientos de instalación incluyendo conexiones de múltiples receptores y televisores y antenas grandes, y delinea una estrategia detallada para localizar averías en cualquier sistema de TV por satélite. El manual es una referencia completa que explica en forma clara conceptos tales como los formatos transmisor- receptor medio y la polarización circular. Es una herramienta muy valiosa que presenta todos los detalles necesarios para instalar, hacer funcionar y mantener un sistema de satélite.

TABLA DE MATERIAS: Vistazo a la Teoría de Comunicación por Satélite. Funcionamiento de los Diferentes Componentes. Elección del Equipo de TV Vía Satélite. Cómo Instalar Correctamente un Sistema de Televisión vía Satélite. Sistemas de Distribución y de TV vía Satélite con Múltiples Receptores. Sistemas y Antenas Grandes. Localización de Averías. Profesionalismo. Apéndices.

341 páginas / 8-1/2 x 11" / 300 ilustraciones, fotografías y tablas / apéndices, glosario / 0-917893-03-4

TAMBIÈN EM PORTUGUESE

THE HOME SATELLITE TV INSTALLATION VIDEO TAPE

The perfect companion to our best-selling *Home Satellite TV Installation and Troubleshooting Manual.* Designed to make owning and operating your satellite TV system a pleasure, it demonstrates how to install a system as well as how to keep it tuned up for best reception. An excellent instructional tool, the video demonstrates professionally accepted satellite TV installation techniques, details how to perform a site survey, detect and avoid terrestrial interference, install the pole and antenna mount, assemble the various components, complete the necessary electrical connections and track the arc of satellites. Illustrated with NASA footage, this high-quality video follows the sequence of the *Manual* so viewers can benefit from the carefully coordinated text and visual aid.

43 minutes / VHS or Beta / 0-917893-04-2

CINTA MAGNETICA DE VIDEO
SOBRE LA INSTALACION DE LA TELEVISION VIA SATELITE

El complemento perfecto de nuestro libro de mayor venta. Está diseñado para hacer que poseer y operar un sistema de TV por satélite sea un placer. Demuestra cómo instalar un sistema y cómo mantenerlo a punto para conseguir la mejor recepción. Este video es una excelente ayuda didáctica, demuestra técnicas de instalación aceptadas por los profesionales, explica en detalle cómo efectuar una exploración de emplazamiento, cómo detectar y evitar la interferencia terrestre, cómo instalar el mástil y el soporte de la antena, cómo armar los diferentes componentes, cómo efectuar las conexiones eléctricas necesarias y cómo seguir el arco de los satélites. Este video de alta calidad incluye tomas de la NASA y sigue la misma secuencia del Manual para que los espectadores se beneficien de la coordinación entre las imágenes y el texto.

45 minutos / VHS o Beta / 02-0-917893-04-2

THE AFFORDABLE STANDARD REFERENCE BOOK
1995 WORLD SATELLITE YEARLY
Footprints, Programming and Technical Use

This affordable book, written in the concise style that has become the trademark of Dr. Frank Baylin, provides the information required to easily determine the satellite programming available and the equipment necessary to receive these satellite signals from any point on our globe. It is organized into four easy-to-use sections that are separated by tabs.

SECTION I outlines required:
- METHODS to receive satellite audio and video signals
- to interpret satellite footprints
- to size antennas and select equipment
- to aim an antenna
- AS WELL AS the latest in audio and video compression
- an overview of scrambling methods and broadcast standards

SECTION II presents:
- a complete listing of footprint maps and other information about over 350 worldwide satellites - past, present and planned
- easy-to-read and accurate footprints of all active satellites

SECTION III lists:
- programming available on each active satellite
- world video standards and scrambling systems

SECTION IV lists:
- Addresses and telephone numbers of satellite manufacturers, service companies, programmers and major satellite system operators

780 pages / 8-1/2 x 11" / illustrations / glossary / index

INSTALL, AIM and REPAIR YOUR SATELLITE TV SYSTEM 2nd Revised Edition

by Frank Baylin

This booklet, a shortened version of *The Home Satellite TV - Installation and Troubleshooting Manual*, explores how to install a satellite TV system, aim the dish at the arc of satellites as well as how to trouble-shoot and repair the system if a problem arises. It also covers the periodic maintenance work that is required to keep a system tuned up and aligned onto the arc of satellites.

64 pages / 8.5 x 11" / 48 figures & photos, 3 tables / 0-917893-16-6

LEARN ALL THE FACTS ABOUT DBS

MINIATURE SATELLITE DISHES

The New Digital Television – 2nd ed.

This affordable book, written in the concise style that has become the trademark of Dr. Frank Baylin, explores all aspects of this exciting new field. Topics covered include:

- The background of DBS. How does it relate to large-dish TVRO systems
- How these satellites and receive systems function and installation methods
- The major players. Learn details of their systems including their satellites and future plans.
- A comparison of the competing programming packages.
- An exciting new strategy to upgrade large-dish TVROs to dual-band, DBS/TVRO operation.
- A study of the encryption systems. Are they secure? Will the pirates be held in check?
- Learn to connect DBS systems to home entertainment centers
- What the future holds in store.

128 pages / 6 x 11" / illustrations

The Future is Here Today!

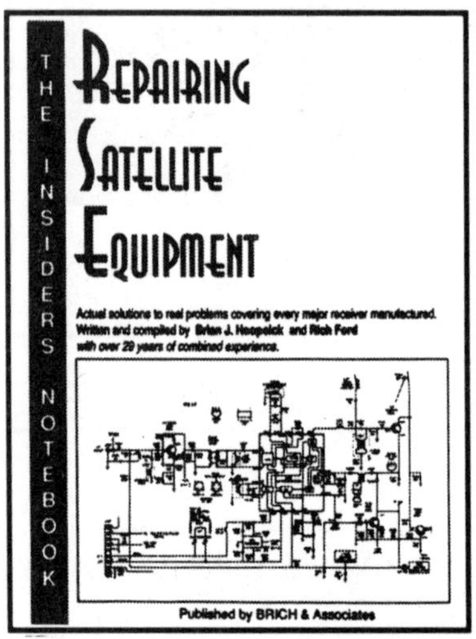

REPAIRING SATELLITE EQUIPMENT

"The Insider's Notebook" by Rich Ford & Brian Hoopsick

Learn the "how-to" of repairing nearly every brand of consumer home satellite TV receive ever used in North America, including current models. This 130-page book in a 3-ring binder, is not theoretical but offers practical, real-world repair solutions. It is organized in a convenient format of PROBLEM in the left column, SOLUTION in the right. Actual repair procedures indicate components or components that need replacement.

Avoid shipping customers' receivers to repair centers and paying hundreds of dollars

REPAIR THAT OLD SATELLITE RECEIVER FOR PENNIES!

CHAPARRAL – CHANNEL MASTER – DRAKE – GENERAL INSTRUMENT – HOUSTON TRACKER – LUXOR – NORSAT – PANASONIC STS TEE-COM TOSHIBA UNIDEN – EQUUS – KENWOOD – MTI – ECHOSPHERE – FUJITSU JANEIL – MA/COM – PICO SCIENTIFIC ATLANTA – PROSAT – ZENITH

Annual Updates Availlable

THE SATELLITE BOOK – 4th Edition

A Complete Guide to Satellite TV Theory and Practice

Available Only From Our UK Office

Edited by John Breeds

This book, written specifically for the European market, presents a complete overview of satellite television written by 22 authors and edited by John Breeds.

CONTENTS: Geostationary Satellites. Tools for the Trade. Working with Ladders. Wall Fixing Systems for Dishes. Cables for Satellite TV Installations. The SCART Connector. Carrier Level Meters. Equipment Installation: A Guideline. Customer Care. Microwave Basics. Ferrite Polar Selectors. Signal vs. Noise. Frequency Modulation. Satellite Antennas. Polyrod Lens Feed. Flat Plate Antenna. Link Budget Analysis. Introduction to SMATV. Intermediate Frequency Distribution Systems. British Telecom's Role. The Future for EUTELSAT. The ASTRA Satellites. The MAC Transmission Standard. The VideoCrypt Encryption System. Eurocrypt for MAC.

292 pages / 21.1 x 29.7 cm / 300 illustrations, photographs, and tables / glossary / index / 1-872567-02-9

WHAT'S on SATELLITE?
Video Activity Report on all Satellites

This accurate and detailed report, published three times a year by Design Publishers, lists the video services on active transponders for all the world's communication satellites. This information allows you to keep abreast of new programming, satellite launches, rapidly changing new services and satellite relocations. Includes:

- Location of each satellite
- Name, transponder number, polarity, bandwidth, center frequency and language of each service.
- Any associated audio subcarriers and/or SCPC (single channel per carrier) frequencies.
- Auxiliary radio services which are transmitted either by means of auxiliary audio subcarriers or independent SCPC carriers
- Estimated beam-center EIRP levels for each satellite TV service

One-Year Subscription – $95.00 U.S.

First Class Air Postage: USA – $8.00, Other Countries – $30.00

3 Issues – January, May, and September

FUTURE TALK

COMMUNICATIONS, TECHNOLOGY and SOCIETY in the 21st CENTURY

by Dr. Joseph N. Pelton

Future Talk, an insightful 222-page book by Dr. Joseph Pelton, explores his vision of the future of telecommunications. He presents a thoughtful study of topics such as the telepower revolution, the global electronic machine, living at super speed, the future of work, "smart energy," money goes electronic, life in the telelcity, electronic education, freedom, justice and telepower, redefining the world with telepower, telewar and the global brain.

222 pages / 6 x 9 inches / selected bibliography / published November 1992

1995 INTERNATIONAL SATELLITE DIRECTORY

The **COMPLETE REFERENCE** source of worldwide communication satellites, manufacturers, service providers, and users of satellite services. This brilliant new edition contains more than 25,000 entries, conveniently organized in 10 tab-indexed chapters.

CONTENTS: International Organizations. Government Regulators and Administrators. U.S. State Agencies. **Satellite Operators.** International Systems: Inmarsat, Intelsat & Intersputnik. Regional Satellite Systems: Arabsat, Astra, Eutelsat, Palapa, etc. Domestic Satellite Systems. Planned Satellite Systems: Alascom, Aussat, Brasilsat, Galaxy, GTE, Spacenet, etc. Planned Systems: ACTS, Cygnus, etc. **Manufacturers of Space Equipment. Manufacturers of Satellite Ground Equipment. Commercial and Domestic Systems. Users of Satellite Services.** Satellite Programmers. Transponder Usage. Location of Programmers by Interest. Broadcasters. Telephone Companies. Cable Companies. **Providers of Satellite Services.** Transmission Services.
Transponder Brokers. Teleports. Videoconferencing. VSATs. **Services for the Satellite Industry.** Distributors. Technical and Consulting Services
Associations. Legal Services. Insurance Services. Research Centers.
Educational and Technical Centers. Publishers and Publications. Financial Institutions. **Geosynchronous Satellites.** List of Satellites. Satellite Database, EIRP Maps, Detailed Information on Satellite Spacecraft. **Index.**

Over 1250 pages / 8-3/8 x 10-7/8" / index / ISSN 1041-4541

NEW, POWERFUL, EASY-TO-USE & COMPREHENSIVE

SATELLITE TOOLBOX SOFTWARE – VERSION 1.2

Designed by Dr. Frank Baylin

PROFESSIONAL, EASY-to-USE, MENU-DRIVEN SOFTWARE

Worldwide Satellite Programming
Antenna Aiming & Sizing
Receive-Site Analysis
Graphs, Tables & Reports

This remarkable software does it all, and with surprising ease:

SATELLITE DATABASE: Lists of all the world's communication spacecraft and the programming they relay. Select any of the world's satellites to view all the video and audio programming available as well as transponder bandwidths, center frequencies or polarization formats. Or enter site latitude and longitude and print a list of those satellites viewable from your receive site.

AIMING SECTION: To activate simply enter your site coordinates. The software ***automatically calculates magnetic variation***, compass heading, azimuth bearing as well as all the aiming angles for a polar or modified polar mount to any satellite within view or any group of satellites that you select. A datafile of major cities worldwide with latitudes and longitudes is provided. In addition, a similar database on every city, town and village in ITU regions I, II and III is available at a small additional cost.

ANTENNA SIZING: This component makes antenna sizing, choice of LNB noise temperature and other system parameters simple. Choose between NTSC, PAL or other broadcast formats to calculate G/T, S/N, C/N, etc... Print detailed analyis reports with your business name as the source. Save an unlimited number of files for each customer site anywhere on the globe.

ANALYSIS: This component allows you to easily calculate virtually every parameter relating to satellite reception including antenna gain, side lobe gain, focal distance as well as path loss, slant distance, G/T, C/N, video and audio S/N and much more. Create graphs or tables of how any one variable affects the outcome of any other. How would changing antenna size or LNB noise temperature effect audio signal-to-noise ratio? Answer a multidude of other pertinent questions.

SYSTEM REQUIREMENTS

BASIC SYSTEM

MSDOS 3.0 or higher version
A minimum of 640 kb of RAM
and
3 MB of harddrive space

ADDITIONAL DATABASES

ITU1 – 2.09 MB

ITU2 – 2.43 MB

ITU3 – 3.42 MB

MAIN PROGRAM........ $95

DATABASES:

ITU1, 2 or 3 .... $70
Any 2 databases ... $105
All 3 databases $120

YOU WILL BE ASTONISHED BY THE BEAUTY, POWER and SIMPLICITY of this SOFTWARE

The program runs without a hitch and the documentation is excellent. In light of the wide range of calculations and data, the program ... would be a bargain at twice Baylin Publication's price – *Satellite Retailer Magazine*

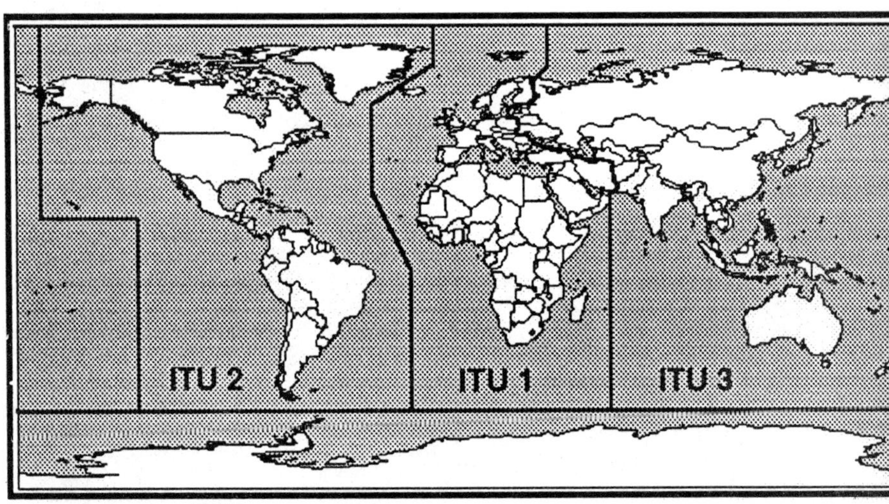

TOOLS for SATELLITE INSTALLATIONS

COAXIAL CABLE CUTTER & RIPPER . This lightweight tool cuts d strips coaxial cable to prepare it r fittings. Adjust- able to fit RG-58, G-59 and RG-6 coax. **$26.00**

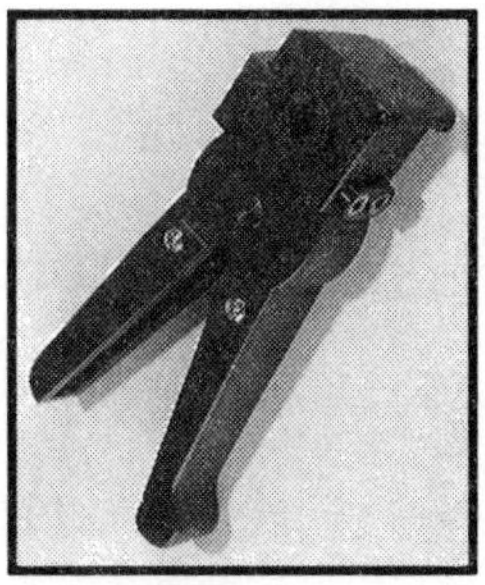

2. SILVA POLARIS COMPASS Lifetime guaranteed compass with magnetized Swedish steel needle, sapphire bearing. Dial reads in 2° increments. **$17**

BIRD FINDER The perfect tool for ghting satellites. Built-in compass d inclino- meter on sturdy tripod. uminum construction - 36 oz. **$174**

4. SILVA COMPASS/INCLINOMETER #15CL Compass & inclinometer in one, with sighting mirror for 1° accuracy. Has rotating capsule, sight line and adjusting screw for declination readings. Azimuth and quadrant scale. $59

. TWIN SUNTO OMPASS/INCLINOMETER ght through or prism views for easurements of both azimuth and evation angles. Jewel bearing with curacy to 0.25°. Threaded tripod cket. Clinometer scales in degrees percentages. **$155**

5. DIGITAL COMPASS Works as easily as point-click-read. Has pistol sights, information display, memory and timer control, and bearing trigger. Memory holds 9-bearings **$167**

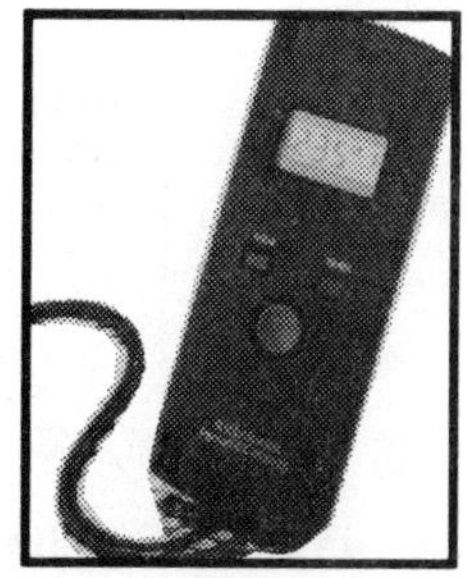

6. ANGLE FINDER (LARGE) Level with magnetic base to measure angles from 0 to 90 degrees easily and accurately in any quadrant. **$23**

7. ANGLE FINDER (SMALL) Accurate to 0.5°. Easy, accurate reading from 0° to 90°. **$20**

8. SATELLITE TUNING METER Bulz-I meter operates in the 70 to 1.4 GHz range. Useful for peaking dishes. For single and block downconversion receivers. **$139**
With audio output **$159**

9. SCOTCH LOK CONNECTORS 100 connectors per box. Self-stripping, insulated and moisture resistant with either 2-wire or 3-wire conductor. **$28**

10. INCLINOMETER Sight through to see slope, percent or degree scales for measuring elevation angles with precision and accuracy. Cross hairs extended by optics for ease of surveying or installing satellite dishes. **$105**

11. COMPASS View-through, liquid-filled instrument with sapphire bearings and rapid settling time. Precision designed for professional use. **$95**

12. COMPASS - SILVA 80 Employs glass optics and is liquid dampened. Scale has sapphire bearing yielding an accuracy of 0.5 degrees. Water resistant. Has internal light. **$118**

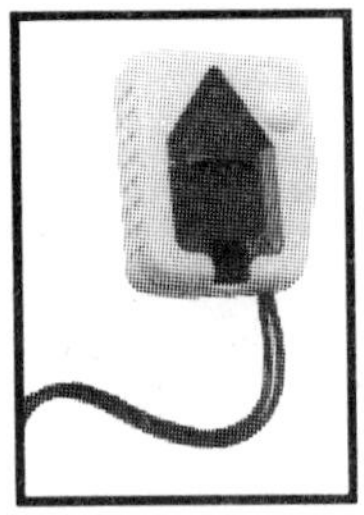

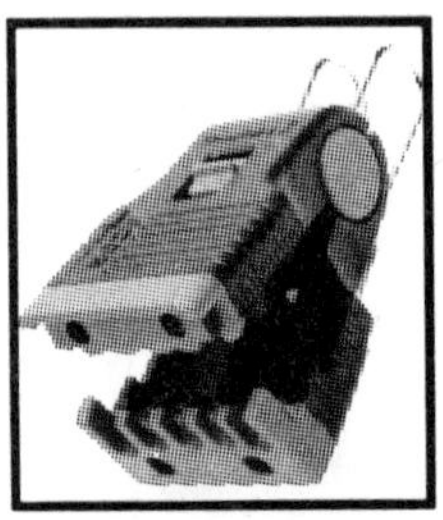

13. KWIK-STRIP Coaxial cable stripper. Cuts and strips RG- 58/U, RG-59/U or RG-6 coaxial cable. Replaceable precision ground cutter blades **$24**
Extra blades **$3**

15. TWO-WAY RADIOS 49 MHz integral FM, simplex operated, voice activated for hands-free use. Range 1/2 mile. PRC-1Y **$125**
Full duplex model PRC-3 **$205**

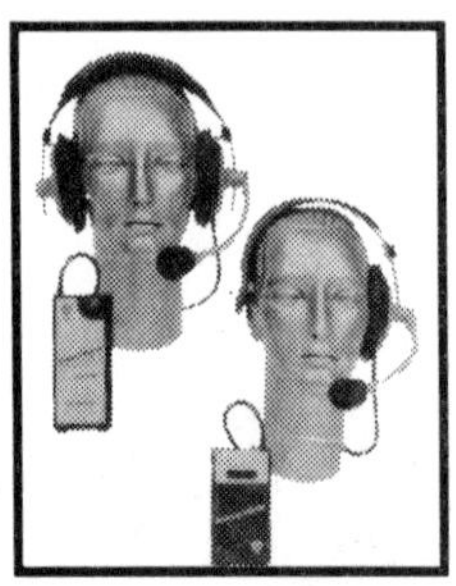

16. FOCUS FINDER CENTERING TOOL 1- or 2-piece ball fits either C-band or Ku-band feeds. Ball inserts into feed to easily measure focal distance. Telescoping tube to 60". For C- and Ku-band **$54**

17. LASER FOCUS FINDER. Beam of red laser light targets center of dish. Small, portable tool. Fits both C-band and Ku-band feeds. **$199**

18. DIGITAL INCLINOMETER 360° Digital read-out. 0.1° accuracy. Reads degrees of slope, % of slope, inches per foot (pitch) & electronically simulated analog bubble display from distance of10-feet. Large LCD display. Length 16cm. **$124** With magnetic base **$130**
24, 48 and 72 inch rails. **$39**, **$65** and **$95,** respectively.

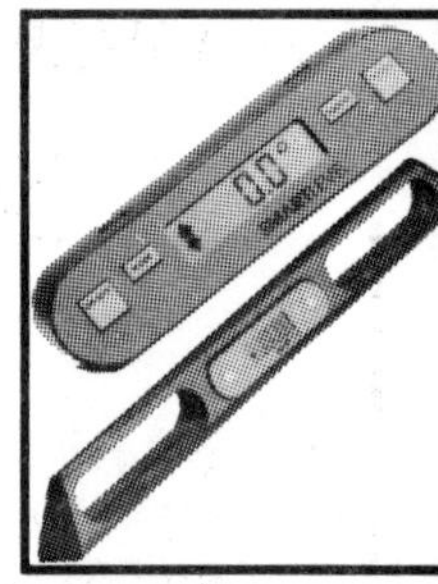

18b. SATLOOK SPECTRUM ANALYZER This unit features a built-in satellite receiver (900 to 1750 MHz), a 4.5" b/w monitor, an audible peaking tone and rechargeable internal battery. Switching on LNB voltage, 13 or 18 Vdc. Ideal for DSS With carrying case. **$799**

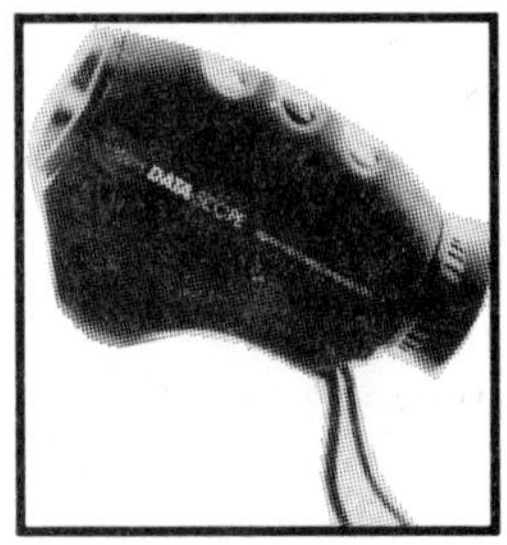

19. DATA SCOPE DIGITAL COMPASS/RANGE FINDER Monocular, fluxgate compass and electronic rang finder in one! Compass: 0.5° accurac stores 9 bearings in memory; calculates change between bearings. Rang Finder: determines range from knowr height & relative size. Includes chronometer.**$479**

20. DISH CRANE For lifting satellite dishes.
Model for 10′ dishes. **$360**
Model for 12′ dishes. **$460**

21. E-Z TRENCHER Cuts 7-inch deep, 100-foot trench in approximately 5 minutes.
Model J-100 cuts 1.5" x 7" **$1155**
Model J-2000 cuts 2.5" x 13" **$2600**
Model TS-1000 cuts 4" x 7" **$1176**

22a. WAYFINDER DIGITAL NUMERIC COMPASS. LCD display, water resistant digital compass with backlighting. Easily calibrated to zero out effect of nearby metal. For boating, provides heading average in rough conditions. **$135**

ORDER INSTRUCTIONS for SCPC RECEIVER and TOOLS

TOOLS MUST BE PURCHASED ONLY FROM OUR BOULDER, COLORADO OFFICE

1. **COAXIAL CABLE CUTTER & STRIPPER** $26.00
2. **SILVA POLARIS COMPASS** 17
3. **BIRDFINDER** 174
4. **SILVA COMPASS/INCLINOMETER - #15CL** 59

4A. **TWIN SUNTO COMPASS/INCLINOMETER** 155

5. **DIGITAL COMPASS** 167
6. **ANGLE FINDER (Large)** 23
7. **ANGLE FINDER (Small)** 20
8. **SATELLITE TUNING METER** 139
   with audio tone 159
9. **SKOTCH LOKS** 28
10. **INCLINOMETER** 105
11. **SIGHTING COMPASS** 95
12. **SILVA COMPASS - 80** 118
13. **KWIK-STRIP COAX STRIPPER** 24
    extra blades 3
15. **TWO-WAY RADIOS**
    PRC-1Y 125
    PRC-3X 205
16. **FOCUS FINDER TOOL (C/Ku-band)** 54
17. **LASER FOCUS FINDER** 199
18. **DIGITAL INCLINOMETER** 124
    with magnetic base 130
    24" rail 39
    48" rail 65
    72" rail 95

18b. **SATLOOK SPECTRUM ANALYZER** 799

19. **DATA SCOPE** 479
20. **DISH CRANE** (10 ft. antennas) 360
    (12 ft. antennas) 460
21. **E-Z TRENCHER**
    MODEL J-100 1155
    Model J-2000 2600
    Model TS-1000 1176

22a. **WAYFINDER DIGITAL COMPASS** 135

25. **BRUNTON TRANSIT–HIGH TECH VERSION** ... 204
    Tri-pod 99
    Ball & Socket Head 39
    Case 88
26. **PORTABLE ACTUATOR** 189
28. **SERVO MOTOR TESTER** 93
29. **PORTABLE TEST RECEIVER** 1310
30. **AVCOM PSA-65A** 2945
    PSA-37D (900 to 1450 MHz) 2545
    Case for each 109
31. **AVCOM PTR-25A TEST RECEIVER** 1445
    Case 109
32. **PANASONIC**
    EY6040 Portable drill 75
    EY6207 Portable drill 339
    EY6220 Drive & driver 146
    EY6990 Hammer drill & driver 275
    Case for each 109
33. **JAWS POWER**
    Large 42
    Small 36

SHIPPING CHARGES for TOOLS

Exact air or sea shipping charges will be added to credit card orders, the preferred method of payment. As a high estimate, add 20% for overseas air, 10% for overseas sea, 10% for US/Canada air, 5% for US/Canada sea.

Total Cost for this Page $_____
Shipping $_____
Total for Tools $_____

TRANSFER TOTAL TO OPPOSITE PAGE

BOOK PRICES

Baylin Publications
1905 Mariposa
Boulder, CO 80302
Telephone: 303-449-4551
FAX: 303-939-8720

Distributed from England by:
Swift Television Publications
17 Pittsfield
Cricklade, Wiltshire SN6 6AN, U.K.
Telephone: 44 (0) 793-750620
FAX: 44 (0) 793-752399

ITEM	PRICE $	PRICE £	NUMBER	TOTAL
World Satellite TV & Scrambling Methods – 3rd edition	$40	£29		
Ku-Band Satellite TV– 4th edition	30	25		
Satellite Toolbox Software (5-14" or 3-1/2" diskette)	95	59		
– ITU Region 1, 2 and 3 Databases	See page 306 for prices			
Update to Satellite Toolbox Databases	20			
Wireless Cable and SMATV	50	35		
Satellite & Cable TV Scrambling – 2nd edition	20	21		
Home Satellite TV Manual, English – 4th revised edition	30	25		
Install, Aim & Repair Your Satellite TV System – 2nd edition	10	12		
Satellite TV Installation Video – English	40	27		
Home Satellite TV Manual ❑ **Spanish** ❑ **Portuguese**	30	25		
Satellite TV Installation Video – Spanish	40			
Hidden Signals on Satellite TV – 3rd edition	20	23		
The "How To" Book of Satellite Communications – 2nd edition	25	22		
1995 World Satellite Yearly	90	59		
What's On Satellite? (3 issues) Shipping: US $10; other $26	95	60		
Future Talk	20			
1995 International Satellite Directory	260			
Repairing Satellite Equipment – The Insider's Notebook	80	50		
European Scrambling Systems – IV	55	32		
Miniature Satellite Dishes – The New Digital TV – 2nd edition	20	14		
The Satellite Book – 4th edition	50	32		
	Total for this page			
	Merchandise Order Total – from previous page			
	Sales Tax – Colorado Residents only (3.7%)			
	Shipping (see below for detailed costs)			
	Grand Total			

Method of Payment:
- Check Enclosed ❑
- Cash ❑
- Credit Card ❑
- COD ❑

Shipping Method:
- Air Mail ❑
- Surface ❑
- UPS Blue ❑
- UPS Red ❑

Visa, Mastercard & Amex - U.S. Office
Visa, Mastercard, Eurocard & Access - U.K. Office

SHIPPING INSTRUCTIONS

to UNITED STATES: Add $4.00 for each item ordered. UPS regular, 2nd day or overnight may be requested at additional charge. UPS COD orders are an additional $5.00. MasterCard, Visa and Amex cards accepted.

to CANADA: Add $5.00 per item ordered. Please remit funds in U.S.$ or by credit card. Orders are shipped via book rate mail. UPS shipping can be requested, but no COD is available to Canada.

WORLDWIDE (from our US office): Remit payment in U.S. funds as checks drawn on U.S. banks, money orders, Master, Visa or Amex credit cards, or cash. Shipments are by surface mail unless additional air mail charges can be added onto credit card billing.

from UK OFFICE: Add £2.50 postage per book within UK. Add £5 per book to Europe; beyond Europe add 30%. Optional insurance £4 extra. Remit payment in UK £ Sterling drawn on UK bank, Eurocheque, Postal Order, cash, Visa, Mastercard, Access or Eurocard.

ORDER TODAY
(Quantity Discounts Available)

Name:____________________

Company:____________________

Address:____________________

____

____

Country:____________________

Telephone:____________________

FAX:____________________

Credit Card No.:____________________

Expiration Date:____________________

Signature:____________________